AN INTRODUCTION TO
EXPERIMENTAL AEROBIOLOGY

ENVIRONMENTAL SCIENCE AND TECHNOLOGY

A Wiley-Interscience Series of Texts and Monographs

Edited by ROBERT L. METCALF, *University of Illinois*
JAMES N. PITTS, *University of California*

PRINCIPLES AND PRACTICES OF INCINERATION
Richard C. Corey

AN INTRODUCTION TO EXPERIMENTAL AEROBIOLOGY
Robert L. Dimmick and Ann B. Akers

AN INTRODUCTION TO
EXPERIMENTAL AEROBIOLOGY

Robert L. Dimmick
Research Bacteriologist and Lecturer in Aerobiology

and

Ann B. Akers
Associated Research Biochemist and Consultant

ASSOCIATE EDITORS

Robert J. Heckly
Research Bacteriologist and Lecturer in Immunology

H. Wolochow
Associate Research Bacteriologist

University of California
School of Public Health
Naval Biological Laboratory
Oakland

WILEY—INTERSCIENCE
A Division of John Wiley & Sons

New York   London   Sydney   Toronto

Library of Congress Catalog Card Number: 75-84963

SBN 471 21558 9

Printed in the United States of America

10 9 8 7 6 5 4 3 2 1

SERIES PREFACE

Environmental Sciences and Technology

The Environmental Sciences and Technology Series of Monographs, Textbooks, and Advances is devoted to the study of the quality of the environment and to the technology of its conservation. Environmental science therefore relates to the chemical, physical, and biological changes in the environment through contamination or modification, to the physical nature and biological behavior of air, water, soil, food, and waste as they are affected by man's agricultural, industrial, and social activities, and to the application of science and technology to the control and improvement of environmental quality.

The deterioration of environmental quality, which began when man first collected into villages and utilized fire, has existed as a serious problem since the industrial revolution. In the last half of the twentieth century, under the ever-increasing impacts of exponentially increasing population and of industrializing society, environmental contamination of air, water, soil, and food has become a threat to the continued existence of many plant and animal communities of the ecosystem and may ultimately threaten the very survival of the human race.

It seems clear that if we are to preserve for future generations some semblance of the biological order of the world of the past and hope to improve on the deteriorating standards of urban public health environmental science and technology must quickly come to play a dominant role in designing our social and industrial structure for tomorrow. Scientifically rigorous criteria of environmental quality must be developed. Based in part on these criteria, realistic standards must be established and our technological progress must be tailored to meet them. It is obvious that civilization will continue to require increasing amounts of fuel, transportation, industrial chemicals, fertilizers, pesticides, and countless other products and that it will continue to produce waste products of all descriptions. What is urgently needed is a total systems approach to modern civilization through which the pooled talents of scientists and engineers, in cooperation with social scientists and the medical profession, can be focused on the development of order and equilibrium to the presently disparate segments of the human environment. Most of the skills and tools that are needed are already in existence. Surely a technology that has created such manifold environmental problems is also capable of solving them. It is our hope that this Series in Environmental Sciences and Technology will serve not only to make this challenge more explicit to the established professional but also to help stimulate the student toward the career opportunities in this vital area.

Robert L. Metcalf
James N. Pitts, Jr.

FOREWORD

In the present volume, the authors have attempted to bring together certain aspects of our knowledge concerning biologically active entities small enough to be rendered airborne and to remain airborne for extended periods of time. Some of the results obtained when such airborne materials contact a susceptible host are also discussed.

The concept of airborne contagions or *miasmas* is as old as recorded medicine, yet crowded urban man is just now awakening to the fact that the very air he breathes may be dangerous to survival. Tons of noxious chemical materials are vaporized and aerosolized daily over metropolitan centers. Our bays, rivers, lakes, and even the offshore oceans are being increasingly polluted with biological and chemical wastes, all of which are capable of being aerosolized by wind and wave action. At home and work man uses an ever increasing variety of sprays that, until proven otherwise, must be considered incompatible with a wholesome environment. It is important to know how biological and chemical aerosols behave and interact, as well as the source and extent of aerosol pollutants. Further, it is vital to know and recognize the effects of such aerosols on man and susceptible animals and plants.

In addition to studies related to respiratory disease transmission, an important role of aerobiological research has been in prophylaxis and therapy of man's respiratory ailments. Questions relative to the respiratory route as a means of immunization as well as drug therapy are today in their infancy, and will remain so until the aerosol behavior of appropriate products has been adequately defined.

Despite the limitation that only principles involved in studies of airborne microbes constitute the major portion of this book, the technology and concepts are of general application to allied fields. The authors have provided one of the first attempts to make the *art* of aerobiology into the disciplined science it must become if man is to live in harmony with his atmospheric environment.

S. H. MADIN

Professor of Public Health
School of Public Health
University of California
Berkeley, California

To William Firth Wells and David W. Hendersen

PREFACE

The word aerobiology, if interpreted in the fullest sense, would seem to define any relationship between the atmosphere and living entities. This is not, however, the definition usually employed by those who consider themselves aerobiologists. Rather, the context is narrowed to include only the study of airborne microbes and their relationship to plants and animals—more specifically, a study of diseases of man that may be transmitted via the respiratory route. Studies of airborne particulate allergens should, perhaps, be included in the definition. *Experimental* aerobiology is the study of aerobiological principles in the laboratory or the field under controlled conditions.

The primary purposes underlying studies in experimental aerobiology are to (a) expand our knowledge of the source of airborne microbes; (b) understand how microbes remain alive while airborne; (c) establish relationships between environmental factors (for example, sunlight, water vapor) and viability and infectiousness; (d) define mechanisms by which pathogens penetrate to sites where disease can develop; (e) relate characteristics of disease to specific respiratory involvement; and eventually, (f) be able to describe the entire process both qualitatively and quantitatively. Although aerobiologists have accumulated reams of data, there is at present no basic theory upon which to construct a *science*, so the present state-of-the-art remains in the data-gathering stage.

To pursue such studies, the aerobiologists must have knowledge of select portions of mathematics, chemistry, physics, meteorology, engineering, electronics, microbiology, physiology, biochemistry, genetics, molecular biology, immunology, epidemiology, medicine and perhaps even cybernetics, as well as having a capability for *gadgeteering*. No single person can be expected to be competent in all these fields, nor can the entirety of these fields be covered in one book.

As a result of this diversity, the scientist who wishes to do meaningful aerobiological research finds it necessary to read literature in fields outside his own specialty without knowing where to start, or how much of what he reads is pertinent to the general problem. As is true in so many sub-branches of science, a *jargon* has evolved—sufficient in size to warrant a glossary. However, we have used words and phrases as they have appeared in print, for to do otherwise would be misleading.

Words sometimes imply situations that are not actually possible, though they seem reasonable at first thought. We use *airborne infection* when we mean airborne contagion (transmission by the airborne route) regardless of whether the focus of infection may be a site other than the lung. We sometimes refer to *cloud chambers* (a phrase with an entirely different meaning in physics) when we mean aerosol chamber, or to *droplet nuclei* when we mean particles containing pathogenic microbes.

Not only are some words or phrases unclear, but a few concepts, if not meaningless, are at least suitable subjects for considerable argument. We often speak of *aerosol decay*, or even *particle decay*. Aerosols and particles do not decay, the particles only become fewer in number. *Biological decay* is only a name for an observed, but as yet undefinable phenomenon—a phenomenon that encompasses the definition of life itself. And finally, to confuse the gathering of useful literature, the chemical industry refers to products in pressure cans as *packaged aerosols*.

Although experimental aerobiology is an art, it is an important art, for it encompasses the study of the most frequent, natural route of entry of disease organisms into the animal system. Of additional importance is the fact that microbes, usually by airborne routes, spoil our food supplies and other materials, interfere with industrial processes, infect wounds and incisions, and contaminate such varied objects as surgical tools and interplanetary vehicles. The airborne microbe is one class of substances that has not been included in most studies on air pollution.

In this book, we have attempted to bring together a sufficient body of knowledge to provide a starting point for future investigators. The book raises more questions than it provides answers. Because the scope is so large, we have assumed the reader to be only partially acquainted with specific concepts and have attempted to describe principles in a simplified manner. The engineer, for example, may find certain chapters unworthy of his attention, whereas chapters on virology or hospital sepsis may require him to refer to outside sources in order to understand the content. The reverse might be true for the immunologist or epidemiologist who wishes to set up experimental units in his own laboratory. We have tried to supply sufficient background knowledge so that scientists in any field will be able to start investigation without extensive literature search, to prevent initial gross errors in assembly and usage of equipment and methodology, and thus to encourage more students to apply their resources to aerobiological problems.

Effective work with pathogens cannot be conducted without elaborate and expensive equipment, so we have provided a chapter on problems of safety and construction methods. Such equipment is required to gain knowledge of real relationships between the pathogenic microbe and the animal—knowledge that can be translated into epidemiological terms. On the other hand, the study of nonpathogenic microbes in equipment that can be simple and inexpensive is no less important than studies with complex tools. Such activities include studies leading to an understanding of processes lethal to microorganisms, the evolution of new devices and techniques, and the possible use of simple chambers as teaching devices in general courses in microbiology. To implement these purposes, we have provided some examples of how a simplified methodology might be employed by workers with restricted budgets. These techniques are covered in the first and second parts of the book.

The third part is a review and analysis of applications of aerobiological techniques in the laboratory and the field. Some well-known and some lesser-known findings are listed, and many are given critical evaluation. The reader will be introduced to problem-areas that need additional investigation, and he will discover that, despite the considerable effort expended to date, there are few agreements as to factual content, applicability of equipment and methodology, or theoretical knowledge revealed by the research.

Additionally, this book is an attempt to provide liaison between the aerobiologist, who is interested in theories and techniques, and the *air hygienist*, the public health officer, the veterinarian, the epidemiologist, and the dentist—all of whom are concerned with practical problems. They have ignored each other's needs and knowledge for too long. In the field, for example, mechanical samplers are taken for granted, whereas the experimentalist knows that mechanical samplers are deficient in many respects. On the other hand, the animal has been shown to be the only sampler capable of providing data in certain field circumstances, but the experimental aerobiologist has not devoted sufficient attention to using the animal as an auxiliary sampler, or to providing answers as to why this situation should exist. We hope this book will contribute to a mutual exchange of information between investigators in the several categories.

Finally, we hope this book will stimulate others to write on the subject in a more specific manner. There is sufficient challenge in areas covered by any chapter to provide impetus for a more comprehensive treatise.

We have not attempted to unearth all the pertinent literature, nor to include all areas applicable to aerobiology research. We leave such inclusions to future writers who, hopefully, will have a greater body of knowledge on which to base their statements and conclusions, and who will undoubtedly provide well-documented theories that will make aerobiology a science rather than the art it is now.

We owe a debt of gratitude to the contributors and to numerous others involved in the tedious and frustrating work of assembling a book of such diversity. We extend our thanks especially to Mr. John Schutz for photographs and to Mrs. Marsha Harris for illustrations. Most importantly, we wish to acknowledge the support and encouragement of The Office of Naval Research, represented in this instance by Dr. Roger Reid and Dr. Robert F. Acker, and the additional support of The Bureau of Medicine and Surgery; without such support this book could not have been written.

Robert L. Dimmick
Ann B. Akers

Berkeley, California

March, 1969

CONTENTS

Chapter 5. Atmospheric Ions and Aerosols   100

Chapter 6. Approaches to the Bioassay of Airborne Pollution   113

PART II: AEROSOL CHAMBERS

Chapter 7. Stirred-Settling Aerosols and Stirred-Settling Aerosol
Chambers   127

PART III: ANALYSIS OF CONCEPTS AND RESULTS

PART 1

TECHNIQUES OF AEROBIOLOGY

1

MECHANICS OF AEROSOLS

Robert L. Dimmick

NAVAL BIOLOGICAL LABORATORY, SCHOOL OF PUBLIC HEALTH,
UNIVERSITY OF CALIFORNIA, BERKELEY, CALIFORNIA

The activities discussed in this book include the creation, transportation, containment, sampling, and biological assay of airborne microbes. Preliminary to such discussions is at least an elementary understanding of behavior of small particles in air. The engineering aspects of aerobiology are also based on a few mechanical principles governing behavior of small particles in air. I prefer to speak of the gross process as mechanics of aerosols rather than physics of aerosol particles, because the former implies practical aspects of a system rather than just descriptions of theoretical behavior of particles.

The term *aerosol* will mean an artificially generated or manipulated collection of particles suspended in a given body of air; this collection will be restricted to particles slightly less than 0.5 μ to slightly more than 20 μ in diameter. For reasons discussed in later chapters, the most important size range is between 1 and 10 μ. The body of air will be construed to be part of the aerosol, and the air-particle mixture will be an aerosol (in the aerosol state) as long as we can detect a single particle in the designated air. A *cloud* will refer to a visible aerosol with defined boundaries in free air. We shall not distinguish between dust (dry) and mist (wet), because biological material is inherently hygroscopic and, in the airborne state, always contains some moisture.

The literature on physical properties of airborne particulates is vast. Fortunately we do not have to consider all of the physical aspects of airborne matter. We need only consider those properties establishing

3

the number and sizes of particles that appear in our various devices or assay menstrua (including animals) at the times we wish to make a measurement. The student who wishes to study the problem beyond the scope of this book is referred to general references at the end of the chapter.

INFLUENCE OF GRAVITY

Consider a spherical particle located in a body of still air. Since the particle has mass, it will be acted upon by the acceleration of gravity and will start to fall. The mass is a function of the density and size of the particle, so velocity is directly related to these properties. As velocity increases, frictional forces of the air tend to oppose the gravitational forces. When these two forces are equal the particle will have attained a terminal, or maximal, velocity.

The equation describing terminal velocity is known as Stokes' law, and may be written

(1-1) $$\mathbf{v} = \frac{1}{18} \frac{\rho d^2 g}{\eta}, \text{ cm/sec}$$

where ρ = particle density, gm/cm^3
d = diameter, cm
g = acceleration of gravity, cm/sec^2
η = viscosity of air, gm/cm sec.

More succinctly, since viscosity changes very little within temperatures and pressures of biological significance,

(1-2) $$\mathbf{v} = 3.2 \times 10^5 \rho d^2, \text{ cm/sec}$$

Thus, a 1-μ particle (10^{-4} cm diameter) of unit density would fall at the rate of 3.2×10^{-3} cm/sec. To be accurate, we should allow for the time it takes the particle to reach this velocity, sometimes called relaxation time, but the interval is very short and not pertinent to our needs.

For particles at the upper end of our size range, 10 to 20 μ, Stokes' law holds to better than 1%. Below this size, particle diameter begins to approach the mean free path of gaseous molecules, a value of about 6.5×10^{-6} cm at ambient temperatures and pressures. The result is that particles tend to slip between air molecules and fall faster than Stokes' law predicts. The correction (usually referred to as Cunningham's correction) may be approximated by adding 0.08 μ to the diameter if we are calculating velocity, and by subtracting this value from any size

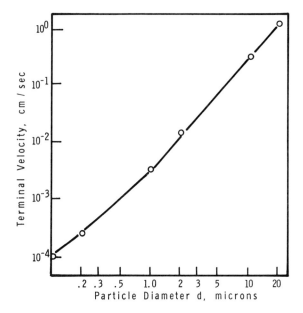

FIGURE 1-1. Terminal velocity of fall of particles in still air. Cunningham's correction applied for particles $< 2\,\mu$.

calculated by measuring rate of fall. The rate of fall of the 1-μ particle, calculated by this simple correction, is 3.48×10^{-3} cm/sec.

Obviously, the smaller the particle, the greater the correction becomes, but particles $0.5\,\mu$ and below fall so slowly that it is difficult to measure the true rate of fall. Also, the range of observed densities of airborne microorganisms is 0.9 to 1.3 gm/cm³ (Orr and Gordon, 1956) and is most commonly found to be 1.1 gm/cm³; thus, for most applications, density can be considered equal to 1.1 gm/cm³. Figure 1-1 shows the relationship between particle size of density 1.1 gm/cm³ and terminal velocity of fall.

INFLUENCE OF BROWNIAN MOTION

Airborne particles are in constant collision with air molecules, some of which may possess sufficient energy to move the particle (Brownian movement). Let us neglect gravity forces for the moment. A single particle will then be subject to erratic and random motion, which is difficult to study or predict; we can, however, talk about average behavior of numbers of particles. Over a period of time, some displacement or diffusion of particles at the edges of clouds will have occurred. Einstein derived an equation for the root-mean-square value of this displacement. In simplified terms it is:

(1-3) $\bar{X} = 5.0 \times 10^{-6} \sqrt{t/r}$, cm/sec

where t = time, sec
      r = radius, cm.

Table 1-1 compares displacement by Brownian movement to rate of fall for particles of various sizes.

Table 1-1 can be used to illustrate two points. First, it is evident from the ratio $\bar{X}/v$ that diffusion by Brownian movement is at least an order of magnitude less than the rate of fall and can be neglected in usual laboratory calculations of behavior of single particles. Second,

Table 1-1 Root-mean-square displacement ($\bar{X}$) cm/sec and terminal velocity (v) cm/sec for several particles sizes of density 1.1 gm/cm³.

Diameter, μ	$\bar{X}$	v	Ratio ($\bar{X}/v$)
20	1.5×10^{-4}	1.3	1×10^{-4}
10	2.2×10^{-4}	3.3×10^{-1}	7×10^{-4}
2	5.0×10^{-4}	1.3×10^{-2}	4×10^{-2}
1	7.1×10^{-4}	3.3×10^{-3}	2×10^{-1}
0.2	1.6×10^{-3}	1.3×10^{-4}	1×10^{-1}
0.1	2.2×10^{-3}	3.3×10^{-5}	7×10

it is possible to answer a practical question that will arise later: What size range of particles may be held in the alveoli as a result of Brownian movement? Although there is considerable room for argument, a reasonable value for the mean diameter of the human alveolus is 15×10^{-3} cm. If the distance the particle travels is assumed to be the radius of the alveolus (7×10^{-3} cm), and the holding time is assumed to be 2 sec, then, for the particle to remain in the alveolus, displacement velocity would have to be at least 3.5×10^{-3} cm/sec, a value approximately that of a particle 0.1 μ in diameter. Table 1-1 shows that particles of about 1 μ and larger, however, would fall this distance in 2 sec and we have answered the question. That is, particles less than 0.1 μ or greater than 1μ diameter would tend to be held in the alveoli to a greater extent than particles within that range. For our purposes, fallout is more important than Brownian movement; i.e., above diameters of 0.2 μ gravity forces are more important than Brownian motion. Beekmans (1965) analyzed particle retention and found essentially the same values.

We defined an aerosol as "a collection of particles," as if particles were independent. Particles may contain an electric charge, and London-van der Waals forces (forces of molecular cohesion) do cause adherence if particles contact each other or contact a surface. We assume that once contact has been made there are no forces we commonly employ that will dislodge or deaggregate them. In the laboratory we confine aerosols in various ways; therefore, we should examine the effect of this containment. Since particles are moving about in a random way as a result of Brownian movement, we should expect them to collide. Reasonably, such collisions are a function of the square of the concentration, since both particles are in motion, so

(1-4) $$\frac{dn}{dt} = -Kn^2, \text{ number/sec}$$

where K = a coagulation constant.

Equation 1-4 would be valid only if the particle size remained constant, which it would not. That is, if we assumed that once two particles collide there is no force to separate them, then even an initially monodisperse aerosol would soon consist of particles of different sizes. Larger particles formed by aggregation would have a lessened density but a greater terminal velocity than the primordial particles. It seems reasonable to suggest that at some limiting concentration the process would be inconsequential. Since the rate of coagulation will obviously decline as time passes, we would like to know what concentration of particles we can start with to assure that coagulation, as a result of Brownian movement, does not appreciably influence aerosol decay. The problem is extremely complex, but K has been derived from theoretical considerations. The value is large for small particles but approaches a limit that does not change much for particles from 1 μ in diameter, or larger; an average and useful value for K is 3.1×10^{-10} cm^3/sec. Using this value for K in equation 1-4 and a 1-sec time for dt, we can construct Table 1-2 for several values of n_0, the initial concentration. Note that coagulation is a second-order process; the longer the time, the less the rate of coagulation.

It is evident from Table 1-2 that as long as the particulate concentration does not exceed about 10^6/cm^3, coagulation by Brownian movement is negligible. We rarely exceed initial concentrations of 10^9 particles per liter of air, so that coagulation as a continuous process influencing physical loss of particles can be neglected in most studies.

Table 1-2 Values of the change in numbers (dn) per second for various initial particulate concentrations $(n_0)/$ cm.^{3}

n_0	dn/sec	t for 10% dn[a]
10^6	3.1×10^2	9 hr
10^7	3.1×10^4	50 min
10^8	3.1×10^6	5 min
10^9	3.1×10^8	30 sec

[a] From table by Hayakawa (1964).

INFLUENCE OF ELECTRIC CHARGE

Airborn particles can have an electric charge, as the well known experiment of Millikan demonstrated.

Mercer (1964) describes the mean charge per particle, $\bar{q}$, emitted from an atomizer as

$$(1\text{-}5) \qquad \bar{q} = 8.2 \times 10^{-7} d^{3/2} \sqrt{N}, \text{ ions/particle}$$

where d = diameter, μ

N = number of ions of one charge (+ or −) per cm^3 of fluid.

Distilled water has an ionic concentration[1] of approximately $10,^{14}$ Substituting in Equation 1-5 for a 2-μ particle

$$\bar{q} = 8.2 \times 10^{-7} \times 2^{3/2} \times \sqrt{10^{14}}$$
$$= 24 \text{ charges.}$$

Particles 1 μ in diameter would contain about 8 charges. As the ionic strength of the fluid increases, however, Equation 1-5 no longer applies because the charge dissipates with greater frequency; at a strength of 10^{19} the number of charges are also about 8. Small insoluble particles

[1] $(C_{H_+}) = (C_{OH_-}) = 1 \times 10^{-7}$ gm eq./liter or 1×10^{-10} gm eq./cm^3

Total ions $= 2 \times 10^{-10} \times \dfrac{6 \times 10^{23}}{18} = 0.7 \times 10^{13}$

atomized from distilled water could have larger numbers of charges than larger particles of high ionic strength. Whether bacteria or virus could be considered to be "insoluble" or not is a moot question; to my knowledge the electrical charge on biological particles has never been studied with the same vigor as inert particles have been studied.

The above is of greatest importance if the particle is in an electric field. The force developed is a function of the number of unit (electron) charges, the strength of the unit charge, and the strength of the applied field. We could substitute this force for the gravitational force in Stokes' law and determine a terminal velocity for a particle under given conditions. In a charging field of 2,000 volts/cm, a 1-μ particle could attain a maximal velocity of 2 cm/sec if the particle accumulated about 100 unit charges. We would seldom encounter fields this large or charges this high in usual laboratory experiments, but the magnitude indicates that electrical forces could be important.

Charged particles can collide at frequencies greater than expected from a consideration of Brownian movement alone. Unless a special effort is made to increase the charge on particles while they are being dispersed, or unless one deliberately holds particles in an electrical field, the effect is mostly confined to coagulation. On the surface of particles there is a continuous exchange of water molecules in equilibrium with the water vapor of the air. As a result, at relative humidities above 50%, the charge dissipates rapidly. At humidities below 30%, the charge may remain for periods long enough to be significant. Unfortunately, observed behavior in this instance does not agree in all aspects with theory, so we will not attempt to develop even approximate formulas, but rather state that the effect of undesired electrostatic precipitation depends on actual situations or conditions.

It is evident that one could collect charged particles by passing them between two charged plates. This effect is utilized in several types of samplers.[2] If equipment used to contain, transport, or collect aerosols has an appreciable electrostatic charge, then an undesirable number of particles could be lost. There is some evidence that air passing through bends in tubing can induce opposite charges on the walls. Although the effect of electrostatic precipitation is often small, and usually unpredictable, one should particularly avoid utilizing nonmetallic tubing to connect ungrounded metallic chambers, and avoid the use of plastic chambers. For example, a plastic chamber about 4 ft in diameter, tested at the Naval Biological Laboratory, acquired a static charge sufficient to attract cotton balls from a distance of 1 ft if relative humidity condi-

[2] Whitby aerosol analyzer. Thermo-Systems, Inc., St. Paul, Minn.

tions were 40% or less. Common sense dictates that one should not use a chamber of such unusual qualities. There is some rather unconvincing evidence that plastic petri dishes, with nutrient medium, can acquire a charge sufficient to cause sampling bias when they are used to collect bacteria.

INFLUENCE OF VAPORS

Vapors may be adsorbed on particle surfaces until an equilibrium is established. The mole fraction of the adsorbed vapor is a function of the surface tension of moist particles. A vapor barrier or "buffer zone" may form around the particle that can either aid or hinder ultimate contact with other particles, or may cause the particle size to change. Substances that lower the surface tension of water are selectively adsorbed on the surface (Gibbs' adsorption equation) and may increase the mass of the particle. Proteins and carbohydrates, usual constituents of nutrient media, lower the surface tension of water. Other organic substances that can vaporize and that are water soluble have surface tensions less than the surface tensions of droplets the size of bacteria, and could be selectively dissolved; hence, the particle could contain a higher concentration of vapors than expected solely on the basis of solubility laws. The importance of this effect to aerobiology is open to question, and there is a dearth of information on the subject.

Hayakawa (1964) reported an interesting coagulation effect. He used ammonium chloride aerosols and found that significant coagulation occurred in the presence of either formic or acetic acid vapors. The effect was not detectable at vapor concentrations less than 1,000 ppm, a far higher concentration than one expects in nature. But formaldehyde and other vapors are often used to decontaminate equipment. In insufficiently ventilated chambers residual vapor concentrations could occur, and unless sensitive physical measurements were conducted, coagulation by vapor action might be mistaken for biological effects.

INFLUENCE OF RADIATION AND TEMPERATURE

Two other processes that are essentially diffusive in nature should be mentioned in passing. One is photophoresis, that is, radiant energy can cause particles to move by producing differential thermal effects. Easily measured effects are noted only on particles much smaller than 0.5 μ. Brock (1968) has defined another phenomenon called *photodiffusiophoresis* caused directly by "molecular diffusion velocities."

If a particle is warmed more on one side than on the other as a result of irradiation, the escape velocity of molecules on the warmer side will be greater than on the cooler side, and the particle will move. Transparent particles may act as lenses to heat the side opposite the irradiation source and to cause the particle to migrate toward the source.

If a particle is in air where a temperature gradient exists, molecules will strike the particle from the warmer side with greater velocity than from the cooler side, and the particle will move; i.e., particles tend to migrate toward cold surfaces. This is the principle of the thermal precipitator. Airborne microbes can be collected by this method, but the technique has not been widely used.

One would expect the kinetic effect of warm molecules to be large, since one commonly finds quantities of dust collected on cold household surfaces, near warm pipes, etc., but these are aggregates of very small particles. Terminal velocity is proportional to the temperature difference across the particle, but if the particle diameter is appreciably greater than the mean free path, γ, of air molecules, then velocity is proportional to $\gamma^2 d$, whereas if the particle is smaller than that, the velocity is proportional to $d^2\gamma$. Figure 1-2 illustrates this situation with other parameters constant, and shows that at temperature differentials

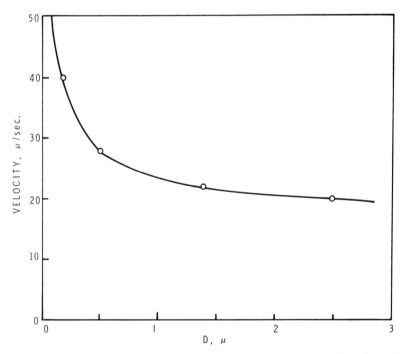

FIGURE 1-2. Terminal velocity of particles of diameter d, (μ) in a thermal gradient of 20°C, 760 mm pressure.

normally encountered, migration of particles greater than 1 μ is inconsequential.

INERTIAL PRECIPITATION; IMPACTION

We have been speaking of aerosols as if the particles were suspended only in still or tranquil air. Almost all air is in motion. Even in closed containers, thermal gradations cause sufficient differences in air density to create air currents with velocities approaching, and often exceeding, those of Stokes' velocity of fall. It is convenient to think of a particle as being suspended in an indefinite volume of air and to talk of the motion of this particle with respect to that volume rather than to some external point. The motion of the air volume is with respect to an external reference point and may be said to be part of an air stream. The particle gains and loses kinetic energy as a result of the air movement, and the particle velocity with respect to the volume can be described by Equation 1-1. If g is replaced by a, which will be any acceleration other than g, then

(1-6)
$$v = \frac{1}{18} \frac{\rho d^2 a}{\eta}, \text{ cm/sec}$$

Some acceleration will be imparted to the particle as a result of velocity changes of the air volume, but the acceleration we are more concerned with is that caused by directional changes. In this case acceleration is defined by:

(1-7)
$$a = \frac{V^2}{R}, \text{ cm/sec}$$

where V = air velocity, cm/sec
R = radius of curvature, cm.

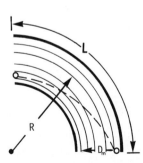

FIGURE 1-3. Particle crossing air streams in the bend of a pipe. D_m is the distance the particle has travelled across the air streams.

We may now ask a very practical question. Referring to Figure 1-3, would a 1-μ particle of density 1 migrate from near one wall of a $\frac{1}{4}$-inch[3] pipe to the other wall, and thus be removed if we were conducting air through a 2-inch bend in the pipe at the rate of 10 liters per minute? First, find the linear velocity of air.

(1-8)
$$V = \frac{F, \text{ cm}^3/\text{sec}}{\pi(D/2)^2, \text{ cm}^2}$$

where V = linear air flow, air velocity, cm/sec
F = volume air flow, cm^3/sec
D = pipe diameter, cm.

Substituting,

$$V = \frac{10^4, \text{ cm}^3/60, \text{ sec}}{3.14 \times (0.3)^2, \text{ cm}^2}$$

or approximately 600 cm/sec.

By equation (1-6) solve for particle velocity:

$$v = \frac{1 \times 1 \times (10^{-4})^2(600)^2/5}{18 \times 1.8 \times 10^{-4}} = 0.22 \text{ cm/sec}$$

Referring again to Figure 1-3, time in the pipe bend is:

(1-9)
$$t = \frac{L}{V} = \frac{\frac{1}{2}\pi R}{V}, \text{ sec}$$

where L = length of the bend,
and so,

$$t = \frac{\frac{1}{2}(3.14)(5.08)}{600} = 0.013, \text{ sec}$$

Distance particle will travel is simply

$$D_m = vt = 0.22 \text{ cm/sec} \times 0.013 \text{ sec} = 0.0029 \text{ cm}$$

Such a particle would not be precipitated under these conditions; it would, however, migrate out of its volume and enter new air streams.

Performing the same calculation for a 10-μ particle, we find $v = 22$ cm/sec and distance traveled = 0.29 cm. This size particle would be impacted onto the walls with an efficiency of about 50%.

---

[3] One has to learn to untilize commercially available equipment which is specified in inches, etc.

Again, the question is not quite so simple. Consider the following interesting dimensional identity:

(1-10) $\dfrac{\eta,\ \text{gm/cm sec}}{\rho_a,\ \text{gm/cm}^3} = v,\ \text{cm/sec} \times r,\ \text{cm} = \text{cm}^2/\text{sec}$

where ρ_a = density of the air
and r = particle radius

The ratio $n/\rho_a = 0.15$, and whenever vr approaches 0.15, the air flow around the particle is found to be turbulent—the particle moves erratically. Equation 1-10 is commonly written:

$$\text{(Reynolds number)}\ \ \text{Re} = \frac{\rho_a vr}{\eta}$$

which is dimensionless, and when Re approaches 1, turbulent motion can occur. Considering the 10-μ particle above, at $v = 22$ cm/sec, Re = 0.1–indicating a minimal turbulent process as the particle crosses air streams; but a 20-μ particle under these conditions would yield a Reynolds number of greater than 1 and the particle motion would be unpredictable. There is an additional source of turbulence to be considered; i.e., turbulence caused essentially by shearing forces between stream lines. This relationship is expressed by another Reynolds number which we will call Re_f:

(1-11) $$\text{Re}_f = \frac{\rho_a VD}{\eta}$$

where V = air velocity cm/sec
D = pipe diameter, cm

When this number begins to approach 2 to 3 $\times$ 10³, air turbulence may be expected. In the case of flow through the bend in the pipe, above, Re_f equals about 2.4 $\times$ 10³, definitely within the turbulent region. Hence, additional numbers of small particles could be impacted and some larger particles might even negotiate the bend. Secondary flow patterns, which are difficult to analyze, can occur under certain circumstances. If the pipe size were changed to $\frac{3}{8}$ inch, then Re_f would be about 1.8 $\times$ 10³ and flow would be nearly laminar. Bends employed in the usual aerobiology apparatus, and usual circumstances, remove only the very large particles. It is advisable, however, to avoid the use of sharp bends at high flow-rates. The process should be tested rather than analyzed by theory (see Chapter 9).

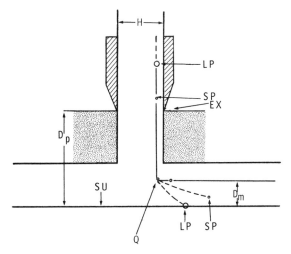

FIGURE 1-4. Idealized concept of jet impaction. H = width of jet; LP = large particle; SP = small particle; D_m = stopping distance; S_u = surface; D_P = distance, slit to surface; Q = point of change of motion.

The principle of imparting a velocity to particles across the lines of the air flow so that they will impact onto surfaces is employed in the most common types of samplers. Details of their construction and application will be discussed later, but we are armed now to consider the principles involved. We can attain high velocities (more than 5×10^3 cm/sec) by drawing air through a slit (area less than 0.1 cm^2) and we can achieve rapid change of air direction by having a surface, either liquid or solid, near the slit as shown in Figure 1-4.

If we return to our concept of a particle acting in a limited volume of air, we can see that, by suddenly changing the direction of the air, the inertial force of the particle will cause it to move out of the volume or to cross the air streams, just as in the bent pipe example. The distance it moves, D, will be equal to the terminal velocity, v, times the time t, which can be written,

(1-12)
$$t = \frac{D}{v}, \text{ sec}$$

Also, the time the particle is in the air stream is given by equation 1-9. The times are equal; hence,

$$\frac{L}{V} = \frac{D}{v}, \quad \text{and} \quad v = \frac{DV}{L}$$

so

$$\frac{DV}{L} = \frac{1\rho d^2(V^2/R)}{18\eta}$$

from Equation 1-6.

Transforming, the Equation becomes

(1-13) $D_m = \dfrac{\rho d^2 V}{18\eta} \times \dfrac{L}{R}$ = Sinclair's "stopping distance"

L/R is dimensionless and is equal to 1.56; i.e., the ratio of 1/4 circumference to the radius.

If we think of the bend as becoming smaller and smaller, as in a true right-angle, we see that both L and R approach the size of the particle; the vector forces become principally those of the original velocity vector and L/R approaches 1, a value equivalent to the empirically derived constant of Ranz and Wong (1952). Furthermore, if we divide through by D, the equation becomes dimensionless and equal to their inertial parameter, ψ.

To make this transformation I have assumed that the dimensions of slit width (in Figure 1-4) and distance between slit exit and the impaction surface are equal. Mercer (1964) shows that, as the ratio approaches 1, the impaction parameter becomes maximal. He also notes that the impaction parameter law holds for square or rectangular jets, but not for round jets; at equivalent dimensions, ψ is smaller for round jets than square ones.

It is important, however, to visualize what happens during the act of impaction rather than to justify this simplified approach. Papers in the included references discuss precise mathematical concepts in more detail. In Figure 1-4 I have shown stream lines of air and two particles, a large one (*LP*) and a small one (*SP*), in the same stream line. In this example, the stopping distance for the large particle exceeds the distance (*D*) for that stream line and is shown impacted, whereas the stopping distance for the small particle was less than *D*, and the particle remained in the air but crossed the stream lines. D_m is related to slit width, H, and is the maximum distance any particle would have to travel. The surface S_m is some finite distance D_p from the slit exit. This distance can be from 1 to 4 times H before the stream lines begin to deviate markedly. Biological efficiency, however, is influenced by H (Goldberg and Shechmeister, 1951). As a result of this separation, D_p, there is an ill-defined volume of air, more or less turbulent, that is shown in the figure by a grey area. Air from the slit expands into this volume so that the change of motion at point Q is not a precise right angle for all stream lines. In fact, the entire flow is turbulent, and measurements of behavior are based on mean particle size and mean impaction distance; results only approach the theoretically predicted behavior.

It is evident that large, heavy particles with a stopping distance greater than D_m will be impacted, regardless of their position in a particular air stream, and almost all small particles with short stopping distances will pass the surface. It is also evident that there exist particles of such size that the chance of impaction is equal to the chance of passage; this size is termed the characteristic diameter for specified jet dimensions under specified conditions; the principle is employed in particle size analysis.

The characteristic diameter may be determined by counting and sizing the impacted particles, and plotting the cumulative data versus particle size, as shown in Figure 1-5. The same shaped curve is found if the inertial parameter, ψ, of Ranz and Wong, (1952), is plotted versus efficiency. The ideal jet would impact only a small range of sizes included by the curve shown in Figure 1-5, whereas in reality a spread of ±10% of the characteristic diameter is considered to be very satisfactory. Again, an exact analysis of all the factors involved and their extent of involvement is neither easy nor practical, but one point is evident. That is, it is the Stokes' velocity, or rather the apparent particle size as measured by the terminal velocity (inertial parameters), that determines whether a particle will fall through a given distance (cross stream lines) to be impacted or to fall out in small spaces such as an alveolus.

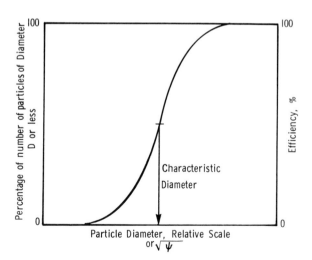

FIGURE 1-5. Cumulative impaction efficiency of a jet operated at a constant flow rate.

A final example of impaction as related to air filters will be considered, although applications will be discussed in Chapter 10. We will not attempt complete analysis here, but instead refer the reader to the review article by Chen (1955). Consider the cross-section of a fiber of diameter Y and the stream lines around it as shown in Figure 1-6. The streams tend to bend around the fiber so that streams approaching the periphery are diverted. In the outer streams, particles that have a stopping distance less than D_m will not be impacted on the fiber. For a given condition there is an effective fiber diameter, y, that is smaller than the real diameter.

The air flow we are discussing here is much less than flow through impactor jets, so it might be supposed that small particles are not trapped easily by fibers. In fact, increasing the flow rate decreases the impaction efficiency of particles about 1 μ in diameter, because at slower rates some particles can approach close enough for "Brownian impaction," or electrical and van der Waal's forces can cause their capture. Since the stopping distance is so short, particles are simply swept around the fiber at higher velocities; in effect, y approaches 0. Particles that equal the size of the fiber or larger are easily captured. An approximate relationship of y/Y to the parameter ψ is shown in Figure 1-7. In this case, ψ should be thought of as the extent to which impaction forces operate. Of course, filters are collections of fibers. Within the interstices, particles can be caught in relatively still air, to settle by Stokes' velocity and be captured by forces noted above. Considering these principles, an ideal filter ought to be composed of large fibers with the spaces filled in with small fibers to provide maximal air velocity but minimal, internal air spaces. This is indeed the principle of absolute filters.

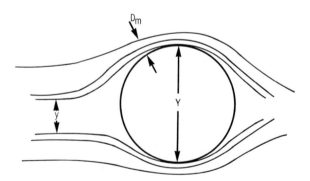

FIGURE 1-6. Streamlines of flow around a fiber. Y = fiber diameter; y = effective fiber diameter; D_m = stopping distance of a particle.

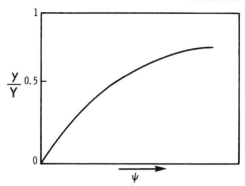

FIGURE 1-7. Relationship between the impaction parameter, ψ, and the ratio of effective diameter, y, to a real fiber diameter, Y.

ACOUSTICAL FORCES

Sound waves can cause particles to move and, if the intensity is high enough, will enhance the probability of collision. There are two mechanisms involved: sound waves have both frequency and intensity (amplitude). The velocity imparted to particles as a result of momentary changes in air pressure can move them at greater velocities than Brownian motion, and since different sized particles will be moved at different velocities, their chance of collision is improved. Furthermore, standing waves or low pressure nodes can occur, and particles tend to migrate toward such points. Distance between particles is decreased in the nodes and the chance of collision is increased. To study acoustical effects it has been necessary to employ sound levels approaching 0.2 watt per cm^2; at this level personnel would need protection, so we can safely say that acoustical forces are not important for our purposes.

SUMMARY

We are concerned with airborne particles that might contain organisms. An aerosol is a collection of particles in air. The most important behavioral aspect of aerosols results from the inertial property of particles. This property establishes the speed a particle will fall, and causes the particle to cross streams of turbulent air to travel for defined distances. The effect of a change in direction of air streams is to impart a momentary acceleration that can cause a particle to collide with other particles or surfaces.

Sometimes the inertial property limits our ability to study aerosols, it causes some particles to be deposited in the lungs when we breathe, and in general it sets the pattern whereby particles gain access to many areas we may prefer they not enter, such as food containers, open wounds, bodies of water, and vehicles intended for use on other planets. Sometimes this property is advantageous because (a) it allows us to collect airborne particles by impacting them on surfaces where they can be examined or incubated to test for presence of life, (b) it provides a means whereby we can filter air required to be free from particles, and (c) it causes many particles to be removed when air is drawn through the nose and throat.

Other actions that can cause particles to move include the thermal energy of molecules (Brownian motion), electrically charged fields, acoustical energy, and radiant energy. If we consider particles in the range of 0.5 to 20 μ, and if we consider only a concentration of particles such that random collisions are infrequent, then these actions, or forces, are so much less than inertial forces that they can be neglected in usual circumstances.

GENERAL REFERENCES

DallaValla, J. M. 1948. *Micromeritics. The Technology of Fine Particles* (2nd ed.). Pitman Publishing Corp., New York and London.

Davies, C. N. (Ed.) 1961. *Inhaled Particles and Vapours*. Symposium Publications Division, Pergamon Press, New York.

Davies, C. N. (Ed.) 1966. *Aerosol Science*. Academic Press, London and New York.

Drinker, P. and Hatch, T. 1936. *Industrial Dust. Hygienic Significance, Measurement and Control*. McGraw-Hill Book Co., New York and London.

Fuchs, N. A. 1964. *The Mechanics of Aerosols*. Pergamon Press, Oxford and The Macmillan Co., New York.

Green, H. L. and Lane, W. R. 1964. *Particulate Clouds: Dusts, Smokes and Mists* (2nd ed.). E. and F. N. Spon, Ltd. London.

Gregory, P. H., and Monteith, J. L. (Ed.) 1967. *Airborne Microbes*. Seventeenth Symposium of the Society for General Microbiology. Cambridge University Press, London.

Orr, C., Jr. 1966. *Particulate Technology*. The Macmillan Company, New York.

Rosebury, T. 1947. *Experimental Airborne Infection*. Williams and Wilkins Co., Baltimore, Md.

Wells, W. F. 1955. *Airborn Contagion and ,Air Hygiene*. The Harvard University Press, Cambridge, Mass.

White, P. A. F., and Smith, S. E. (Eds.) 1964. *High Efficiency Air Filtration*. Butterworths and Co., Ltd., London.

REFERENCES

Beekmans, J. M. 1965. Deposition of aerosols in the respiratory tract. I. Mathematical analysis and comparison with experimental data. *Can. J. Physiol. and Pharmacol.*, **43**: 157–172.

Brock, J. R. 1968. Some new modes of aerosol particle motion: Photodiffusiophoresis. *J. Phys. Chem.*, **72**: 747–749.

Chen, C. Y. 1955. Filtration of aerosols by fibrous media. *Chem. Rev.*, **55**: 595–623.

Goldberg, L. J., and Shechmeister, I. L. 1951. Studies on the experimental epidemiology of respiratory infection. V. Evaluation of factors related to slit sampling of airborne bacteria. *J. Infect. Diseases*, **88**: 243–247.

Hayakawa, I. 1964. The effects of humidity on the coagulation rate of ammonium chloride aerosols. *A.P.C.A. Journal*, **14**: (9) 339–346.

Mercer, T. T. 1964. Aerosol production and characterization: Some considerations for improving correlation of field and laboratory derived data. *Health Phys.*, **10**: 873–887.

Orr, Clyde Jr. and Gordon, M. T. 1956. The density and size of airborne *Serratia marcescens. J. Bacteriol.*, **71**: 315–317.

Ranz, W. E., and Wong, J. B. 1952. Jet impactors for determining the particle size distribution of aerosols. *Arch. Ind. Hyg. Occup. Med.*, **5**: 464–477.

Sinclair, David. 1950. In *Handbook on Aerosols.* Superintendent of Documents, U.S. Government Printing Office, Washington, D.C.

2

PRODUCTION OF BIOLOGICAL AEROSOLS

Robert L. Dimmick

NAVAL BIOLOGICAL LABORATORY, SCHOOL OF PUBLIC HEALTH,
UNIVERSITY OF CALIFORNIA, BERKELEY, CALIFORNIA

Almost every human activity tends to create airborne bacteria from one source or another. When we cough or sneeze, we produce biological aerosols. Surprisingly, for periods close to half an hour after taking a shower, humans shed bacteria into the air at concentrations much higher than at other times, and even the wearing of clothes does little to alleviate the extent of dispersion (Williams, 1966). Fortunately, the predominant mass of such aerosols is in particles too large to reach the innermost spaces of the lung—they are impacted in the upper respiratory tract (Drinker and Hatch, 1963; Hatch and Gross, 1964).

Because some particles are small enough to reach possible sites of infectivity, the potential for respiratory disease is always present, particularly in the case of "carriers" (Williams, 1966). Our problem in the laboratory is to attempt to duplicate in terms of size, biological content, and concentration, natural aerosols which arise from a variety of known and unknown sources. In some respects, this is easy; in others, it may be impossible. Extensive studies have been made of dispersion principles and methods from an engineering standpoint (Orr, 1966), but few of these studies are directly applicable to the above problem.

If we divide 1 cm^3 of liquid into 1-μ particles, we change the surface area from 6 cm^2 to roughly 3×10^8 cm^2. It is evident that a considerable amount of energy would be required to do this. Many kinds of energy can be applied to break the volume into pieces: electrical energy of suffciently high potential will cause a water stream to break into small particles; explosive forces will disrupt liquids or solids; sonic (vibrational) energy can be applied to contained liquids to produce clouds; shear forces can be applied by propelling liquids through jets, by dispersing liquids from spinning disks, or by exposing the liquid surface to high velocity air streams. We might even drop the volume and let it splash, observing from this procedure that whatever energy we use must be applied equally to the entire volume, either instantly or to parts of it constantly. The high velocity air stream into which liquid can be fed is the most simple and effective method and is the one we will be primarily concerned with.

JET SPRAYERS

The simplest atomizer, or sprayer, is like the right-angle arrangement shown in Figure 2-1a, which we will use as a model to describe how droplets are formed. The immediate effect of the air stream is to deform the emerging liquid, as pictured in Figure 2-1b. The distorted portion

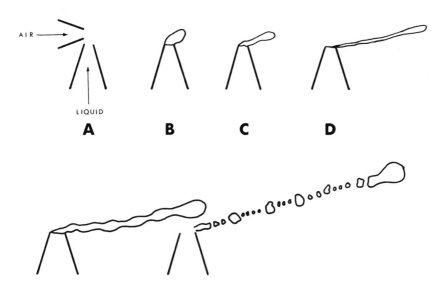

FIGURE 2-1. Steps in the breakup of a liquid stream by an air jet.

then becomes elongated (2-1c) and drawn into a thread (2-1d). Turbulence causes waves on the surface of the thread as it elongates (2-1e), and finally the thin portions separate into beads (2-1f) of a variety of sizes (Green and Lane, 1964). A more efficient arrangement is to locate the air jet centrally to the liquid. The dispersion principle is essentially the same in such peripheral-type atomizers, and also in spinning disks, except that thin sheets of liquid are formed, rather than threads.

Factors that influence the ultimate size, and size distribution, of particles emerging from a variety of different types of jets have been studied by numerous scientists, but the work of Nukiyama and Tanasawa (1939) is the most useful. They derived an empirical formula that describes the output of jets in terms of a surface mean diameter. Their formula includes the effects of viscosity, surface tension, liquid density, and air and liquid flow on the expected diameter. Of more immediate interest is the very practical analysis of jet design and operation published by Frazer and Eisenklam (1956). The results of their studies and those of Lewis, et al. (1948), show that the distribution of particle sizes from either twin-fluid peripheral jets (Figure 2-2) or right-angle jets (Figure 2-3) is quite broad, and that mass median diameters of such aerosols

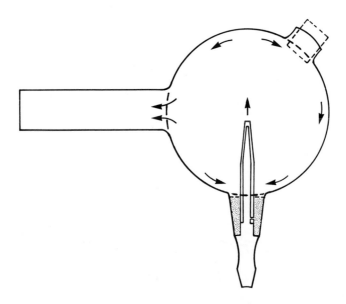

FIGURE 2-2. Cross section of an all-glass, Wells-type atomizer as used by DeOme et al. (1944).

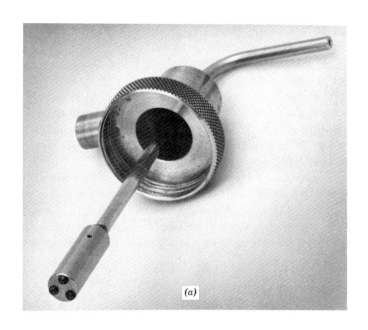

(a)

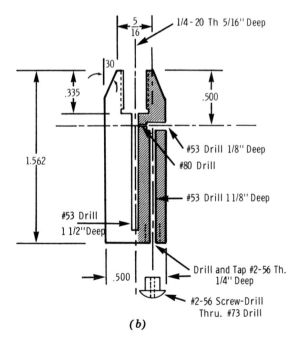

5/16

1/4 - 20 Th 5/16" Deep

30

.335

.500

1.562

#53 Drill 1/8" Deep

#80 Drill

#53 Drill 1 1/8" Deep

#53 Drill
1 1/2"Deep

.500

Drill and Tap #2-56 Th.
1/4" Deep

#2-56 Screw-Drill
Thru. #73 Drill

(b)

FIGURE 2-3. The Collison atomizer. (a) Module that can be used with screw-cap jars. (b) Cross section of the atomizer head. Only one of three jets shown.

vary from 50 μ to more than 150 μ. Mercer, *et al.* (1968), found the median diameter to be about 39 μ. In fact, more than 95% of the mass output from such atomizers is composed of particles larger than we wish to include in our studies.

We have a small dilemma. One possible answer is to increase the air to fluid volume ratio (the major factor that influences particle size) by making orifices smaller or increasing air flow. If we utilize smaller orifices, there is a reasonable chance the opening will clog with the biological fluids we employ; furthermore, the output will be low (Fuchs and Sutugin, 1966). If we increase air flow, we must increase air exchange in whatever chamber is used to contain the aerosol, and this could prevent particulate build-up or could exceed desired flow rates in ducts. Some jets of increased efficiency are illustrated by Frazer and Eisenklam (1956), but even these designs do not approach the size range we need.

ATOMIZERS (NEBULIZERS)

Another answer is to remove larger particles by impacting them on baffles, or to position jets in small containers that act as baffles. In fact, we would do this even if smaller particles could be created without baffles because we must confine infective aerosols within essentially gastight equipment. It is convenient to have the spray device external to ducts or chambers, and this can be done only by confining the jet and delivering the aerosol through tubing. Such devices are often termed "nebulizers"; the three most commonly employed by aerobiologists are the DeVilbiss nebulizer (The DeVilbiss Company, Somerset, Pa.), the Wells-type atomizer (DeOme *et al.*, 1944) (Figure 2-2), and the Collison-type atomizer (Green and Lane, 1964) (Figure 2-3). These units operate with pressure drops across the jet from 5 to 30 psi, but practical pressure required is 6 to 12 psi. All are reflux atomizers; i.e., impacted fluid is returned to a reservoir and reatomized. The recently described Lovelace nebulizer (Mercer *et al.*, 1968) may prove to have utility for the aerobiologist, though no biological data are yet available. Dautrebande and Walkenhorst (1964) designed a disperser to produce particles < 0.5 μ by a stripping action.

The true size distribution of particles directly produced by jets is difficult to predict, for this depends on (*a*) dimensional factors and operating conditions, (*b*) the possibility of particles colliding, (*c*) potential creation of satellite droplets from large masses of splashing liquids, (*d*) viscosity and density of the fluid, and (*e*) concentration of solutes. The apparent distribution depends especially on the method used to measure particle size. For example, since aqueous particles evaporate, it is practically impossible to obtain a representative sample from air within, or just emerging from, such atomizers and to observe particles under the microscope without changes occurring. The lifetime of pure water particles near 1 μ in diameter is less than $\frac{1}{2}$ sec in conditions where air is not saturated, whereas 15 μ particles might exist for 200 sec; hence, not only size, but distribution of sizes would change.

Distance of the sampler from the jet would influence the size of the particles collected. Lane and Edwards (cited in Green and Lane, 1964) investigated the problem using dibutylphthalate, a liquid with a low vapor pressure. They found the mass median diameter of particles emerging from the Collison atomizer to be about 3 μ with very few particles greater than 15 μ. Their findings also showed that splashing against walls or baffles did not add to the break-up of liquid to form additional small particles, but this may not be true with aqueous solutions because of viscosity differences. Mercer *et al.* (1968), using an ingenious, adjustable baffle principle previously reported by Wright (1958), show that maximum output occurs when jet-to-baffle distance is about 0.015 inch. Again, a biological effect, if any, has not been investigated.

DROPLET EVAPORATION

If the liquid to be dispersed contains dissolved solids, particles will not completely evaporate, but will be reduced to a new size. Green and Lane (1964) have reported the equation for evaporation to be:

$$(2\text{-}1) \qquad \bar{d}_f = \bar{d}_0 \sqrt[3]{\frac{s\rho_0}{100\rho}}$$

where s = gm of solute/100 gm solution
ρ_0 = density of solution
ρ = density of solute
$\bar{d}_0$ = average diameter, initial
$\bar{d}_f$ = average diameter, final.

Because of the complex nature of bacteriological medium we cannot always know ρ, but a suitable estimate would be the measured density (1.1) of particles formed in equilibrium with most values of relative humidity (RH) as reported by Orr and Gordon (1956). We will assume an average value of 2% for total dissolved solids in common spray fluids and that the density of such solutions is 1.0. The evaporative factor,

$$\sqrt[3]{\frac{s\rho_0}{100\rho}} = \sqrt[3]{\frac{2 \times 1}{100 \times 1.1}} = \sqrt[3]{0.02 \times 0.9}$$

then becomes approximately 0.26.

This solution for the evaporative factor is an approximation—not because of the arithmetic—but because the formula is not entirely realistic. The diameters should reasonably be considered as average diameters, because we do not know what conditions actually exist in particles of evaporated biological material. Problems of supersaturation, surface tension, internal particle pressure, selective adsorption, hygroscopicity and selective crystallization—all should influence the final size; and all interfere with the possibility of using a simple equilibrium calculation such as Raoult's law to arrive at a more theoretically satisfying answer. Cox (1963) discussed these factors in detail. Kethley et al. (1957) are the only ones to have investigated this problem from the viewpoint of the aerobiologist. They exposed several "biological fluids" to known humidities and determined the water content at equilibrium. They then calculated the expected new size of a 13-μ particle, which was the average size of particles initially formed in their atomizer, and compared calculated to measured sizes at various humidities; agreement was close in most instances, and their reported average diameters of "dry" particles ranged from 1.8 to 4.2 μ, which is very close to the size reported by Mercer et al. (1968).

Equation 2-1 does not consider the influence of relative humidity on particle size at equilibrium. In effect, the ratio ρ_0/ρ is equivalent to water activity, and hence to relative humidity. Substitution in Equation 2-1 yields

$$(2\text{-}2) \qquad \bar{d}_f = \bar{d}_0 \sqrt[3]{\frac{s}{100} \text{RH}}$$

Table 2-1 shows data from Kethley et al. (1957) compared to values calculated by Equation 2-2. The difference ratio is about 1.5, and this should serve as an empirical correction factor for beef extract fluids, but not for other materials. The correction factor is not exact because the observed moisture contents at equilibrium were not directly related to water vapor pressure—except in the range of 30 to 80% RH. Within this range, Kethley's percentage estimates of water content of beef extract are the same as values reported by Gillespie and Johnstone (1955) for phosphoric acid aerosols. This may or may not be the case with

Table 2-1   Comparison of two methods of calculating an equilibrium particle size

Relative Humidity	Diameter calculated by equation 2-2	Reported[a] diameters	
		A	B
20%	$1.1\,\mu$	$1.8\,\mu$	$1.80\,\mu$
40	1.4	2.0	1.85
60	1.6	2.1	2.0
90	1.8	2.5	2.5

[a] Data from Kethley et al. (1957).
A observed
B calculated
Data based on average initial diameter of 13 μ and 0.3% beef extract in water as atomizer fluid.

most other materials, so we are left with the disconcerting conclusion that, until more is known about small particles in equilibrium with water vapor, we have no generally applicable evaporation factor; each case is unique. We shall use Equation 2-1 for calculations to follow, considering it as a suitable starting-point from which more exhaustive studies could be initiated.

SIZE DISTRIBUTION (VARIANCE) OF PARTICLES

The best available data of size distribution of droplets initially formed by jets, and where actual counts are given, are those of Lewis et al. (1948). If we use their data and assume that jet enclosures remove all particles above 20 μ, we can construct Table 2-2. Referring to Figure 2-4, we find the mass median diameter for initially formed particles to be about 10 μ. This value agrees approximately with data shown by Green and Lane (1964) for typical distributions, but not at all with other data from unenclosed jets.

Several distributions of particles from unbaffled jets are shown by Frazer and Eisenklam (1956). All show, on the basis of volume, that considerably less than 1% of the particles is in the range of 20 μ or below, but smallest particles greatly outnumber the largest particles found (Friedlander, 1965). Lane and Edwards (cited in Green and Lane, 1964) measured the mass output of a Collison atomizer and found it to be about 0.17%, based on total flow through the jet. The value is

Table 2-2   Cumulative distribution of particle sizes[a]

| Original particle size | Number of particles | Initial values | | | d_f |
		No.	Relative mass	Cumulative mass	
1μ	3.9 × 10⁵	41.4%	<1%	<1%	0.26 μ
5	3.4 × 10⁵	36.1	10	11	1.3
10	1.6 × 10⁵	17.0	37	48	2.6
15	4.0 × 10⁴	4.25	34	82	3.9
20	1.0 × 10⁴	1.09	18	100	5.2

[a] Data from Lewis et al. (1948).
d_f is the size the particle would attain in equilibrium with air at about 50% relative humidity.

not directly applicable to a Wells atomizer, but it is of interest, since it was found not to change appreciably when several different operating conditions were employed. Evidently, increasing the air flow in a refluxing nebulizer also increases the liquid flow, but the mass output remains relatively constant. Fraser and Eisenklam (1956) state that increased air flow decreases the particle size, but Mercer et al. (1968) indicate that the effect is small (e.g., 3.8 μ at 10 psi; 2.4 μ at 30 psi). Since that the effect is small e.g., 3.8 μ at 10 psi; 2.4 μ at 30 psi). Since

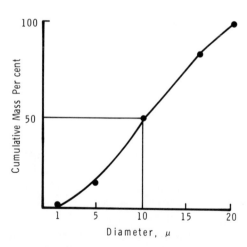

FIGURE 2-4. Cumulative distribution of particles less than 20 μ initially formed by a twin-fluid jet. Data from Lewis et al. (1948).

the mass-fraction containing 20 μ or less is so small, slight changes in the initial mass median diameter, or the variance of the distribution, can have a large comparative effect on the particle size output of a nebulizer. That is, if the mass variance could be increased by use of less efficient jets, and without changing the mass median diameter, the mass containing smaller particles would be increased. Hence, jets that could produce a wide distribution of sizes are probably more effective for our purposes than those of decreased efficiency. This is speculation, however, for I know of no comprehensive study of jet design specifically related to aerobiological needs. The answer, for the student, is to measure the output parameters of the disperser he intends to use.

Atomizer Output

The most useful and direct way to measure the output of cells from a nebulizer is to stain cells, disperse them from distilled water, collect them in an impinger, and measure the concentration by photometric means (nephelometry). If an accurate measurement of the percentage recovery of viable cells is required for an experiment, this value—an effective output—must be determined for the particular nebulizer employed.

At the Naval Biological Laboratory we have employed photometric techniques to measure the output of several Wells-type nebulizers and found values from 0.02 to 0.05 ml/min. Because these measurements were conducted with cells suspended in distilled water, the value deviates slightly from total mass output values.

In many experiments, a highly accurate determination is not required, and we usually accept a value of 0.025 ml/min for dispersers of the same general type operated at 10 psi. If the density of the microbial cell is 1.1 and the flow of liquid through the jet is 30 ml/min (an observed value), then efficiency, as calculated by Lane and Edwards (cited in Green and Lane, 1964), becomes 0.09; i.e., 10% less than their reported value of 0.10:

$$\frac{\text{aerosol output (volume} \times \text{density), gm}}{\text{liquid throughput, gm}} \times 100 = \frac{0.025 \times 1.1}{30} \times 100$$

$$= 0.09$$

(Cells assumed to be in pure distilled water.)

Another way to determine the output for purposes of calculating expected aerosol concentrations is to measure the change in volume or mass of the nebulizer fluid. Values (apparent output, O_a, ranging from 0.13 to 0.20 ml/min have been found for various refluxing units. Of course, most of this loss is caused by the evaporation of water. Consequently, solid content as well as particle density/ml increase in the atomizer fluid, but the effect is not important for spray times less than 5 minutes.

The true percentage output (observed mass output/volume loss $\times$ 100) varies from 12 to 20%; if initial survival (recovery) is to be determined, the true percentage output must be measured. The equation for calculating the number of cells dispersed then becomes

$$\frac{\text{number cells}}{\text{ml}} \times \frac{\text{atomizer output, ml}}{\text{min}}$$

$$\times \text{ time, min} \times (O_a) = \text{number of cells dispersed, } N_0$$

We can summarize at this point by saying that there are few definitive data on the actual size and distribution of particles initially formed from biological suspensions refluxed in nebulizers. We can safely say that, with reflux atomizers, the mean size formed and capable of emerging is between 5 and 15 μ, or a final equilibrium size (d_f) of 1.3 to 4.9 μ. In a properly operating reflux atomizer the mean size varies from 1.2 to 2.8 μ. Whether a more precise estimate is needed depends on the objectives of the experiment.

DISTRIBUTION OF CELLS IN PARTICLES

We should now consider the distribution of particulate matter in droplets emerging from the disperser. It is obvious that droplets might contain numbers of cells or virus particles ranging from none to the number that could be packed into a selected droplet volume. Since we have a variety of sizes, the distribution of which is not known precisely, and a variety of numbers of cells per particle, we can talk only about average values. Let us assume that bacteria in the disperser fluid were distributed according to the Poisson distribution. To calculate the expected probability of a cell being in a droplet, we need to know the volume of the droplet and the number of cells per milliliter. Figure 2-5 shows volumes of representative particle diameters, from which suitable values can be obtained. The volume times the number of cells per milliliter yields the expected probability, and from published tables of Poisson distributions we can obtain the expected number of cells within given particle sizes. Representative data are shown in Table 2-3. Let us assume that the cumulative distribution of particle sizes shown in Table 2-2 is representative of distributions we would usually encounter. Indeed, the number distribution is approximately log-normal, as Mercer et al. (1968) found, except for the smallest particle size. I suspect the reason

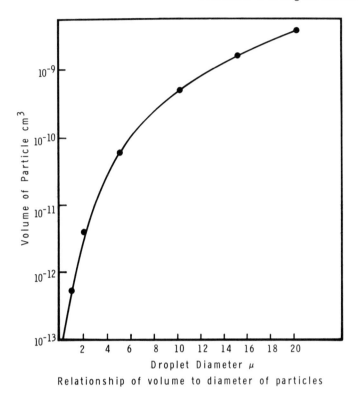

Relationship of volume to diameter of particles

FIGURE 2-5. Relationship of volume to diameter of particles 1 to 20 μ diameter.

that the smallest reported particles are fewer in number than expected is that, by present methods, they were more difficult to collect and count in the presence of larger particles. No $0.1\text{-}\mu$ particles were reported, yet they should have been formed in countable numbers.

We can combine the Poisson with the reported distribution by multiplying the percentage of a give size by the probability of having one, two, or more cells in the particle, and construct the somewhat unorthodox distribution curves shown in Figures 2-6, 2-7 and 2-8. The curves tell us, in a subjective way, how the distribution of numbers of cells in the various particles change as a function of concentration of cells in the atomizer fluid.

One sees that as the concentration of cells in the disperser fluid is increased, higher numbers of cells are being contained in smaller particles. Since smaller particles outnumber the larger, the effect for particles containing cells is to shift the distribution toward the smaller particle

Table 2-3 Expected probabilities of finding N (and only N) cells (assumed volume 1 μ^3) in particles of sizes initially formed, and in the equilibrium sizes, as a function of cell concentration in the disperser fluid

Cell conc. No./ml	Initial size, μ	1	5	10	15	20
	Equilibrium size, μ	0.26	1.3	2.6	3.9	5.2
	N					
5×10^8	0	0.999	0.97	0.77	0.43	0.10
	1	...	0.03	0.20	0.36	0.25
	2	...	0.0004	0.026	0.15	0.27
	3	...	...	0.002	0.04	0.19
	4	...	...	0.0001	0.008	0.09
	5	...	...	...	0.002	0.04
	>6	...	...	...	...	0.01
1×10^9	0	0.999	0.94	0.60	0.18	0.015
	1	0.0004	0.06	0.30	0.31	0.063
	2	...	0.002	0.08	0.26	0.13
	3	...	...	0.01	0.15	0.19
	4	...	...	0.001	0.06	0.19
	5	...	...	...	0.02	0.16
	>6	...	...	...	0.008	...
2×10^9	0	0.99	0.87	0.35	0.03	...
	1	0.009	0.11	0.37	0.11	0.002
	2	...	0.007	0.19	0.19	0.007
	3	...	0.0003	0.06	0.21	0.02
	4	...	...	0.01	0.18	0.04
	5	...	...	0.003	0.12	0.08
	>6	...	...	0.0006	...	...
1×10^{10}	0	...	0.522	0.004	...	...
	1	...	0.34	0.022	...	...
	2	...	0.11	0.06	...	...
	3	...	0.02	0.113	...	...
	4	...	0.004	0.155	...	...
	5	...	0.0005	0.175	...	...
	>6	...	...	0.308	0.997	1.00

sizes. Note that the particle sizes listed in the figures are those at equilibrium, but the calculated number of microbes per particle is based on sizes initially formed.

These curves can provide semi-quantitative values. For example, if we atomized at a concentration of 10^9 cells/ml, we should find about

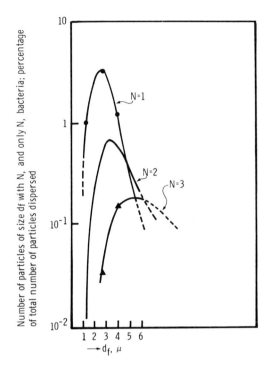

FIGURE 2-6. Percentage of particles of size distribution shown in Table 2-1 expected to have 1, 2, or 3 bacteria if concentration in atomizer were 5×10^8 cells/ml.

8% of all dispersed particles to be both 2 μ in diameter and to contain only 2 cells. Also, the modal values of these curves tell us that only 1 cell is contained mostly in particles about 2 μ; only 2 cells in particles about 2.8 μ; only 3 in 4-μ particles; and only 4 in 5.5-μ particles. It is obvious from these figures that it would be very difficult, and somewhat meaningless, to determine precisely the number of cells per particle, but we can easily arrive at a fair estimate.

Let us assume that N is contained only in the modal size at the modal percentages. For $N \gtrless 3$, this is essentially true because the percentages of particles containing 3 or more, is 0.1% or less. Table 2-4 shows the modal size for $N = 2$ and $N = 3$, and the percentages related to $N = 1$ as equal to 100%. A plot of the values for different concentrations of atomizer fluid is shown in Figure 2-9. Since these modal sizes really contain a distribution of numbers, percentages are only approximate and the expected error is shown between dotted lines in the figure.

We can use Figure 2-9 to predict the average number of cells per

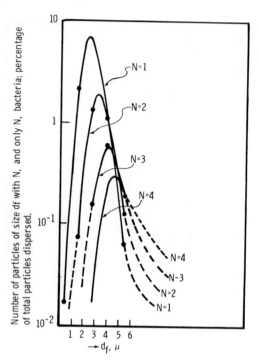

FIGURE 2-7. Percentage of particles of size distribution shown in Table 2-1 expected to have 1, 2, or 3 bacteria if concentration in atomizer were 1×10^9 cells/ml.

average particle. Suppose we spray a suspension containing 10^{10} cells/ml. For every 100 particles containing 1 cell, we would expect to find (by Figure 2-9) another 60 containing 2 cells, another 20 containing 3 cells, and another 7 containing 4 cells. We should find, then, 308 cells in 187 particles—or a ratio of 1.6 cells per particle. If we compared evaluations of the initial numbers of viable cells (immediate recovery) by means of slit samplers (1 colony per particle) and impingers (1 colony per cell), we should find a ratio of about 1.6.

An interesting corollary to this reasoning now becomes apparent. Suppose that the size of particles formed in the atomizer, and eventually airborne, was completely unknown to us. The ratio of the numbers found with the two samplers, divided by cell concentration, would permit us to estimate a mean probable initial particle volume; from Figure 2-5 or similar calculations, we could determine a central size which would yield that ratio. Assuming an observed ratio of 1.6 and a measured concentration of 10^{10}, we find the particle volume to be 1.6×10^{-10}

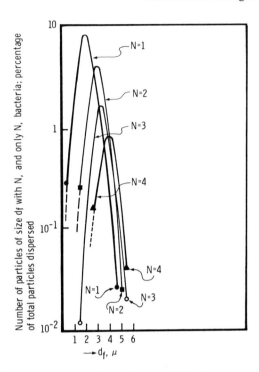

FIGURE 2-8. Percentage of particles of size distribution shown in Table 2-1 expected to have 1, 2, or 3 bacteria if concentration in atomizer were 2×10^9 cells/ml.

Table 2-4 Expected modal size and percentage of particles containing N cells as a function of cell concentration in the disperser fluid

Conc.	N	$N_1 = 1$	$N_2 = 2$	$N_3 = 3$	$N_4 = 4$	$(N_2/N_1)\%$	$(N_3/N_1)\%$	$(N_4/N_1)\%$
5×10^8	Size, μ	2.7	3.5	4.5				
	% of total particles	3.5	0.65	0.18		18	5.0	
1×10^9	Size, μ	2.0	3.0	4.0	4.3			
	% of total particles	7.0	1.9	0.6	0.29	27	8.5	4
2×10^9	Size, μ	1.8	3.0	3.1	4.0			
	% of total particles	10.0	3.9	1.6	0.8	39	14.0	8
1×10^{10}	Size, μ	1.2	1.8	2.2				
	% of total particles	12.0	7.2	2.3		60	19.0	

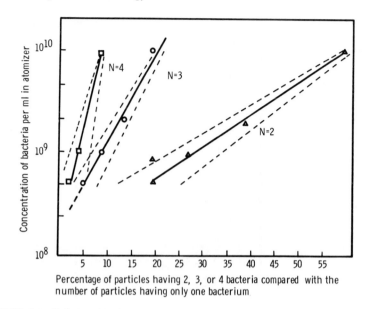

Percentage of particles having 2, 3, or 4 bacteria compared with the number of particles having only one bacterium

FIGURE 2-9. Relationship between number of bacteria per particle (regardless of size) and concentration of bacteria in atomizer. Estimated variance is shown by dotted lines.

$(1.6/1 \times 10^{10})$ and the mean size (Figure 2-5) to be 6.8 μ. An observed ratio at a cell concentration of 2.5×10^{10} was 1.4. The mean volume was 5.4×10^{-11}, which corresponds to an initial particle size of 4.8 μ or an equilibrium size, d_f, of roughly 1.3 μ. The mass median diameter of an equivalent aerosol, by the stirred settling method, was found to be 1.32 μ (Dimmick et al., 1958). Kethley et al. (1957) reported a ratio of 1.1 at a cell concentration of 1.4×10^9/ml, which indicates an initial central size of 11.5 μ, but the nebulizer was not the same as above.

Both Hayakawa (1964) and Mercer et al. (1968) show that particle size varies with concentration of dissolved solids in the atomizer fluid. A reasonable approximation is:

$$(2\text{-}3) \qquad \bar{d}_f = k \sqrt[3]{S/\rho}$$

where $\bar{d}_f$ = average particle diameter
S = concentration of solute, weight percent
ρ = density of dry particle (roughly 1.1)
k = empirical constant depending on nature of solute.

No values of k are available at this time, and unless derived under standard conditions they would not be universally applicable. The investigator, if interested, can determine k for his own situation by plotting at least two determinations of $\bar{d}_f$ versus $\sqrt[3]{S}$ and drawing a straight line between the points.

I have presented this exercise principally to show how data can be analyzed, rather than to establish definitive values of constants. That the analysis agrees fairly well with observed facts may be fortuitous since the theory is somewhat tenuous. If one has available a more appropriate estimate of a number-size distribution than employed here, one should certainly use it to apply to one's own problem, with the above analysis as a guideline procedure.

PRACTICAL ASPECTS

By a circuitous route we have arrived at the concept of an atomizer (a) that embodies a peripheral-type, twin-fluid atomizer in a container or "baffle-housing" to strip off large particles, (b) that utilizes air and liquid flow in desired amounts, and (c) that has reasonably consistent and measurable parameters conforming to approximate theory; a rather unhappy situation from the viewpoint of the purist, but adequate until proven otherwise. Since an efficiently operated right-angle jet, the Collison atomizer, performs just as well as the Wells-type, we conclude there is little difference between them. The principal factor allowing us to obtain particles in the size range below 10 μ is the baffle, or container, in which jets are operated.

At the Naval Biological Laboratory we have utilized hundreds of Wells-type atomizers built from glass. They have three deficiencies of a practical nature: they break easily, are very difficult to build to consistent tolerances, and they are hard to clean. Figure 2-10 shows the design for a metallic jet insert that overcomes some of these deficiencies. The advantage, if any, over the Collison atomizer is that the container is less complicated.

Some workers (Rosebury, 1947) have rejected reflux atomizers on the basis that the act of constant impaction against the walls, together with shear forces involved, will kill sensitive microbes. Indeed this may occur, but the extent of loss of viability by refluxing is far less than other, usually unpredictable, factors that influence the livelihood of airborne microorganisms. Figure 2-11 shows a relationship between the duration of refluxing and the number of cells of the species *Serratia marcescens* recovered near the exit of a reflux atomizer. For a period of almost 2 hr there was little change in the numbers recovered. The slight increase during the first hour was probably caused by concentration of particu-

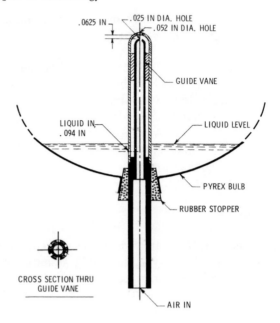

FIGURE 2-10. Schematic cross section through stainless steel, peripheral air jet.

lates in the fluid, as mentioned above, or by disruption of cellular aggregates. After 2 hr there was a marked decline in numbers that could be attributed to a decrease in the number of *viable* cells in the atomizer fluid. Vedros and Hatch (personal communication) found that a strain of meningococci, thought to be more sensitive in many respects than

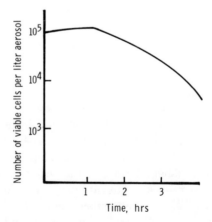

FIGURE 2-11. Immediate viable recovery of cells of *Serratia marcescens* from a continuously operated Wells-type atomizer.

S. marcescens, survived 6 hr of refluxing without measurable trauma, whereas Hatch (personal communication) found stored *Escherichia coli* to be very sensitive, although freshly grown cells were not. Most bacteria will withstand pressures as high as hundreds of thousands of pounds per square inch without dying, and it seems unlikely that forces involved in dispersion alone could cause death. Other biophysical factors involved in the causes of death of bacteria will be discussed in later chapters. Obviously, the influence of atomization should be measured with each species dispersed, and the measurement related to other factors such as growth medium, temperature, air pressure, etc., for which no guiding principles are available at this time.

Although the work-horse of the aerobiologist is the reflux atomizer, other types of dispersers have been used enough to deserve mention. It is obvious that we are not going to obtain true monodisperse aerosols from reflux atomizers without considerable effort, and if we want to gain quantitative knowledge of retention in the respiratory tree we need aerosols having fairly narrow distributions of particle size. Additionally, we would like to have a disperser that would allow us to vary particle size for the purpose of calibrating collection devices. Spinning disks have been used by industry for many years to produce bulk powders; these devices have a high output and efficiency, and can produce particles both smaller and more uniform than jets. The modified air-drive top, as fully described by May (1949), approaches the ideal. Whereas the jet produces threads of liquid, the spinning top produces a uniform sheet of liquid that breaks into two sizes. The smaller particles (termed "satellite particles") are about one-tenth the size of the larger particles, and in May's instrumentation they are effectively removed by suction.

Were it not for the fact that the spinning top is difficult to build and is extremely sensitive both to operational faults and effective maintenance, it would find more acceptance. Currently, it is the only instrument that will serve in experiments where a useful quantity of particles with a narrow distribution of sizes is required, and one is sometimes forced to learn to operate it.

Schwendiman *et al.* (1964) and Lippmann and Albert (1968) utilized a high-speed electric motor attached directly to the disk. This eliminated many stability problems associated with air-driven disks. They found a direct relationship between particle size and fluid concentration in the range of 1 to 8 μ diameter. Output was low (80 particles/ml), but this concentration could be used to standardize samplers. No biological data were reported.

A vibrating reed device that produces very uniform droplets has recently been described by Wolfe (1961), but reported behavior indicates that the sizes and number of particles are not in the desired range.

DISPERSION OF POWDERS

Aerosols can be produced directly from dry material. In studying this problem, I learned two important facts: (1) one cannot grind material in ball mills beyond a certain degree of fineness. The reason for this is that as grinding continues, the mass of the smallest particles that are forced together begins to equal the mass of larger particles that are ruptured, and a state of quasi-equilibrium is reached; (2) once small particles are formed by any method, they must be kept apart. In fact, the energy required to separate two small particles that have coalesced is almost as great as the energy needed to break them apart initially (Derr, 1963). Hence, if one had a bulk powder of 1-μ particles, it would require about the same amount of energy to make an aerosol of 1-μ particles as it would to make the same aerosol from a powder composed of indefinite-sized particles. Various methods have been used to prevent agglomeration (Orr, 1966) or to grind by mechanical means, but none are suited to laboratory needs. For example, the Tanner mill (Tanner, 1957) requires a 5-horsepower motor and liquid cooling.

The principle of grinding and dispersing at the same time is embodied in a disperser described by Dimmick (1959). In this device, air jets tear particles from the surface of tightly packed material and impart a sufficient spin to particles to break them into smaller ones. A series of cavities prevent larger masses from being ejected until particles are small enough to move with air streams that are conducted by as short a path as possible into the aerosol container.

The device, operated within an enclosure, will produce particles almost as small as those dispersed by the Wells atomizer, and do so more or less continuously. However, it suffers from the same faults as the spinning tops; it is difficult to build and to use. Additionally, because of the hygroscopic nature of biological material, "bone dry" air or nitrogen must be used. As a result, particles have a high electrostatic charge and they impact in great numbers on any nearby surface. This effect undoubtedly influenced the reported size-reduction efficiency of the device. Biological recovery of *S. marcescens* from the powder dispenser was found to be high enough to be useful. About 40% of the cells (on a per milligram basis) was lost as a result of packing the powder for dispersion; of that, 2% could be recovered by air samplers within 1 minute after dispersion, and about 8% by collecting powder electrostatically as it emerged from the jet.

Explosive forces have been used for dry material; e.g., the CO_2 pistol (Gordieyeff, 1957), and the principle of the rupture of a diaphragm by compressed gas have been reported (Payne, 1948). Unless such aerosols are stripped of the larger particles (as is done in refluxing atomizers), the mean size is too large for laboratory use.

ULTRASONIC GENERATORS

Ultrasonic generators will produce aerosols, and there is one generator manufactured for the purpose of producing an "ultrasonic fog"[1] and another (Andrews, 1964) used for aerosol therapy.[2] However, the fact that ultrasonic energy is commonly used to disrupt cells—some generators are reputed to sterilize surgical instruments—leads one to think this method of producing viable airborne bacteria for laboratory purposes is of little value. No comprehensive studies of the method have been reported. Mitchell and Pilcher (1959) produced aerosols of polystyrene latex by ultrasonic methods, but reported no details. It is important to note, however, that enough viable airborne bacteria are produced by these instruments to represent a potential, but as yet unmeasured, hazard if used with infectious material. The possible utility of ultrasonic generators to aerobiology remains to be determined.

In summary, there is no effective, or natural, manner in which living biological material can be converted *in toto* into particles ranging from 1 to 10 μ in sufficient quantities for use in detailed studies of infectivity of pathogens or survival of airborne cells. It is quite practical, however, to enclose simple atomizers or jets in a container so that larger particles will be impacted or removed by settling, and the emerging aerosol will contain sufficient particles in the desired size range to permit effective aerobiological research in the laboratory. Since these methods probably do not simulate "natural" conditions of dispersion, we shall eventually have to turn our attention to mechanisms found in nature and attempt to duplicate the process in the laboratory. Until we create a theoretical background sufficient to allow us to predict biological behavior with present devices, the building of new devices seems futile.

REFERENCES

Andrews, A. H., Jr. 1964. Ultrasonic aerosol generator: A new dimension in aerosol therapy. *Presbyterian-St. Luke's Hospital Med. Bull.*, Chicago. October.

Cox, C. S. 1965. Protecting agents and their mode of action. In *A Symposium on Aerobiology, 1963* (R. L. Dimmick, Ed.), pp. 345–368. Nav. Biol. Lab., Nav. Supply Center, Oakland, Calif.

Dautrebande, L., and Walkenhorst, W. 1964. Deposition of microaerosols in human lung with special reference to the alveolar space. *Health Phys.*, **10**: 981–993.

[1] Edison Instruments Co., Rahway, N.J.
[2] DeVilbiss Company, Toledo, O.

DeOme, K. B., and Personnel, U.S. Nav. Med. Res. Unit #1. 1944. The effect of temperature, humidity, and glycol vapor on the viability of air-borne bacteria. *Am. J. Hyg.*, **40**: 239–250.

Derr, J. S. 1965. Models for deagglomeration and fracture of particulate solids. In *A Symposium on Aerobiology, 1963* (R. L. Dimmick, Ed.), pp. 227–265. Nav. Biol. Lab., Nav. Supply Center, Oakland, Calif.

Dimmick, R. L. 1959. Jet disperser for compacted powders in the one-to-ten-micron range. *Am. Med. Assoc. Arch. Ind. Health*, **20**: 8–14.

Dimmick, R. L., Hatch, M. T., and Ng, J. 1958. A particle-sizing method for aerosols and fine powders. *Am. Med. Assoc. Arch. Ind. Health*, **18**: 23–29.

Drinker, P. and Hatch, T. 1963. *Industrial Dust*. McGraw-Hill Book Co., New York and London.

Fraser, R. P., and Eisenklam, P. 1956. Liquid atomization and the drop size of sprays. *Trans. Inst. Chem. Engr.*, **34**: 294–313.

Friedlander, S. K. 1965. The similarity theory of the particle size distribution of the atmospheric aerosol. In *Aerosols, Physical Chemistry and Application*, K. Spurny (Ed.). Publishing House of the Czechoslovak Academy of Sciences, Prague, Czechoslovakia.

Fuchs, N. A., and Sutugin, A. G. 1966. Generation and use of monodisperse aerosols. In *Air Science* (C. N. Davies, Ed.). Academic Press, London and New York.

Gillespie, G. R., and Johnstone, H. F. 1955. Particle size distribution in some hygroscopic aerosols. *Chem. Engr. Prog.*, **51**: 74F–80F.

Gordieyeff, A. V. 1957. Studies on dispersion of solids as dust aerosols. *Am. Med. Assoc. Arch. Ind. Health*, **15**: 510.

Green, H. L., and Lane, W. R. 1964. *Particulate Clouds, Dusts and Smokes* (2nd ed.). E. and F. N. Spon, Ltd., London.

Hatch, T. F., and Gross, P. 1964. *Pulmonary Deposition and Retention of Inhaled Aerosols*. Academic Press, New York and London.

Hayakawa, I. 1964. The effects of humidity on the coagulation rate of ammonium chloride aerosols. *J. Air Pollution Control Assoc.*, **14**: 339–346.

Kethley, T. W., Cown, W. B., and Fincher, E. L. 1957. The nature and composition of bacterial aerosols. *Appl. Microbiol.*, **5**: 1–8.

Lewis, H. C., Edwards, D. G., Goglia, M. J., Rice, R. I., and Smith, L. W. 1948. Atomization of liquids in high velocity gas streams. *Ind. Engr. Chem.* **40**: 67–74.

Lippman, M., and Albert, R. E. 1968. A compact electric-driven spinning disk generator. *Am. Med. Assoc. Ind. Hyg. Assoc. J.*, **28**: 501–506.

May, K. R. 1949. An improved spinning top homogeneous spray apparatus. *J. Appl. Phys.*, **20**: 932–938.

Mercer, T. T., Tillery, M. I., and Chow, H. Y. 1968. Operating characteristics of some compressed air nebulizers. *Am. Ind. Hyg. Assoc. J.*, **29**: 66–78.

Mitchell, R. I., and Pilcher, J. M. 1959. Measuring particle size in air pollutants, commercial aerosols and cigarette smoke. *Ind. and Engr. Chem.*, **51**: 1039–1042.

Nukiyama, S., and Tanasawa, A. 1939. An experiment on the atomization of liquids. *Trans. Soc. Mech. Engr.*, Japan, **5**: 1–4.

Orr, C., Jr. 1966. *Particulate Technology*. The Macmillan Co., New York.

Orr, C., Jr., and Gordon, M. T. 1956. The density and size of airborne *Serratia marcescens*. *J. Bacteriol.*, **71**: 315–317.

Payne, R. E. 1948. The measurement of the particle size of sub-sieve particles. Third Nat. Conf., Am. Instr. Soc. Reprinted in Bull. #1244: 1–14. Sharples Corp., Philadelphia, Pa.

Rosebury, T. 1947. *Experimental Airborne Infection.* Williams and Wilkins Co., Baltimore, Md.

Schwendiman, L. C., Potsma, A. K., and Coleman, L. F. 1964. A spinning disk aerosol generator. *Health Phys.,* 10: 947–953.

Tanner, H. G. 1957. New type of mill for refined chemicals. *Ind. Engr. Chem.,* 49: 170–173.

Williams, R. E. O. 1966. Epidemiology of airborne staphylococcal infection. *Bacteriol. Rev.,* 30(3): 660–674.

Wolfe, W. 1961. Study of the vibrating reed in the production of small droplets and solid particles of uniform size. *Rev. Sci. Instr.,* 32: 1124–1129.

Wright, B. M. 1958. A new nebulizer. *Lancet* (July), 24–25.

3

RELATIVE HUMIDITY: MEASUREMENT AND MEANING

Robert L. Dimmick / George F. Marton

NAVAL BIOLOGICAL LABORATORY, SCHOOL OF PUBLIC HEALTH,
UNIVERSITY OF CALIFORNIA, BERKELEY

DETERMINATION OF RELATIVE HUMIDITY

Except for the assay of whatever "life" is, measurement of the moisture content of air is the most difficult parameter an aerobiologist must determine. Moisture content may, indeed, be the most important environmental factor influencing survival of airborne microbes, a subject to be discussed in more detail in another chapter. Many persons will be surprised to learn that the commonly used wet-dry bulb determination of relative humidity (RH) is empirical and is useful only under the conditions initially established to determine the relationship between two temperatures (Zwart, 1966). Values are constantly being revised. To arrive at suitable knowledge, we must first examine what we mean by RH.

We will start by assuming that water vapor in the air acts as a perfect gas. It really doesn't, but the difference is small and we want to keep the formulas shorter than the sentences. The density of the water vapor in air at a given temperature can be written as:

$$(3\text{-}1) \qquad \rho_v = \frac{p_v}{RT} M_w = \frac{p_v}{R_a T} \frac{M_w}{M_a} = 0.622 \frac{p_v}{R_a T}$$

where ρ_v = density of water vapor in moist air, or absolute
          humidity
          p_v = water vapor pressure
          R = universal gas constant
          M_w = molecular weight of water
          M_a = molecular weight of dry air
$R_a = R/M_a$ = gas constant for dry air
          T = absolute temperature.

The density of moist air is the sum of the density of dry air and vapor:

(3-2) $\rho = \rho_a + \rho_v$

where ρ = density of moist air
      ρ_a = density of dry air in moist air.

The total pressure of the moist air is:

(3-3) $P = p_a + p_v$

where p_a = dry air pressure
      P = total pressure of moist air, i.e., atmospheric pressure

Therefore,

(3-4) $\rho_a = \dfrac{p_a M_a}{RT} = \dfrac{P - p_v}{RT} M_a$

The ratio of the molecular weight of water or water vapor and dry air is:

(3-5) $\dfrac{M_w}{M_a} = 0.622$

Hence, substituting equations (1), (4) and (5) in (2):

(3-6) $\rho = \dfrac{P - p_v}{RT} M_a + \dfrac{p_v}{RT} M_w = \dfrac{P - 0.378 p_v}{RT} M_a = \dfrac{P - 0.378 p_v}{R_a T}$

Since density and partial vapor pressure are the actual variables, to know the real moisture content of air we must extract all of the water vapor from a known volume of air, measure the mass of water extracted, and measure either the change in pressure as a result of this extraction or the mass of dry air remaining—all, of course, at constant temperature. Absolute standards are obtained in this way. Once pressures or densities have been measured carefully, they can be employed to calculate whatever value we need. As one result of such measurement, we can define a relative measure of moisture content called percent RH:

$$(3\text{-}7) \qquad \%RH = 100 \frac{p_v}{p_s} = 100 \frac{\rho_v}{\rho_s}$$

where p_s = saturation vapor pressure over a flat water surface
ρ_s = density of water vapor in moist air at saturation over a flat water surface.

Air-conditioning engineers talk about the humidity ratio:

$$(3\text{-}8) \qquad r = \frac{m_v}{m_a} = \frac{\rho_v}{\rho_a}$$

whereas meteorologists discuss specific humidity:

$$(3\text{-}9) \qquad q = \frac{m_v}{m_a + m_v} = \frac{\rho_v}{\rho}$$

where m_v = mass of water vapor in a given amount of moist air
m_a = mass of dry air in a given amount of moist air.

Air-conditioning engineers, meteorologists, and physicist have created no less than a chaotic situation with respect to nomenclature. We have not found moisture content expressed as grains per bushel, but there may be someone who is planning to incorporate bushels into some chart for purposes of maximum obscurity. The serious student is advised to have at hand a rather complete set of conversion tables before going to the literature.

For our purpose, indeed for almost all practical purposes, the method of extracting and weighing water is too slow and too elaborate—we must turn to indirect methods that measure effects of moisture content rather than the water mass. Spencer-Gregory and Rourke (1956–57) provide a detailed explanation of standard, generally applicable methods.

Wet-Dry Bulb Methods

One of the first thoughts that comes to mind is the fact that as long as the vapor pressure of air is less than that of water at the same temperature, water will evaporate and cooling will take place. As the water temperature drops, its vapor pressure will drop until some kind of equilibrium occurs. The equilibrium temperature compared with the ambient temperature would be a measure of the water vapor pressure of the air. The practical application of this process, called wet-dry bulb psychrometry, is an easy and reasonably effective method for achieving

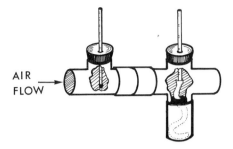

AIR
FLOW

FIGURE 3-1. A simple wet sleeve hygrometer for use in measuring the relative humidity of aerosol chambers during the filling process.

a good estimate. It is the most widely used measurement and it may, indeed, be the one most widely misused. There are wet bulb depression tables (Wexler, 1965; MacPhee, 1966), psychrometric tables (Zimmerman and Lavine, 1945), charts, and well-constructed slide rules[1] that allow one to convert the depression of temperature, expressed by the difference in readings of a dry and a wet bulb thermometer, into RH values. These tables etc. are quite accurate, but are of little value unless the measurement of wet bulb depression is obtained under the same conditions used to construct the tables. For example, air velocity past the w,et bulb, which is equipped with a wet sock or sleeve, should be approximately 900 ft/min, satisfying the requirement of turbulent flow with a minimum Reynolds number of Re = 2,300.

For our use, thermometers should be mounted in line in a tube through which air can be drawn, with the wet bulb mounted downstream from the dry bulb, and with the end of the wet sock hanging into a water reservoir below the air passage. Thermometers, or thermisters if used, should be calibrated and capable of being read to 0.2 F. A commercial thermister-psychrometer is available at the present time.[2]

The wet-dry bulb psychrometer shown in Figure 3-1 can be constructed from standard copper fittings and is suitable for general use. One problem enountered with an arrangement where air must be drawn some distance through pipes is that the air might expand, or change in temperature. It is not advisable to draw air from the chambers through small tubes and then into the unit shown in Figure 3-1. It is usually necessary to fill a static chamber with preconditioned air while at the same time

[1] Psychrometric Computer CP-165A/UM, Felsenthal Instruments, Chicago, Ill.
[2] Atkins Technical Inc., Box 14405, University of Florida Field Station, Gainsville, Fla.

withdrawing contained air. The tubes used for these two purposes are the best location for the psychrometers. When the two dry bulb and wet bulb temperatures are identical, the chamber can be considered to be "filled" with air of the indicated humidity.

A potential source of error is created when an aerosol, in contrast with air, is drawn through the psychrometer. Airborne particles can contaminate the sock, which ideally should be wetted only with distilled water. Soluble contaminants will change the vapor pressure of the water and, consequently, the temperature depression caused by evaporation. This is a serious problem, both with the wet sock and the sensing devices listed below. Filters could be used to remove particles, but they adsorb moisture and cause pressure changes, both of which lead to erroneous readings.

Electronic Dewpoint Instruments

Other methods of determining RH will be discussed briefly in an approximate order of practical importance. The most recent development, and one that shows promise of becoming a more accurate and efficient method than the wet-dry bulb, is the electronic measurement of the dewpoint.[3] A former method of determining dewpoint was to cool a stainless steel mirror, attached to a thermocouple, with a cold solution such as dry ice and alcohol. Either visual or photoelectric observation was used to determine the moment moisture formed on the mirror.

The new devices cool the mirror by running current through a semiconductor block that essentially decreases in temperature as a function of increasing current[4] (Peltier effect). The current is regulated by a photocell that monitors the presence or absence of water on the mirror. The result is that a moisture film of precalibrated thickness is maintained on the mirror surface; hence, the mirror is held at the dewpoint temperature. In a recent publication, Jury et al. (1967) report a response time of 1.3 sec. One can either monitor the dewpoint temperature (i.e., the temperature of the mass of the mirror) directly, or monitor the current and relate it to the dewpoint.

Although electronic dewpoint instruments are fairly expensive, they have the advantage of providing continuous recorder output, and they are ideal for controlling and measuring air input to dynamic chambers. They also provide frostpoint measurements when the air temperature

---

[3] The dewpoint, or saturation temperature, is the temperature to which air has to be cooled, at constant pressure, to be saturated. It is not the same as wet bulb temperature except under saturated conditions.
[4] Bendix Corporation, Southfield, Mich.; Cambridge Systems, Newton, Mass.; Climet Instruments, Inc., Sunnyvale, Calif.

is below freezing. The most serious fault is that the mirror can become contaminated by aerosol particles that interfere with accuracy and response time. Properly operated, electronic dewpoint instruments could provide a laboratory with primary standards for calibration of other instruments.

Membrane and Fiber Hygrometers

Protein membrane and hair "hygrometers" serve useful special purposes. The operational principle depends upon the change of volume or length of a dry protein body with absorbed moisture. Hair hygrometers provide only an approximate value of the RH, whereas membrane hygrometers can be calibrated to a claimed accuracy of ±1%.[5] Either type can be placed inside chambers and viewed through ports, or they can be placed in small boxes through which air can be drawn to provide continuous monitoring of, for example, a laboratory air source. Some are available with a small pen recorder attachment. We have found that continued usage at very low humidities appears to denature the sensing elements, and that the sensing elements will eventually become contaminated to the extent that reliability is lost. Hygrometers can be used effectively to monitor ambient humidities where changes occur slowly over limited ranges (20 to 80% RH) and where possibility of contamination is minimal (e.g., rooms, wards, sheltered outdoor situations).

Hygroscopic Substances

Pure, inorganic hygroscopic salts attain a moisture equilibrium with the vapor pressure of the air. If solutions of moist salts are heated continuously, moisture will be lost until only crystals remain, and if cooled they will absorb moisture until all crystals dissolve and equilibrium occurs. This is not the same equilibrium that occurs if the salt solution in contact with crystals were confined in only a small, fixed volume of air and allowed to equilibrate. In the latter case, both air and solution change until an equilibrium is reached. If heat is supplied by passing an electric current through a solution saturated with respect to ambient temperature, additional moisture will be driven off until the ionic mobility (conductance) of the solution decreases to a level such that, with current flow now restricted, a new equilibrium is attained. The temperature of this new equilibrium state can be related to the dewpoint of the air.

The salt-temperature-moisture equilibrium is the principle of the Fox-

[5] Lab Standard Hygrometer, Serdex Inc., Boston, Mass.

boro Dewcell[6] (registered trade mark) instrument. Like the psychrometer, conversion charts for the instrument are empirical. Furthermore, it is not sensitive at the high and low ends of the RH scale and, for our purposes, is readily contaminated. If used according to the manufacturer's directions, it is reasonably accurate and convenient; the output can be recorded and it will serve in a manner similar to the psychrometer.

Almost any substance that adsorbs or absorbs moisture in equilibrium with water vapor pressure can be used as an electrical resistance element. In fact, it is difficult to find a substance that does not act this way to some extent. The resistance of a piece of paper, measured by a high-impedance vacuum-tube voltmeter, will depend on RH. There are few, substances, if any, that exhibit a linear relationship of resistance to moisture content over more than 20% of the RH range—different substances yield linear responses over different portions of the total range. A number of commercial instruments utilizing this principle are available, but none have been found more suitable to our purpose than the first three types mentioned.

Potential Methods

Almost the same principle can be employed by using various materials as the dielectric in the capacitor of an electronic oscillator circuit. Air can be used as the dielectric—the frequency of the circuit changes as a function of water content of the air. The idea is a good one, and nearly linear response can be attained over an RH range of 10 to 90%. It is expensive, however, to construct and maintain instruments with frequency stability suitable for the purpose. Also, dimensional stability of the capacitor plates is critical. These problems may have been overcome, because a novel capacitor-type hygrometer[7] has been announced recently, although aluminum oxide serves as the dielectric material.

Electrolysis of water in solution, or water adsorbed on various materials, is another principle that can be used in some instances. Electrolysis of phosphorous pentoxide held between two electrodes is the principle of the Beckman hygrometer.[8] A recently announced humidity meter[9] utilizes porous glass to adsorb water, which is then electrolysed; the reported accuracy is 1% at a 60%-response-time of 120 sec.

Water vapor absorbs radiation in certain bands of the infrared range.

---

[6] The Foxboro Company, Foxboro, Mass. Also, Honeywell Inc., Minneapolis, Minn. has recently made a portable instrument available.

[7] Hygrometer, Model 1000, Panametrics, 221 Crescent St., Waltham, Mass.

[8] Beckman Instruments, Fullerton, California.

[9] Bulletin HM 101 F, American Standard, New Brunswick, N.J.

Where long light paths can be attained, this principle should represent another method by which absolute values could be obtained. It has been tested and used in outdoor situations, but at present no suitable laboratory instrument exists.

In summary, there is no highly accurate, convenient, rapid, and at the same time inexpensive method of determining the moisture content of air containing hygroscopic particles, and the situation is currently the most exasperating technical problem the aerobiologist faces.

IMPORTANCE OF RELATIVE HUMIDITY

The question of how accurately we must measure moisture content, in view of the above, is answered by saying that we do not know. Some of the earliest observations about the livelihood of airborne bacteria indicated that survival varied markedly under different RH conditions, and that there seemed to be ranges of maximal sensitivity. In many instances, the change of sensitivity appeared to exceed the ability to differentiate between one RH value and another (Dunklin and Puck, 1948; Cox, 1966). Until we have more accurate methods of measuring humidity, we will not know how accurate we have to be.

That is not the only quandary related to moisture content of the air. Since the instruments we use are generally calibrated in terms of percent RH, we customarily report our data in those terms. But this may not be the factor directly causing death of a dried cell. Surprisingly, only a few investigators, in agreement with Webb's initial concern (Webb, 1959), have suggested that it is the absolute moisture content that influences survival of airborne microbes, and that a particle in equilibrium with RH1 at T1 is not the same as one in equilibrium with RH1 at T2. Perhaps we should report our results in some way that includes the effect of temperature.

To determine whether a change of temperature at constant RH produces an effect caused only by biological reactions to physicochemical changes occurring in the constituents of the particle, or whether changes were the result of biological reactions to temperature *per se,* is indeed difficult. If it is the moisture content of particles that is important, we could express this according to Scott (1956) as:

$$a_w = \frac{p}{p_0}$$

where p = vapor pressure of the solution,
p_0 = vapor pressure of the solvent, and
a_w = water activity.

This is derived from Raoult's law

$$\frac{p_0 - p}{p} = \frac{n_1}{n_1 + n_2}$$

where n_1 = moles of solute and
n_2 = moles of solvent.

At equilibrium,

$$\frac{p}{p_0} = \frac{n_2}{n_1 + n_2}$$

for ideal solutions. Using our previous notations from equations 3-1 and 3-7,

(3-10) $$a_w = \frac{p_v}{p_s}$$

This actually means that water vapor activity times 100 is numerically equal to the corresponding percent RH at equilibrium. Furthermore, a_w appears not to be temperature-dependent. We could say, in a sense, that a_w takes the place of the mole fraction of a solute in a perfect solution. If it is not related to temperature, however, we intuitively feel it is not suited for our purpose.

Another measure of the effect of water content on biological systems is the osmotic pressure. It can be related to water activity according to:

(3-11) $$\Lambda = -\frac{RT \ln a_w}{\bar{V}}$$

where Λ = osmotic pressure,
R = gas constant,
T = temperature, and
$\bar{V}$ = partial molar volume.[10]

We could, therefore, report results as effective osmotic pressure; that is, droplets would have attained equilibrium with, or be isotonic to, a solution having the calculated osmotic pressure. Since we may continue assay of viability as long as 24 hr, or even longer in drums, we could presume that the equilibrium condition is a realistic situation related to the moisture effect that acts on long-lived cells.

Webb (1959) has utilized another relationship between moisture and

---

[10] The partial molar volume is the change in volume that would occur if 1 mole of pure solvent were removed from a large volume of the solution. For water, it is approximately 18.

temperature. By modifying a formula for the rate of change of mass as a droplet evaporates, he derived the expression

(3-12)
$$\frac{d_m}{d_t} = A[100 - (\%\text{RH})]\frac{p_s}{T}$$

where A = constant, assumed equal to 1,
m = mass of particle, and
t = time.

At equilibrium, however, $d_m/d_t = 0$, so if we use this expression to relate moisture and temperature, we are assuming that something happens to the biological mechanisms of the cell during evaporation of the droplet to influence the effect of moisture on bacterial survival for an extended time thereafter. In other words, if the initial drying rate determines the way cells behave, then subsequent RH changes should have no effect on the established death rate—a statement we know to be wrong. Webb treated biological decay as typical thermodynamic data and showed that he could obtain a satisfactory Arrhenius plot with any of the several decay constants he observed from individual runs with washed bacterial cells. Data he presented were convincing; however, the oxygen effect (Hess, 1965) was not well known at that time, so we do not know how this might have influenced his results. Nevertheless, we have used Webb's function to adjust data obtained under conditions of day-to-day limited RH and temperature changes, and found that otherwise scattered data points could be rearranged to form coherent patterns.

A conversion factor such as Webb's is exactly what is needed to allow investigators to compare data on an equivalent basis. We are unable to explain why this has not been done before, especially since the quibble about d_m/d_t being equal to zero can be avoided easily by substituting Equation 3-7 in Equation 3-1 to yield absolute humidity:

(3-13)
$$\rho_v = \frac{p_v}{RT}M_w = \frac{p_s(\%\text{RH})}{100RT}M_w = \frac{p_s(\%\text{RH})}{100R_aT}\frac{M_w}{M_a}$$

$$= \frac{0.622}{100R_a}\frac{p_s(\%\text{RH})}{T} = B\frac{p_s(\%\text{RH})}{T}$$

where B = constant = $\dfrac{0.622}{100 \times 53.34}$ = $1.166\ (10^{-4})\ \dfrac{\text{lb, }^\circ\text{R}}{\text{ft, lb}}$

$$= 2.23\ (10^{-3})\ \frac{\text{gm, }^\circ\text{K}}{\text{cm}^3\text{, mm Hg}}$$

This equation has the same utility as Webb's.

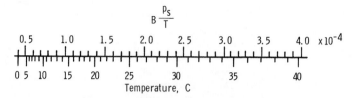

FIGURE 3-2. Relationships of absolute humidity to the factor $B(P_s/T)$, where $T = 273 + t$ (measured, C); P_s = pressure, mm Hg, $B = 2.23 \times 10^{-3}$ (gm, °K/cm³, mm Hg); RH = relative humidity, per cent. To use, multiply observed per cent RH by the value of $B(P_s/T)$ for the observed temperature, C.

For convenience, values of $(Bp_s)/T$ versus values of temperature from 0 to 40°C have been plotted in Figure 3-2. The importance of this type of conversion to practical situations is best illustrated by Waddy's (1957) finding that the incidence of cerebrospinal meningitis in Africa correlated not with RH, but with absolute humidity. In his paper, Waddy expresses regrets that experimental data have not been reported in a way that will allow him to draw additional inferences from his observations. Kingdon (1960), on the other hand, considered RH to be the important measurement.

Of course, another simple way to achieve a basis for comparison of data would be to report the dewpoint or the humidity ratio. It follows that, with any method used, the dry bulb temperature must also be indicated.

Finally, one might rearrange various forms of the Clausius-Clapyron equation to derive some relationship between equilibrium moisture content of droplets and temperature. There seems to be little value in doing this, however, because of the unknown nature of all the factors involved in the final equilibrium. It should be emphasized again that we are really interested in how much water is available to the airborne cell, and how this affects the cell's livelihood. Indeed, any theoretical treatment based only on behavior of ideal solutions or perfect gases, as all the above are, fails to consider such additional variables as:

(a) adsorption of vapor on surfaces;
(b) change of vapor pressure with droplet size;
(c) change of surface tension as a result of adsorption;
(d) denaturation of proteins resulting in changed adsorptive capacity;
(e) effects of micrometeorological changes (Fleagle and Bosinger, 1963);
(f) crystallization of dissolved salts and the resulting competition for water molecules with proteins; and
(g) potential utilization, or production, of water by cellular enzymes.

Effects of micrometeorological changes possibly deserve more consideration than simply listing them as part of the overall picture. Orr *et al.* (1958) have shown that in the mid-range humidities salt particles do not grow at the same rate as they shrink—a kind of hysteresis. Dimmick (unpublished data) measured the electrical resistance of dried films of biological substances at different RH levels and observed that the rate of change in resistance as the humidity decreased was not the same as when humidity was increased. There seemed to be critical moisture levels, but the methods were not sensitive enough to be certain. Considering the complex nature of mixtures that biological menstrua consist of, such critical conditions could occur at more than one interval over the RH range. Slight changes in temperature or vapor pressure during an experiment conducted near these points might cause changes of state that would grossly affect the moisture available to the cell without changing the overall moisture content of the particle to a measurable extent. This problem needs additional critical investigation.

We mentioned that most results have been reported on the basis of RH. Many investigators, thoughtfully, have reported the temperature also. Considering the abysmal lack of knowledge that now exists regarding exact behavior of airborne bacteria, reporting the temperature, RH, and the methods used, as well as estimating accuracy, seems to be the accepted method of communicating aerobiological data in a form useful to the reader. The curious or skeptical student can then transform them into whatever parameter he wishes. We strongly recommend, however, that biological data be compared with some measure of absolute humidity in graphical presentations. Certanly the creation of new and more accurate methods of determining the moisture content of air, allowing meaningful assay of the effects of moisture on airborne bacteria, remains one of the most needed, as well as the most productive, paths of research open to the investigator.

REFERENCES

Cox, C. S. 1966. The survival of *Escherichia coli* sprayed into air and into nitrogen from distilled water and from solution of protecting agents, as a function of relative humidity. *J. Gen. Microbiol.*, 43: 383–399.

Dunklin, E. W., and Puck, T. T. 1948. The lethal effect of relative humidity on airborne bacteria. *J. Exptl. Med.*, 87: 87–101.

Fleagle, R. G., and Bosinger, J. 1963. *An Introduction to Atmospheric Physics*. Academic Press, New York and London.

Hess, G. E. 1965. Effects of oxygen on aerosolized *Serratia marcescens*. *Appl. Microbiol.*, 13: 781–787.

Jury, S. H., Bosanquet, L. P., and Kim, Y. W. 1967. Fast response sensing element for a frost point hygrometer. *Rev. Sci. Inst.*, 38: 1634–1637.

Kingdon, K. H. 1960. Relative humidity and airborne infection. *Am. Rev. Respirat. Diseases*, 81: 504–512.

58   Techniques of Aerobiology

McPhee, C. W. (Ed.) 1966. *ASHRAE Handbook of Fundamentals.* American Society of Heating, Refrigeration and Air-conditioning Engineers, Inc., New York.
Orr, C., Jr., Hurd, F. K., and Corbett, W. J. 1958. Aerosol size and relative humidity. *J. Colloid Sci.,* 13: 472–482.
Scott, W. J. 1956. Water Relations of Food Spoilage Organisms. *Advances in Food Research,* pp. 88–127, Vol. 7. Academic Press, New York.
Spencer-Gregory, H., and Rourke, E. 1956–57. *Hygrometry,* Pitman Publishing Corp., New York.
Waddy, B. B. 1957. African epidemic cerebro-spinal meningitis. *J. Trop. Med. Hyg.,* 60: 218–223.
Webb, S. J. 1959. Factors affecting the viability of air-borne bacteria. I. Bacteria aerosolized from distilled water. *Can. J. Microbiol.,* 5: 649–669.
Wexler, A. 1965. (Ed. in Chief) *Humidity and Moisture,* Vols. I, II, III. Reinhold Publishing Corp., New York.
Zimmerman, O. T., and Lavine, I. 1945. *Psychrometric Tables and Charts,* p. 162. Industrial Research Service, Dover, N.H.
Zwart, H. C. 1966. Errors in humidity measurement. *Instruments and Control Systems.* 39: 95–96.

NOMENCLATURE

ρ = density of moist air
ρ_a = density of dry air
ρ_v = density of vapor in moist air
ρ_0 = density of vapor in moist air over flat water surface or saturated vapor density
P = pressure of moist air
P_a = partial pressure of dry air
P_v = partial pressure of vapor in moist air
P_s = partial pressure of vapor in moist air over a flat water surface or saturation vapor pressure
m_a = mass of dry air

m_v = mass of water vapor
M_a = molecular weight of dry air, 29
M_w = molecular weight of water vapor, 18
T = temperature, °K
R = universal gas constant
$R_a = \dfrac{R}{M_a}$ = gas constant of dry air
n = moles
$\bar{V}$ = partial molar volume
RH = relative humidity
Λ = osmotic pressure
$°R = 460 + °F$

4

ASSAY OF LIVING, AIRBORNE MICROORGANISMS

Ann B. Akers / William D. Won

NAVAL BIOLOGICAL LABORATORY, SCHOOL OF PUBLIC HEALTH,
UNIVERSITY OF CALIFORNIA, BERKELEY, CALIFORNIA

One prerequisite to the study of the behavior of aerosols of bacteria and viruses is a quantitative method for assay. There is no known sampler that will provide a complete description of the biological properties of an aerosol, and there is no absolute means of determining whether all airborne bacteria are recovered from the sampled air. Hence, reliance must be placed on comparative evaluations of several different types of samplers used to measure a single aerosol source.

The number of viable organisms decreases in an aerosol because of physical and biological mechanisms, the effects of which must be separately determined. Physical losses occur because of (1) deposition on surfaces by settling, convective flow, or diffusion; (2) electrostatic attraction. Since the rate of physical loss within a particular chamber is variable, and since sample removal results in dilution of the remaining aerosol, physical loss should be measured each time a sample is assayed.

The total microbial numbers can be determined by direct count under the microscope, by light-scatter measurements, or by tracer techniques (using dyes, spores, radioisotopes, or enzymes); none of the microbes measured by these methods need be viable. Goldberg's (1968) fluorescent tracer technique is an extremely sensitive and useful innovation for determining physical loss in an aerosol. Physical aspects of sampling and methods for determining total particles recovered from microbial aerosols are discussed in detail by Green and Lane (1964) and May (1967).

Ideally, an aerosol sampler for microbiological assay should be capable of "counting" the total number of living airborne particles in a unit volume of air, as well as determining the number of viable units per particle, and the size of the particles containing such units. However, this presupposes that 100% of the airborne cells, living or dead, can be physically removed from the air, and that no loss of viability occurs during or after sampling. The sampling substrate used (whether liquid or solid) should be a medium in which maximal growth of the organism is obtained; the choice therefore depends on the organism studied. The assumption, which may not be necessarily true, is that any medium capable of supporting maximal growth of a test organism under "ordinary" conditions will also support maximal growth of the same cells damaged by aerosolization. Actually, two situations can occur: (1) the "injured" cell may need additional growth factors, or (2) the "ordinary" medium may be, in some way, "toxic" to the collected cells.

In contrast to sampling of particles by physical methods, assay for viability is inherently variable. The concentration of bacteria per unit volume of air is usually measured by the appearance of visible colonies on an agar plate exposed directly to an air stream, or a plate to which portions of liquid from impinger samples are added. However, Cox and Baldwin (1964) found that death resulting from a breakdown of the cell's productive capacity (enzyme systems) could be distinguished from loss of its reproductive power (nucleoprotein integrity). They atomized two aliquots of an *E. coli* suspension, one of which was infected with T7 bacteriophage. Samples of the aerosols were taken and growth of uninfected control cells was compared with extent of lysis of test cells containing phage. More *E. coli* cells at 0.5 hr aerosol age could support phage growth and lysis than could support colony formation. In this case, the "bacterial count" would depend on whether colony-forming units or lysis by phage was the criterion used for viability. An enzyme method for determining total bacterial numbers is described by Anderson and Crouch (1967).

Plaque-forming units per volume can be determined for those viruses and rickettsiae that form plaques on tissue culture. Nonplaque-formers, which cause cytopathogenic changes (CPC) in tissue culture, are assayed by determining the dilution that causes CPC in an arbitrary number of samples (TCID, $TCID_{50}$). However, if a virus or rickettsia will not propagate on tissue culture, concentration is measured in terms of the egg or animal infective dose and the assay is not directly comparable to viable count obtained by plaque or $TCID_{50}$ methods. In addition, infectivity titrations vary, depending on the route of inoculation and animal used.

Methods for sampling bacterial aerosols listed by Tyler and Shipe (1959) and by Anderson and Cox (1967) include sedimentation, filtration, agar impaction, electrostatic deposition, liquid impingement, centrifugation, and thermal precipitation. Sampling methods have been thoroughly reviewed by Bachelor (1960), Wolfe (1961), Wolf et al. (1964), and Noble (1967). This chapter will therefore be confined to a discussion of the most commonly used samplers and the factors which, by affecting the biological properties of the atomized organisms during sampling, result in variable assay data.

ISO-KINETIC SAMPLING

Before considering factors that affect biological properties of airborne microbes during sampling, the problem of obtaining a representative sample should be considered. As an analogy, consider a spherical volume containing a heterogeneous aerosol of particles 1 to 20 μ in diameter. A centrally located, hypothetical, "gravity sink" (rather than normal gravity) toward which all particles would move according to Stokes' law, and which might be turned "on" or "off," is created. It would then be possible to have the system "on" for a period of time such that all of the 20-μ particles had moved to the gravity sink by virtue of their more rapid rates of fall, but many of the smaller particles would not move through the entire volume. If particles collected on the sink were counted and sized at that time, the number of large particles would appear to be increased with respect to their ratio in the aerosol. Thus, unless the sink was "on" until all particles had been collected, the sample would be biased (Davies, 1965).

The reverse situation exists when a sample is removed from an aerosol by applying a vacuum source through an orifice (a substitute for the gravity sink) connected to a chamber. The outrushing air withdraws a disproportionate number of small particles, which travel with the air stream more efficiently than larger particles. The sample is therefore biased toward smaller particle sizes. Theoretically, a representative sample cannot be withdrawn from still air by suction in a finite time. However, if the tubing is kept reasonably large, and the air flow rate slow, the error is not great.

If air is moving, as in ducts and most situations outdoors, sampled air can be removed by directing the sample tubing into the air stream, and sampling at the same velocity as the air stream—hence, the term *iso-kinetic sampling.* In this situation, particles do not cross air streams before entering the sample port; all sizes have an equal "chance" of being sampled, and the sample obtained is representative of the distribution in the original aerosol. If the velocity of sampling air is greater than that of the moving stream, small particles will dominate the sample. If the velocity is slower, large particles will dominate because they are less easily diverted around the sampling orifice, which then acts as a kind of obstruction as do fibers in a filter (May, 1967).

Errors caused by non-iso-kinetic sampling are not great when particles are in the 1- to 5-μ range, but error increases as diversity of size increases. There is no "finagle factor" that will correct for mismatched sampling and air flow velocities—one must adjust flow rates to correspond to the given situation. In the following we optimistically assume iso-kinetic conditions.

IMPINGEMENT

Impingement implies collection of particles in a liquid menstruum rather than on a solid surface. The all-glass impinger (AGI) operated at an airflow rate of 12.5 l/min, and having a 30-mm clearance between orifice and sampler bottom, was chosen as one of two reference samplers by aerobiologists who met during the First International Symposium in 1963. They suggested that data obtained with any specialized sampler should be correlated with at least some results obtained with the standard AGI-30 reference sampler (Brachman *et al.*, 1964). The potential loss of viability incurred as a result of the sampling step is difficult to assess and may not always be constant. The sampling medium, duration of sampling, volume of medium, collection temperature, and the holding time and temperature between the sampling and the assay should therefore be stated in published papers to allow readers to judge results for themselves. An example of variability inherent in liquid impingement samplers is that long intervals required to sample dilute aerosols (less than 600 colony-forming units per liter of air), especially at low relative humidity (RH), cause excessive cooling and evaporation of the collecting fluid that may lead to death of the organisms being assayed.

Capillary Impingers

Most capillary impingers (AGI-30, Shipe impinger, Porton raised impinger) operate by drawing the aerosol through an inlet tube and then through a critical orifice. When the ratio of pressure across the capillary to inlet pressure (1 atm) is 0.5 atm or less, particles in the aerosol impinge into the fluid at sonic velocity, the flow rate is constant, and after the sampler has been calibrated no flow meter is needed (Cown et al., 1957). The air flow rate depends on the pressure ratio; if the ratio is greater than 0.5 atm, recovery efficiency of the sampler is affected and flow rate must be measured (Tyler and Shipe, 1959). For maximum collection of bacterial cells a volume of 20 ml of sampling fluid and a distance of 30 mm between the capillary tip and the bottom of the AGI-30 has been experimentally established. Variation in either of these parameters may lead to variation in sampling efficiency. The assumption that this distance and flow rate is also optimal for collection of phage, rickettsiae, and viruses may not be valid.

Shipe et al. (1959) hypothesized that direct impingement of vegetative bacteria against the bottom of the AGI at near sonic velocity might kill or injure cells to such an extent that the sample would not be an index of the total number of viable cells entering the sampler. Data of Tyler et al. (1959) show that there can be loss of viable count as compared to radioactive count when P^{32} and S^{35} labeled suspensions of Serratia marcescens, Brucella suis and Pasteurella pestis are sampled. Similar results were obtained when a stained-cell count was compared with a viable-cell count. Particles of less than 0.3 μ may be carried by the high velocity of the jet air stream through the impinger fluid without being trapped, or particles of greater than 3 μ diameter might be retained by the capillary intake tube and not be impinged. The Shipe sampler was designed to eliminate the intake tube and minimize splashing of sampler fluid. Data in Table 4-1 are presented to show the variation in bacterial count obtained with the use of four different liquid impingers.

Frequently, two samplers are placed in series and bacteria recovered from the second are assumed to be a result of "leakage" of the first. This conclusion is not necessarily valid since particles not trapped in the first sampler may also penetrate the second sampler. Alternately, since cotton is harmful to some vegetative organisms, the cotton filter used as a total sampler to evaluate the biological efficiencies of other samplers, may not yield valid data. Some observed variations in sampler "efficiencies" are shown in Table 4-2.

The relative effectiveness of different samplers varies with the size of particle being sampled. Data in Table 4-1 illustrate the high variability of mean recoveries when small, heterogeneous droplets were sampled; the larger the droplets the greater the disparity in recovery. This could have been caused by (1) unequal de-aggregation of large droplets when impinged on the liquid surface, or (2) retention of droplets in the inlet tube of the impinger followed by death of the trapped cells. Tyler et al. (1959) found that the percentage of organisms retained decreased

Table 4-1   Recoveries in four samplers of small or heterogeneous aerosol droplets of *Serratia marcescens* and *Bacillus subtilis* spores.

Aerosol	Type of Sampler	Plate count per L aerosol at indicated period (min):			
		2–3	4–5	8–9	16–17
S. marcescens in mixed aerosol (small droplets[a])	Shipe sampler	4,700[b]	4,550	3,500	1,820
	Midget impinger	4,500	3,430	2,950	1,980
	All-glass impinger	3,780	3,250	2,060	1,710
	Capillary impinger	1,560	1,480	1,210	700
B. subtilis in mixed aerosol (small droplets[a])	Shipe sampler	5,340	6,930	4,230	4,800
	Midget impinger	4,100	3,700	4,000	3,600
	All-glass impinger	5,440	4,780	4,230	4,280
	Capillary impinger	5,900	4,700	5,000	4,860
S. marcescens aerosol (heterogeneous droplets[c])	Shipe sampler	20,400	16,600	10,900	5,900
	Midget impinger	12,800	12,300	12,900	4,000
	All-glass impinger	6,800	5,400	4,000	2,800
	Capillary impinger	6,800	6,700	5,000	3,300

[a] Less than 3.0 μ MMD (mass median diameter), Devilbiss No. 40 nebulizer.
[b] Each value is mean of 5 samples. Suspension counts: mixed, *S. marcescens*, 11.7 $\times$ 10[8] per ml; *B. subtilis*, 7.1 $\times$ 10[8] per ml; *S. marcescens* only, 24.6 $\times$ 10[8] per ml.
[c] Greater than 3.0 μ MMD, explosive disseminator in test chamber.
(Shipe, Tyler and Chapman; *Appl. Microbiol.*, **7**:352, 1959. Reprinted with permission of American Society for Microbiology.)

with time and ascribed this phenomenon to a decrease in the number of large droplets remaining airborne.

Air Washing Samplers

The midget impinger, Edgewood bubbler, and Venturi scrubber operate at a lower flow rate than capillary impingers. As a result they are less destructive of vegetative cells, but collection is inefficient (Table 4-1). However, the midget impinger collects large droplets (>3 μ) more efficiently than the AGI-30.

Pre-impinger/Impinger Designs

If particles 5 μ or larger are removed before the sample is conducted into an impinger, there is less likelihood that material will be trapped in the impinger inlet tube. This pre-impingement technique provides a more accurate measure of the fraction of particles small enough to be potentially hazardous, especially when poly-disperse aerosols are sam-

Table 4-2 Loss of bacterial droplets from exhaust of 3 samplers in *Bacillus subtilis* aerosols of small droplet size.[a]

| Sampler type | Sample No. | Spores per L aerosol | | Loss from exhaust |
		Test sampler	Cotton collector in series	
		$\times 10^3$	$\times 10^3$	%
	1	6300	16	0.25
	2	4700	17	0.35
	3	7800	12	0.14
All-glass impinger	4	4100	13	0.32
(typical trial)	5	6400	10	0.16
	Mean	5900	14	0.24
Mean loss in 23 tests: 0.30 ± 0.06[b]				
	1	58.5	11.2	16
	2	28.5	20.0	41
	3	133.5	18.3	12
	4	152.5	18.8	11
Midget impinger	5	184.5	17.0	8
(typical trial)	6	200.5	33.6	14
	Mean	257.0 *(sic)*	19.8	17
Mean loss in 24 tests: 30.0 ± 4.0				
	1	7400	995	12
	2	8000	1070	12
	3	9900	1075	10
	4	9700	1195	11
Venturi scrubber	5	9900	1255	11
(typical trial)	6	9300	1235	12
	Mean	9000	1138	11
Mean loss in 42 tests: 10 ± 0.35				

[a] Aerosol MMD (mass median diameter) < 3.0 μ in plastic sphere aerosol unit.
[b] All variations from means refer to standard deviation of the mean, unless otherwise noted.
(Tyler and Shipe; *Appl. Microbiol.*, **7**:344, 1959. Reprinted with permission of the American Society for Microbiology.)

pled. The pre-impinger, fitted to the input of an impinger, behaves in a manner similar to the respiratory tree; i.e. larger particles are retained in the "upper" portions and only particles below a given size penetrate to more distal areas.

The multistage liquid impinger (May, 1966) simulates the respiratory tract even more closely. Particle discrimination is in three stages: particles that would lodge in the upper respiratory tract are removed in stage 1, those that would impact in the bronchi or bronchioles in stage 2, and those reaching the alveoli in stage 3. In addition, rates of evaporation of sampling fluid are less than in a standard impinger, and impingement is gentle; fewer organisms are killed or injured during prolonged sampling periods than with standard impingers. When compared with other samplers, this multistage impinger yielded an increased number of viable *E. coli* cells. There is no proof that cells are not killed in both samplers being compared, although under some conditions estimated recoveries near 100% have been obtained with *E. coli*.

IMPACTION

Impaction implies collection of particles by solid or semisolid surfaces, such as nutrient agar. A characteristic of impactors is that one colony is obtained for each viable droplet or particle, even though the droplet may contain more than one organism. This is in marked contrast to liquid impingers in which bacterial clumps are usually broken into single-cell units as a result of agitation in the collecting fluid. Because particles may superimpose on the agar surface, the colony count may be considerably less than the actual number of viable organisms in the air sampled. This is particularly true when sampling air of high bacterial concentration. Injury to vegetative cells by air flow during agar impingement may occur, possibly as a result of evaporation and increased concentration of inhibitors in the agar.

Slit Samplers

The most commonly used impaction samplers are the Bourdillion or Decker slit samplers in which a measured volume of air is directed through a slit against a rotating nutrient agar surface. Any variation in the viable particle concentration with time can therefore be measured. Goldberg and Shechmeister (1951) evaluated a number of factors that affect the quantitative recovery of viable organisms sampled with the Bourdillion slit sampler. They found that slit-to-agar distance, slit width, and air velocity are interrelated factors significant for sampling efficiency. For example, with a given air velocity and slit width, a 40% reduction in recovery of *Pasteurella pestis* occurs when the slit-to-agar distance is decreased from 4 mm to 2 mm; with a given slit-to-agar distance, decrease in slit width from 0.028 cm to 0.015 cm also resulted in a reduction in the number of viable colonies obtained. Goldberg and Shechmeister hypothesize that a fracture or shattering of bacterial cells may occur with the use of a slit width less than a critical value and that when the slit-to-agar distance was increased, widening of the air jet before its impaction against the agar may have increased the degree of shattering. Sampling of bacterial aerosols by intense air velocities may also result in an appreciable reduction in the recovery of organisms.

Comparative studies made at the Naval Biological Laboratory have shown that slit configuration, agar plate rotation rate, and relative humidity of the aerosol being sampled influence the collection of viable airborne particles in the slit sampler. The use of two samplers at a given station, one with a low rotation rate to give resolution in time and the other with a rapidly rotating plate to indicate the total integrated number of viable airborne particles passing the sampling station, is suggested. At present there is no general rule that will allow one to predict the efficiency of recovery of a given species under given conditions.

The Stacked Sieve

The Andersen sampler has been selected as an alternate reference sampler, where "concentrations of cells are too low to be adequately sampled by the impinger, or where the number of airborne particles is being determined" (Brachman *et al.*, 1964). The unit consists of a series of six stages, each composed of a perforated plate located above a petri dish containing agar culture medium. The size of the holes is constant for each stage, but of smaller diameter with each succeeding stage. The velocity of the air drawn through the sampler increases with each succeeding stage; hence, viable airborne particles are both sized and counted. The smaller the particle, the greater will be the chance of shattering of the bacterial cell by high velocity impaction, although it is no greater than in any other impactor that would collect the given particles. Dessication of viable particles on lower stages is minimized because of moisture transferred from preceding stages (Andersen, 1958).

Andersen reported that glass and aluminum petri dishes were equivalent, but plastic dishes gave consistently lower counts, retaining aerosol particles on the exterior of the plastic dishes and on the walls of the sampler itself. The phenomenon was hypothesized to be due to an electrostatic charge on the plastic dishes. Dimmick (see chapter 7) found no significant difference in sampling efficiency whether aluminum, glass, or plastic plates were used in the Andersen sampler.

May (1964) reported that for stages 1 and 2 (large particles) valid calibration of the Andersen sampler is impossible because of variations in particle deposition in different areas of a given stage. The intake efficiency for the larger particles was also poor. May, therefore, modified the sampler by adding a stage for impaction of large particles. The jet-to-agar clearance was increased and the pattern and size of holes was altered in stages 1 and 2. Lidwell and Noble (1965) recommend adding a pre-stage in order to adapt the Andersen sampler for use with airborne, bacteria-carrying particles between 12 μ and 19 μ. Dimmick suggested that the flow rate should be high when small particles are expected, and lowered when large particles are anticipated. This avoids using plates of several different volumes, or changing the configuration of the cast pieces.

A series of trials carried out by Leif, W. R. and Hebert, J. E., at the Naval Biological Laboratory, indicated that the Andersen (1962) aerosol monitor, on which a neoprene band had been employed to contain the molten nutrient agar, was not suitable for sampling all viable airborne microorganisms. A series of petri plates was prepared, using the same batch of agar medium in contact with wedges of neoprene material. *Francisella tularensis,* deposited from an aerosol on these plates, did not grow on the medium above the sectors of neoprene. Data have indicated that growth of *S. lutea* and *B. suis* is also suppressed by agar which has been in contact with neoprene band material. Quantitative sampling can be accomplished if the neoprene band is replaced by a stainless steel band to contain the molten agar. These results suggest that neoprene contains a substance that is toxic to *F. tularensis, S. lutea,* and *B. suis.* Since growth of *Bacillus globigii* is not affected by use of a neoprene band, toxicity must vary with the organism sampled. The general principle of testing potential toxicity of materials used on or near sampling medium cannot be stressed too highly.

Effects of atmospheric salt on collection of airborne *F. tularensis* by direct impaction on nutrient media have been compared in Andersen and Decker slit samplers by Leif and Hebert at this laboratory. At salt concentrations of over 10 μg per liter of air, the slit of the Decker sampler became occluded during a sampling period of about 20 minutes. No occlusion of the holes in the Andersen sampler was noted during sampling periods as long as 1 hour.

Green and Green (1968) nebulized a mixed suspension of P^{32}-labeled *Staphylococcus aureus* and S^{35}-labeled *Proteus mirabilis;* the aerosol was collected on an Andersen sampler with nutrient broth in the sampling plates. Loss of viability was calculated from the ratio of bacterial to radioactive count for each organism in the nebulizer suspension and aerosol. The authors report that viability of *S. aureus* was unaffected by aerosolization and assay, whereas viability of *P. mirabilis* declined.

FILTRATION

Filtration implies collection of particles by removal on fibers or fine-structured sieves. Samplers of this type are generally considered to be "total collectors." Spores and other resistant bacteria in viable form are collected completely and efficiently on a variety of filter media such as cotton, glass wool, paper, or molecular filter membranes. However, variations in the degree of "resistance" of microorganisms to air flow through a filter leads to inconsistent results. Except with membrane filters, it is necessary to wash entrapped viable particles off the filter surfaces, or to disintegrate the filter material, and then culture the wash fluid. Accurate assay presupposes that all entrained microorganisms can be washed from the filter; data of Wolochow (1958) show this assumption to be untenable in all instances.

Leif, W. R. and Hebert, J. E. (personal communication), at the Naval Biological Laboratory, investigated the quantitative collection of eight bacterial species on five different filter materials and found that airborne spores of *B. globigii* could be quantitatively recovered from the filter materials tested immediately after collection. A 72-hour delay in assaying the spore content of a filter pad did not materially affect the recovery of airborne spores. However, none of the vegetative forms tested was recovered quantitatively, even when filter materials were subjected to assay immediately after collection.

Molecular Filter Membranes

Bacterial cells deposited from aerosols on the upper surface of membrane or molecular filters (MF) form colonies when the filter is placed on a suitable nutrient surface such as agar or blotting paper saturated with liquid medium. Assay is therefore easier and should be more quantitative than that with other types of filters. However, it has been found that erratic and inconsistent viable counts of several organisms are frequently obtained on the membrane filter (Wolochow, 1958). No one factor seems to be operating: (1) media satisfactory for use in other applications is not necessarily optimal for growing colonies on the MF, (2) weight and wetting time vary within and between filters of different lot numbers, (3) incubation of the MF in plastic or glass petri dishes, and in a water saturated or ambient atmosphere, results in a difference in the maximal number of colonies obtained, (4) the use of too much liquid medium may result in contamination of the medium by growth from the upper side, and (5) since filters may not be wetted uniformly, detergents have been added during the manufacturing process (Cahn, 1967). Detergents are known to have detrimental effects on some microbes.

Webb (1965) reported that the death rate of cells on a millipore filter is affected by the number of cells on the filter. When large numbers of cells were used, considerable protection of those cells was noted, but no mechanism whereby cells can protect each other has been proven.

Although loss of bacterial cells does not seem to occur either through the MF itself or around it, further developmental work is necessary before the MF can be considered to be entirely satisfactory.

PRECIPITATION

Thermal Precipitation

Thermal precipitation implies that particles are gently impacted on cooled surfaces at velocities imparted by the differential pressure of air in a thermal gradient.

Air to be assayed, flowing at 0.3 liter per minute, passes between a heated and a cooled surface separated by 0.015 inch. For assay of microorganisms, the collecting surface is covered either with filter paper saturated with nutrient medium or with nutrient agar (Kethley et al., 1952).

Comparative assays of airborne S. marcescens and Bacillus subtilis spores, collected by both the thermal precipitator and the liquid impinger, were carried out (Orr et al., 1956); no slippage was observed and results were considered comparable. For relatively low cell concentrations, the method is rapid, simple, and direct. However, low viable cell recoveries are obtained at high aerosol concentrations. Another disadvantage of the method is the low air flow rate and the fact that, with sampling times greater than about 5 minutes, drying of the filter paper may occur, leading to desiccation and destruction of sensitive organisms. It is possible that the inlet air might be humidified.

Electrostatic Precipitation

Electrostatic precipitation implies that particles are gently impacted on surfaces at velocities imparted by charge differentials in an electrostatic field gradient. Several types of electrostatic samplers have been developed. All operate on the principle of precipitation of charged airborne particulates onto liquid or solid medium contained in petri plates or on the inside of a cylinder. Since influent air passes along an ionizing electrode maintained at a high potential, viable organisms are subjected to the known lethal action of ozone. In view of the bactericidal effects of air ions discussed by Krueger in chapter 5, it is probable that viable recovery is affected by air ions produced in the precipitator.

Tubular Precipitators

An electrostatic sampler developed by Houwink and Rolvink (1957) for assay of bacterial aerosols was modified by Morris *et al.* (1961) to permit rapid consecutive sampling over a wide range of aerosol concentrations and air volumes. The numbers of viable bacteria obtained from parallel samples with the precipitator/slit sampler and with the syringe or fluid precipitator/impinger were compared. Collection efficiency approaching 100% was obtained when aerosols of *B. globigii* spores, *S. marcescens*, *E. coli*, or *S. lutea* were assayed by using the appropriate precipitator.

Viability of *E. coli* and T3 bacteriophage collected in impingers was found to increase with RH; at 90% RH the recovery of coliphage was greater in impinger fluid than from fluid in a precipitator. Morris *et al.* (1961) attribute the phenomenon to the lethal action of ozone produced in the electrostatic sampler. Coliphage particles were protected when they were sprayed from suspensions containing 0.1% peptone. Another interpretation of these data might be possible in the light of Hatch's (1967) RH-dependent reactivation work with phage mentioned in chapter 11.

One major drawback to the use of the tubular electrostatic precipitator is the necessity for special technical skill in preparation and operation of the equipment.

Large Volume Sampler

Direct air sampling is of particular value when it is desirable to recover respiratory pathogens from natural aerosols. Meningococci have been detected at a concentration of 1 viable particle per 100 ft^3 of air and adenoviruses at 1 tissue culture infective dose per 300 to 3,000 ft^3 of air (Artenstein and Miller, 1966). This indicates the need for sampling large volumes of air. The large-volume sampler (LVS) designed by Litton Systems Inc., Minneapolis, Minn., functions by electrostatic precipitation and is capable of drawing air flows up to 10,000 liters per minute (Gerone *et al.*, 1966). The air passes through a high-voltage corona that charges particulate matter, causing it to precipitate on a grounded disc which rotates at 200 and 300 revolutions per minute. The disc is covered by a thin flowing film of collecting fluid that is varied in composition, depending on the organism being collected. Evaporation of collecting fluid is such that only short periods of sampling are possible. Nonvolatile collection fluids, such as glycerol, have been used, but many microorganisms cannot tolerate these solvents, and viability would be decreased with their use. Contamination of the sampling fluid sometimes occurs, and great variations in the viable recovery data have been reported. Gochenour (1966), in his critical evaluation of the LVS, states that it is at present only a qualitative sampler.

Airborne rabies virus was recently isolated from Frio Cave, Texas, using the LVS (Winkler, 1968). Virus was not detected by either the AGI-4 or Andersen sampler, although sentinel animals became infected. Winkler suggests that the success was a result of the increased capacity of the LVS to concentrate air in a given volume of collecting fluid. A 100-fold greater concentration of particulate matter was found in the LVS compared to the AGI-4.

CONTINUOUS SAMPLING TECHNIQUES

One problem encountered in the field is detection of changes in airborne bacterial contamination during an extended period of time. Epidemiological studies of modes of transmission of airborne disease would be simplified if more were known regarding the bacteriological content of the air in locales where disease has been transmitted by suspected airborne routes. Two time-sequence, impaction-type devices that permit continuous collection of airborne organisms are the slit-incubator sampler (Decker et al., 1958) and the Andersen air monitor (Andersen and Andersen, 1958). Both of these samplers are somewhat cumbersome and difficult to use. An additional drawback is that media is held in special containers—standard petri plates are not used. There have been a number of other complex instruments of like nature tested in various laboratories, but none has been satisfactory.

A continuous-flow liquid impinger applicable for use as a bacterial or viral sampler for particulates larger than 1 μ diameter has been developed by Goldberg and Watkins (1965). They reported that it was possible to concentrate the viable particulate content of 50 ft³/minute of air (at 1 ft³/min) into a terminal recycled fluid volume of less than 2 ml. A novel multi-slit, high-volume sampling device developed by Buchanan et al. (1968) will permit recovery of a high percentage of viable microorganisms from the atmosphere. It operates at a sampling rate of 500 liters of air/minute, with an output of 6 ml fluid/minute, has a high collection efficiency, and functions at a low pressure drop. Neither of these two samplers has been tested sufficiently to allow us to recommend them, pro or con.

BACTERIOPHAGE ASSAY

In general, when a phage aerosol is sampled, the sampling fluid from the impinger is mixed with an appropriate dilution of soft agar, previously inoculated with the host culture, then poured onto the surface of a plate and incubated. Plaques are read at a suitable time thereafter. Harstad (1965) compared liquid impingers, filter paper, and fritted bubblers as collecting devices for T1 phage aerosols. Using dynamic aerosols having a mass mean particle diameter of 0.2 μ, he recovered the largest number of viable phage particles by using liquid impingers (an AGI-4 with a flow rate of 12.5 liters/min or a capillary impinger similar to the AGI-4, but with a flow rate of 2.5 liters/min). However, 30 to 48% slippage occurred. Filter papers were the most destructive to phage. Slippage was greater than 80% with fritted bubblers. Regardless of sampler, phage death was highest at 85% and lowest at 55% RH. Particle sizes were observed to be larger at 85% than at 30% or 55% RH.

The most significant conclusions drawn from Harstad's report are: "Impinger physical slippage varies inversely with relative humidity, presumably because of changes in particle size with relative humidity. However, the particle size differences at the three relative humidities studied were small, indicating that impinger slippage is extremely sensitive to changes in the size of submicron particles." It is interesting that Green and Green (1968) also found that particle size was an important determinant in the assay of airborne P. mirabilis.

Sampling of bacteriophage aerosols by electrostatic precipitation was discussed in the previous section of this chapter. Again, humidity markedly influenced viable recovery from impinger samples.

VIRUS ASSAY

Viruses are usually assayed by modifications of standard aerosol sampling methods. Harper (1961) sampled four different viruses by using standard Porton impingers, whereas Thorne and Burrows (1960) used single and multi-jet impingers, membrane filters, and filter paper samplers to collect foot-and-mouth disease virus. Membrane filters and impingers were found to be of equivalent efficiency, but filter paper was less efficient. For maximum elution of virus from the filter paper, a detergent was added to the buffer solution; however, it is probable that detergent is toxic to many viruses. Encephalomyocarditis viruses, and nucleic acid isolated from these viruses, have been satisfactorily collected in the AGI-30 prior to assay in tissue culture (Akers *et al.*, 1966a, 1966b).

Jensen (1964) states that the AGI is less efficient for virus assay than the Andersen sampler, and reports slippage of less than 10% of the total for bacteriophage aerosols sampled by impaction. Use of the Andersen sampler results not only in the collection of virus, but also in sorting out of the sizes of droplets (with the correction for mass: see chapter 7) in which the virus is contained. Where the amount of virus is extremely small, samples from each stage can be concentrated by centrifugation. Collection medium can be inoculated into tissue culture, chick embryo, chorioallantoic membrane, or animals, depending on the growth requirements of the virus assayed. The fact that "dead" virus is capable of producing a positive hemagglutination reaction, allowed Guerin and Mitchell (1964) to determine both "fallout" and "decay" of airborne influenza PR8 virus.

Dahlgren *et al.* (1961) found the slit sampler to be only 75% as efficient as the AGI-30 when used to assay aerosols of T3 bacteriophage. Kuehne and Gochenour (1961) reported that comparable results are obtained, using either the slit sampler or liquid impinger for the assay of airborne Venezuelan equine encephalomyelitis virus.

MEDIA

The nature and composition of collection medium, dilution fluid, and plating medium constitute important parameters influencing survival characteristics of aerosolized microorganisms. Different classes of microorganisms (bacteria, fungi, viruses, and rickettsia) require different media for propagation. Some cells in an aerosolized population may be so sensitive to nutritional requirements that they are unable to produce countable colonies under conditions that exist during sampling. Generally, enriched-complex media are used to provide the most favorable environment for these cells to recuperate and grow, but limited media occasionally elicit better growth response than enriched media.

Collection medium which will preserve the capacity to form colonies on agar, while not permitting replication to occur in the liquid, should be chosen for impingers. This idealistic state is seldom achieved. Unless dilution and plating are carried out immediately after removing each impinger sample, some cell multiplication may occur. Therefore, a maximal colony yield is not necessarily a sign of an efficient impinger.

Additives to impinger fluid, such as agents to prevent foaming and to increase density, may be deleterious to sensitive organisms. When dilute aerosols are sampled for times longer than 1 minute, evaporation and concentration of the impinger fluid may lead to increased "toxicity" of some additives; injured cells are selectively susceptible to such increased toxicity. It is not valid to assume that because an uninjured cell will grow in the presence of a given additive, an aerosolized cell will be equally tolerant of the additive.

Serial dilution of cells from impinger fluid may result in inconsistent and variable data. Dilution not only changes cell concentration, but the dilution medium usually differs from impinger fluid. If cells respond to these environmental changes as they do to the changing conditions employed by Heckly et al. (1967), who demonstrated rhythmic responses to environmental shifts, erratic results are to be expected. An example of such data is presented by Won and Ross (1966a) who found disproportionate counts among triplicate plates, as well as between decimal dilutions, when P. pestis aerosol samples were serially diluted in 1.0% peptone. This discrepancy was not observed for cells aerosolized and held at RH values below 50% prior to dilution in peptone water. In contrast, cells maintained at 87% RH and then collected and serially diluted in heart infusion broth, yielded the expected proportionality of counts as a function of dilution, and produced higher recoveries of viable cells as well. Furthermore, the use of heart infusion broth as a diluent permitted the detection of cells from an aged aerosol of P. pestis, whereas the same sample diluted with 1.0% peptone water gave no indication of viable cells. It is likely that the variable results reported by Won and Ross reflect the phenotypic expression of injured cells in growth environments where optimal conditions for rejuvenation are not precisely known or employed. It is noteworthy that the effect of diluent was greatest for cells aerosolized at the most deleterious RH values and therefore, presumably, the most injured while airborne.

Direct impaction or plating on solid medium also constitutes a major and sudden environmental change. As with a liquid medium, the ideal solid medium should contain factors capable of maximizing the survival and multiplication of injured cells. In an in-house report, Dimmick and Hatch noted that diminutive (petite) colonies of S. marcescens and P. pestis appeared on plates from aerosols diluted with secondary air, under certain conditions of relative humidity. As reported by Dimmick (1965), for cells injured by heat, the diminutive characteristic was apparently not transmissible. Significant differences in numbers of survivors were found when cells were assayed on blood agar base (BAB) or BAB plus whole blood. These data are indicative of cellular injury, since the cells demonstrated a need for additional nutrients to repair damaged mechanisms.

Incubation temperature can also affect the composition of the medium required for maximum growth. Brownlow and Wessman (1960) have shown a more exacting requirement by *P. pestis* for essential nutrients, when grown at 37°C as compared to 30°C or below. Hatch and Dimmick (1966) have found that suppression of growth processes of *P. pestis* by storage at 4°C apparently permits some injured cells, on both BAB and BAB plus whole blood, to rejuvenate. The highest number of cells, in this instance, appeared on supplemented medium. When conditions optimal for recovery of damaged cells are required, they cannot be selected on the basis of conceptions derived from established information obtained with undamaged cells of the same species or by inference from results with other species (Harris, 1963).

Under special sampling conditions, e.g. at temperatures near freezing, the duration of sampling must be short enough to avoid freezing of the agar. A more satisfactory alternative is the development of a cryotolerant medium suitable for aerosol assay in sub-zero temperature environments. Won and Ross (1966b) developed a medium-base consisting of 3.0% cornstarch and 3.0% agar. Unlike conventional agar, the base is able to withstand freeze and thaw processes without water expression, and without losing structural integrity.

Aerosolized *E. coli*, *S. marcescens*, and *B. globigii* spores, sampled on both frozen heart infusion broth agar and starch base agar frozen with broth, yielded well-isolated colonies and significantly higher numbers of viable cells than those collected on corresponding unfrozen media at room temperature. The unexpected relationship between the high recovery phenomenon and a cold or frozen surface, is interesting, and corroborates results of Hatch and Dimmick with *P. pestis*. It seems conceivable that cold temperature reduced or arrested metabolic activities, inducing a dormant state that permitted cells to recuperate from stress incurred by the aerosolization process.

Consistently low and erratic growth of airborne *F. tularensis* on media solidified with the freeze-tolerant base, led to modification of the base by reducing the agar concentration to a 2.0% level and adding to it a 1.0% potassium alginate to provide controlled syneresis during incubation. Recoveries of airborne *F. tularensis* at various RH levels on unfrozen, modified, freeze-tolerant medium, were similar to those obtained on unfrozen conventional 1.5% agar base medium. Fewer colonies were always found from initial aerosol samples taken on frozen, starch-alginate base medium than on either the unfrozen control medium or the unfrozen freeze-tolerant medium. It is obvious, therefore, that a medium satisfactory for use in one application is not necessarily optimal under other circumstances.

Wolochow (1958) points out that culture media devised on the basis of other uses, may not be satisfactory for use with the membrane filter. It is conceivable that, as in other sampling situations, some cells in the population being filtered are borderline with respect to nutritional requirements. Under conditions that exist on the filter (in contrast to those on an agar surface), such cells are unable to reproduce to the point of producing countable colonies, although these cells may not be the same ones that fail to grow on agar surfaces.

In conclusion, one must always be aware, as Kethley *et al.* (1957) so aptly point out, that it is possible to vary the total number of "viables" (apparent concentration level) by altering the sampling fluid without appreciably varying the death rate constant. This suggests that the differential of total viable cells collected is attributable not to a failure to mechanically collect organisms, but rather a failure to demonstrate them as being viable.

RELATIVE HUMIDITY

In any discussion of the RH effect on survival of microorganisms, it is assumed that the shift from a particular RH in the aerosol chamber to 100% RH in the sampler fluid does not alter the number of viable organisms in the sample assayed. This is not a valid assumption. Webb (1965) reports that rehydration has a marked effect on survival; one can then infer that rehydration would also affect the viable count obtained on assay. Ion control and protein synthesis studies summarized by Anderson and Cox (1967) suggest that viable bacteria recovered from aerosols are not unchanged rehydrated forms of the original bacterium. No evidence is presented, however, that would permit differentiation between cellular alterations occurring as a result of aerosolization (dehydration from 100% RH in a liquid suspension), storage as an aerosol, and rehydration of sampling.

That an RH shift does in fact alter the capacity to replicate, has been demonstrated by Hatch and Dimmick (1966) and Akers *et al.* (1966a). It is difficult, however, to predict the magnitude or "direction" of error introduced into sampling data by this phenomenon, especially since bacteria, phage, and viruses appear to respond differently to an upward RH shift.

THE ANIMAL AS A SAMPLER

Historically, "sentinel" animals have been used as indicators of the presence of several species of airborne pathogens in naturally contaminated environments. Sentinel mice became infected by airborne *Mycobacterium tuberculosis* bacilli according to studies reported by LeNoir and Camus in 1909. Dermatophytes (*Trichophyton mentagrophytes* and *Microsporum gypseum*) were recovered from lungs of sentinel mice, rats, rabbits, guinea pigs, and monkeys (Lurie and Way, 1957); rabies virus was shown to be present in air by infections produced in sentinel coyotes, foxes, ringtails, opossums, and hamsters (Constantine, 1962, 1967). Furcolow (1961) recovered airborne *Histoplasma capsulatum* from sentinel rodents as did investigators in Mexico, Panama, and Africa. Converse and Reed (1966) recovered *Coccidiodes immitis* from monkeys exposed to desert air and from dogs exposed to air and soil. Brachman *et al.* (1966) demonstrated *Bacillus anthracis* in air by means of sentinel monkeys that became infected.

Experiments by Riley *et al.* (1962), in which air flow from wards housing tuberculosis patients was directed through a chamber of guinea pigs, demonstrated the usefulness of direct air sampling toward providing an understanding of the aerosol transmission of this disease. Their results demonstrated that 1 infective particle is contained in approximately 12,000 ft^3 of sanitorium air. The extreme sensitivity of the guinea pig as a sampler for tubercle bacilli is exemplified in these experiments. Hospital air contains 1 to 10 viable particles/ft^3, or 10^4 organisms/ 12,000 ft^3; using a mechanical sampler it would be almost impossible to detect the "one" tubercle bacillus among 10^4 viable particles.

Based on epidemiologic data, calculations for measles virus suggest that one human infective dose is contained in about 3,000 ft^3 of schoolroom air (Riley and O'Grady, 1961). Couch *et al.* (1966) have used human volunteers as samplers for the presence of respiratory disease agents. Their results showed that individuals infected with respiratory viruses produce airborne virus in quantities sufficient to infect susceptible individuals. Assuming normal breathing by susceptible volunteers and an infectious dose of less than 50 TCID$_{50}$, assay of air samples from rooms occupied by infected volunteers indicated that transmission would be accomplished in from 5 minutes to 24 hours.

Serologic testing of humans in suspect locations or occupational groups was selected as one of the main sources of information in an epidemiologic investigation of an urban Q-fever outbreak in 1959 (Wellock, 1960). A majority of cases resided within a fan shaped area spreading out downwind from the suspected area of origin. Evidence is presented that wind-borne effective dissemination of Q-fever occurred up to a distance of 10 miles. A warm dry northwest wind was prevalent in the late afternoon during the spring and summer of 1959 when the most marked dissemination of Q-fever occurred.

Experimental Laboratory Animal Exposure

Experimental investigations of airborne contagion afford a valuable working basis for epidemiological studies not feasible under field conditions such as those described in the preceding section. In fact it is likely that many models of epidemiological processes, constructed from data obtained by routes of infection other than the respiratory route, are not applicable to real situations. In order to reproduce (as naturally as possible within the framework of safe experimental techniques) conditions underlying aerial transmission of bacteria and viruses, it is necessary to study atmospheres containing the infectious agent in a finely dispersed form, preferably droplets with diameters of 5 μ or less. Since animals infected by the pulmonary route must be considered a potential source of infectious airborne particles, the development of an isolation unit became a necessary adjunct to any study of infectious airborne droplets.

The principles developed by Reyniers and his associates (Reyniers, 1943), although designed for germ-free studies, were uniquely suited to studies involving airborne infection. Various systems adapted from that of Reyniers were developed to measure quantitatively the response of animals subjected to graded doses of bacterial aerosols at given temperatures and relative humidities. Isolation of the animals from their natural laboratory environment must be provided so that variations in the isolated environment can be introduced and controlled. Chambers of various configurations and degrees of sophistication are described in reviews by Rosebury (1947) and Leif and Krueger (1950). Schematic diagrams, photographs and flow sheets of exposure chambers, animal storage units, sampling equipment, and safety devices are detailed in articles by Henderson (1952), Druett and May (1952), Wolochow, Chatigny and Speck (1957), Griffith (1964), Wolfe (1961), Thiéblemont et al. (1965), Akers et al. (1966b), and Beard and Easterday (1965).

In brief, an aerosol exposure system consists of an atomizer (usually a modified Wells atomizer (DeOme 1944)) to generate the aerosol, a mixing chamber in which the aerosol is diluted with secondary air to provide for humidity control, an exposure chamber in which experimental animals are held, and a sampling device (liquid impinger, slit sampler, Andersen sampler) from which the concentration of organisms in the aerosol can be estimated (Figure 4-1). Ideally, a particulate light scatter monitor should be included in the system so that physical fallout can be determined and operational conditions can be monitored. A simplified exposure system requiring no secondary air, and which can be assembled within a negative pressure hood is shown in Figure 4-2 (Akers et al.

FIGURE 4-1. Aerosol exposure unit (ABX): *A*, atomizer; *B*, secondary air; *C*, exposure chamber; *D*, effluent air line; *E*, impinger; *F*, mouse exposure cage. (Akers, Bond, Papke, and Leif. 1966. *J. Immunol.*, 97(3): 380.) Reprinted with permission of the Williams and Wilkins Co., Baltimore, Md.

1966b). The unit is used in a gas-tight hood under slightly negative pressure. The plexiglass cover for the mouse-chamber, *B*, is taped in place. Air pressure within the chamber is balanced by the simple expedient of observing whether a piece of tape covering a ½-inch hole in chamber, *B*, bulges or sags. Flow rate through an AGI-30 impinger,

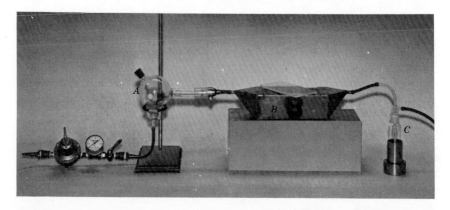

FIGURE 4-2. A simple, dynamic aerosol exposure apparatus: *A*, atomizer; *B*, exposure chamber; *C*, impinger. (Akers, Bond, Papke, and Leif. 1966. *J. Immunol.*, 97(3): 381.) Reprinted with permission of the Williams and Wilkins Co., Baltimore, Md.

C, is fixed at 12.6 liters/minute and air-flow rate through the atomizer, A, is adjusted to keep the tape flat.

For testing the infectivity of aged or persistant aerosols, in contrast to transient aerosols, airborne microorganisms can be held in a rotating drum (Goldberg et al., 1958) prior to animal exposure.

Particle Size

It should be noted that an experimental animal both inhales and swallows microbes as a result of respiratory exposure. Goldberg and Leif (1950) exposed mice to a test aerosol of P^{32}-labeled $P.$ pestis and found that whereas the total retention was over 80% of the calculated dose per mouse, only 30% of that material retained (24% of the total) was in the respiratory tree; 70% was in the gastrointestinal tract. Pappagianis (1953) reported an average of 13% retention of spores in the lungs of mice exposed to an aerosol of $C.$ immitis (unpublished data).

Between species, lung volume is proportional to body weight, whereas alveolar surface area correlates linearly and directly with metabolic activity (Tenney and Remmers, 1963). Alveolar size appears to be a function of animal size (Figure 4-3). Two animals of equal body size will

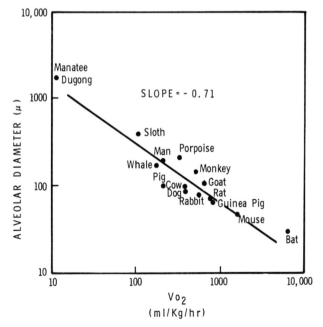

FIGURE 4-3. Logarithmic plot of mean alveolar diameter in microns as a function of metabolic rate per unit of body weight. (Tenney and Remmers, 1963. *Nature,* **197:** 55.) Reprinted with permission of Macmillan (Journals) Ltd., London.

have lungs of the same volume, but if one has a higher rate of metabolism, its alveoli will be smaller. Particle size is therefore an extremely critical factor for respiratory penetration and retention in laboratory animals. With a homogeneous aerosol of 1- to 2-μ particles, average respiratory retention is 15% for the mouse, 50% for the guinea pig, and nearly 100% for monkeys.

Druett *et al.* (1953) reported that the number of anthrax spores necessary to cause infection increases with the particle size, and is related to the fact that the majority of the larger sized particles are deposited in the upper respiratory tract. Similar results were obtained when groups of guinea pigs were exposed to single-organism clouds of *P. pestis* and the results compared with those from animals exposed to particles of 12 μ diameter (Druett *et al.*, 1956). It was noted, however, that over the dosage range covered, the time of death in any of the groups was independent of dosage. Small particles initiated bronchopneumonia followed by septicemia and death, whereas animals exposed to large particles developed a septicemia without pneumonia. Again, this reflects the difference in site of deposition of particles in the respiratory tract, since a large number of particles deposited higher in the respiratory tract appear to move rapidly to the lymphatic system whereas smaller particles are more likely to reach alveolar walls.

Treating Sample Data

The respiratory LD_{50} (also referred to as the LRE_{50}) is defined as the number of airborne organisms that results in death of 50% of a group of animals exposed to a virulent aerosol. Virologists, however, define an LD_{50} as the *dilution* of the original virus suspension resulting in 50% mortality; the LD_{50} titer is reported as a negative power of the base 10 and the larger the exponent, the more virulent the organism. A respiratory end point can be expressed in terms other than mortality. Lurie *et al.* (1952) used the number of primary pulmonary foci as a quantitative index of resistance to experimental tuberculosis as well as an index of the virulence of the infecting bacteria. Dannenberg and Scott (1956) showed that in melioidosis, where a single primary pulmonary lesion is subsequently fatal to mice and hamsters, it is possible to use the lesion-count method to estimate the LD_{50} dose. Their method shows good agreement with usual titration methods.

A suspension being aerosolized should be assayed immediately before and after the spray period to determine the effect of the nebulization process on "survival" of the test organism. The "spray factor" is defined as the ratio of the number of organisms per unit volume in the aerosol to the number of organisms per unit volume in the suspension from which the aerosol was produced. If secondary air is introduced into the exposure chamber, a correction factor must be applied to correct for the dilution of the actual aerosol by this secondary air.

$$\text{Theoretical maximum aerosol concentration} = \frac{\text{(organisms/ml in suspension)} \times \text{(ml suspension aerosolized/min)}[1]}{\text{(air flow in liters/min)}}$$

$$\text{Concentration of aerosol/liter recovered} = \frac{\text{(organisms/ml in sampler)(ml of sampling fluid)}}{\text{(volume of aerosol sampled in liters)}}$$

$$\text{Nominal \% recovery} = \frac{\text{concentration of aerosol recovered}}{\text{theoretical maximum aerosol concentration}}$$

In assigning a numerical value to the LD_{50} dose of a given organism, Guyton's (1947) formula is generally used.

If t = time of exposure of an animal to the aerosol,

c = organism concentration in ml of aerosol computed from the impinger samples,

W = average weight of the animals (gm),

A = respiratory volume of the experimental animal

= 1.25 ml/min/gm for a mouse (Guyton, 1947).

Then

$$\text{organisms inhaled per animal} = (t)(c)(W)(A)$$

and the respiratory LD_{50} = organisms inhaled/animal resulting in 50% mortality.

The LD_{50} can alternately be calculated as organisms retained per animal by multiplying the inhaled dose by the appropriate factor for percent retention in the respiratory tract of the experimental animal used. One assumes that the concentration of an aerosol determined by a particular sampling device is the "effective" concentration to which animals are exposed. It is evident that no sampler will collect a higher concentration of viable organisms than actually exists—it is more likely that the apparent concentration of viable organisms will be less than the true number. If the indicated concentration is less than actually exists, numerical values assigned to the LD_{50} will be low by an amount determined by sampling loss.

Differences in the LD_{50} values of given bacterial aerosols, as reported

[1] The effective output (See Chapter 2).

by various laboratories, may be a reflection of different conditions under which aerosols were generated. If an aerosol is maintained in essentially a dynamic state while animals are exposed, the transit time (i.e., the time from generation of a particle to exposure of an animal or a sampler) can be estimated. To illustrate, a total aerosol flow rate of 1 ft³/minute through an infection chamber of 4×4 inches inner cross section and 24 inches length would have a transit time of the order of 15 seconds. The linear flow of the aerosol within the chamber would be approximately 9 ft/minute and the probability of any large number of particles remaining in the infection chamber for longer than 15 seconds will not be great. On the other hand, if the aerosol is generated by atomizing into a chamber with a volume of approximately 20 ft³ while maintaining a flow rate of approximately 4 ft³/minute, there is a fair probability that a significant fraction of particles in the aerosol would have an aerosol age greater than 15 seconds.

During a relatively short exposure interval, aerosol decay can be expressed as a constant percent loss per unit time, or mathematically as

$$N_t = N_0 e^{-kt}$$

where N_t = viable organisms/liter at time t, the interval (minutes) between start and end of challenge,

N_0 = viable organisms per liter at the time of initiation of animal challenge,

k = exponential decay/minute,

M = breathing rate of the animal in liter/minute.

Inhaled dose for exposure time, t:

$$D = \int_0^t MN_t \, dt = \frac{MN_0(1 - e^{-kt})}{k} = \frac{M(N_0 - N_t)}{k}$$

if $kt < 0.1$, then $1 - e^{-kt} \cong kt$, and

$$D \cong MN_0 t \qquad \text{(Goldberg } et \ al., \text{ 1958)}$$

Since the ability of an organism to be sampled and to propagate itself on an agar plate varies as a function of aerosol age, differences in LD_{50} values may be observed. In addition, if it is assumed that an organism might be better able to propagate itself in the respiratory tree of an animal than on the surface of an agar plate (either directly or after having been entrapped by a capillary impinger) one can understand why a low indicated concentration of an aerosol sometimes has a high "effective" infectious concentration; this is especially apparent if propagation differences change as a function of time.

Host-Parasite Relationship

A common measurement, in fact the one that we are most interested in, is degree of infectivity, or virulence. There are a number of probabilities involved in this measurement, which results in two rather unfortunate circumstances: (1) an absolute dosage (one that will just infect all host animals) cannot be measured, but must be estimated in terms of a 50 percentile measurement—the lethal or infective dosage required to affect half of the exposed animals; and (2) the measurement is inherently uncertain.

Three methods have been evolved to estimate an LD_{50}; two, entirely arithmetical, are based on a short-cut method for maximum likelihood (Reed and Muench, 1938; Kaerber, 1931), and the other (Goldberg, 1960; Goldberg *et al.*, 1954) is a visually adjusted, triplet-probability method utilizing a special graph paper termed a "linearized mortality grid." These methods will be discussed briefly and the reader is referred to the original papers for a more detailed analysis. It is required that at least two graded doses, which infect fewer than all exposed animals, be available.

Example of the Method of Reed and Muench

Make a table, as shown, (Table 4-3). The first four columns are the observed data; the others are calculated. In the example, both dilutions of a "standard suspension" and measured numbers of viable units are used to demonstrate that, whereas the calculations are not identical, results can be interconverted.

Select two values, one just above and one just below the 50 percentile level.

Using dilutions, interpolate between the appropriate two dilutions; in the example, between 10^{-2} and 10^{-3};

$$\frac{\% \text{ mortality above } 50\% - 50\%}{\% \text{ mortality above } 50\% - \% \text{ mortality below } 50\%}$$

$$\frac{82 - 50}{82 - 8} = \frac{32}{74} = 0.432$$

$$\text{Logarithm of } LD_{50} \ titer = \frac{\log \text{ of dilution above}}{50\% \text{ mortality}} + \frac{\text{interpolated}}{\text{value}}$$

$$\text{Logarithm of } LD_{50} \ titer = \overline{2}.0 + 0.432 = \overline{2}.432$$

$$LD_{50} \ titer = 10^{-2.432}$$

Table 4-3   Animal Mortality data

I Measured numbers of viable units	II Dilutions of a "standard suspension"	III Died[a]	IV Survived[a]	V Mortality Ratio	VI Percent
212	10^0	10 ↑	0	10/10	100
21.2	10^{-1}	10	0	10/10	100
2.12	10^{-2}	8	2	8/10	80
0.212	10^{-3}	1	↓ 9	1/10	10
Cumulative mortality data					
212	10^0	29 ↑	0	29/29	100
21.2	10^{-1}	19	0	19/19	100
2.12	10^{-2}	9	2	9/11	82
0.212	10^{-3}	1	11 ↓	1/12	8

[a] Arrows indicate direction of addition for cumulative data.

Using dose, or numbers of organisms/animals: interpolate between the two dilutions; in the example between 2.12 and 0.212:

$$\frac{50\% - \% \text{ mortality below } 50\%}{\% \text{ mortality above } 50\% - \% \text{ mortality below } 50\%}$$

$$\frac{50 - 8}{82 - 8} = \frac{42}{74} = 0.568$$

$$\text{Logarithm of LD}_{50} \; dose = \frac{\text{log of dose below}}{50\% \text{ mortality}} + \frac{\text{interpolated}}{\text{value}}$$

$$\bar{1}.326 + 0.568 = \bar{1}.894$$

$\text{LD}_{50} \; dose = \text{antilog } \bar{1}.894 = 0.78 \text{ organisms}$
Conversion of LD_{50} *titer to* LD_{50} *dose*

$$10^{-2.43} = (10^{-3})(10^{-0.57}) = \bar{3}.57$$

$$\text{antilog } \bar{3}.57 = 0.0037$$

$$(0.0037)(212) = 0.78 \text{ organisms}$$

Example of the Method of Kaerber

This method is used primarily by virologists—the example will be given only for dilutions.

Log LD_{50} = 0.5 + log of highest concentration of organism used

$$- \frac{\text{sum of } \% \text{ of dead animals}}{100}$$

Using data in the preceding table,

$$\text{Log } LD_{50} \text{ titer} = 0.5 + (0) - \frac{100 + 100 + 80 + 10}{100}$$

Log LD_{50} titer = 0.5 − 2.9 = 2.4

LD_{50} titer = $10^{-2.4}$ and LD_{50} dose = 0.78 organism, in good agreement with the value calculated by the method of Reed and Muench.

Of course, 0.78 is an absurd number and the real value is probably 1.0 organism. The example used, which is taken from real data, illustrates that the estimate of the number of administered viable cells contained some indeterminate error—though not enough to be meaningful. In this example, the pathogen was administered via the intraperitoneal route and the varied doses were obtained by diluting the original suspension containing 212 cells per volume. Exposure via the respiratory route cannot always be performed with uniform dilution or increment ratios—the dose for each exposure must be measured and, despite unusual precautions, seldom turns out to have exactly the desired value. One can expose animals, dilute the atomizer fluid, expose again, etc., and assume the doses will decrease in a like manner, or one can expose groups of animals for different time periods. One might also employ the dosage increment that caused the percentage in column VI to change from less than 50% to more than 50%, but with the knowledge that the estimated LD_{50} will be less certain than if dosage increments had been uniform. The problem of obtaining uniform graded doses makes the employment of the Reed and Muench method not ideally suited to infectivity studies with aerosols.

Example of "Linearized Mortality" Graph Method

Figure 4-4 shows a "linearized mortality" graph with data from the previous example. Note that, in this instance, percentages of mortalities (column VI) are obtained directly from columns III and IV. A line of best fit at an exact 45-degree angle is drawn through the points. To insure that this criterion is met, the ordinate and abcissa on the

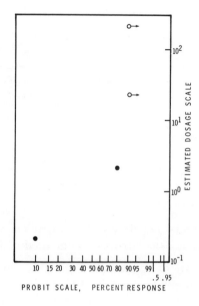

FIGURE 4-4. A "linearized mortality" graph showing data from Table 4-3. The probit scale is obtained from the scale in Figure 4-5, and the dosage scale is from Figure 4-6.

graph must be appropriately related. Axes are shown in Figures 4-5 and 4-6 in proper ratio so the reader may copy and/or enlarge them, and draw his own graph paper. Also shown in Figure 4-5 is a device to aid in determining the line of best fit. These sets for stated group sizes should be traced individually on a transparent sheet to act as an overlay. Sizes relative to the scales shown in the figures must be maintained.

In use, the 50 percentile line on the overlay is aligned to cover a vertical line drawn through a 50 percentile line on the graph, and the overlay is moved up or down along that line until all data points are subjectively located, equally, within the outer lines; the center 45-degree line is then the line of best fit. By lifting the overlay, several points to delineate the line of best fit could be marked on the graph, the overlay removed, and a line drawn through the points. A horizontal line drawn through the junction of the line of best fit and the 50 percentile line will terminate on the right-hand scale at the estimated LD_{50}.

One more step is needed. Figure 4-6 shows a "reactor count" scale which should also be traced on transparent overlay. The scale, which is used to determine the 95% confidence interval of the estimated LD_{50},

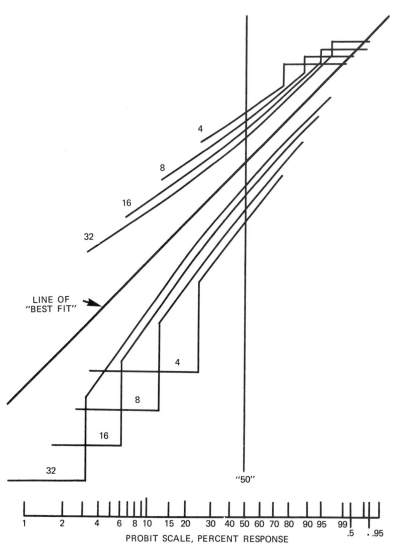

FIGURE 4-5. Ninety-five percent confidence interval overlay for groups of 4 to 32 animals. The probit scale is shown aligned with the 50 percentile axis. (Goldberg, L. J., in *Quantitative Methods in Pharmacology*, p. 189, 1960.) Reprinted with permission of the North Holland Publishing Co., Amsterdam, The Netherlands.

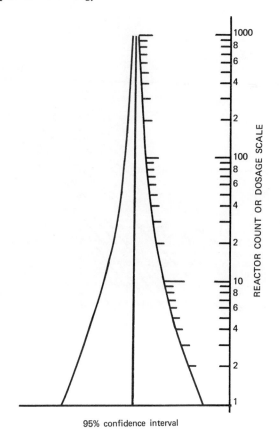

95% confidence interval

FIGURE 4-6. Ninety-five percent confidence interval of the LD_{50} as a function of total number of reacting animals. Note that the reactor count scale and the dosage scale are identical except that the log intervals of the dosage scale can be any value whereas the reactor scale must be as shown. (Goldberg, L. J., in *Quantitative Methods in Pharmacology*, p. 185, 1960.) Reprinted with permission of the North Holland Publishing Co., Amsterdam, The Netherlands.

is placed with the center line on the horizontal line representing the LD_{50} and moved until the total number of animals that have reacted (in groups where less than 90% reacted) is aligned with the right margin. In the included example, where group size was 10, 1 reacted in one group, 8 in another, and in the rest of the groups more than 90% reacted. In this instance the proper number is 9 animals. The points where the outer lines intersect the margin indicate the 95% confidence interval for the LD_{50}. The reverse process can be used: determine what confidence interval is required and continue gathering data until the number of reacting animals meets the requirement. Figure 4-7 shows a completed analysis for the included data; the LD_{50} is estimated to be 1.05 (a value very close to that found by the Reed and Muench method) and the 95% confidence interval is 0.5 to 2.2.

There are some cautions: (A) Since the exact location is subjective (the method is affectionately known as the "Ouija board"), two or more operators may not agree, but, with practice, deviations are well within the 95% confidence interval. (B) Mortality percentages above or below certain values (vertical lines on the best-fit overlay) must be plotted as if they were limited, minimal percentages. In the example, the 10% point is real; a 5% response would have been plotted on the 10% line, but with an arrow as in the response-point for a dosage of 21 cells,

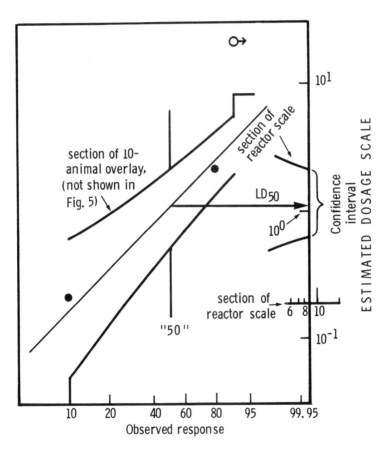

FIGURE 4-7. Completed analysis of infectivity data by the "linearized mortality" graph method.

and the overlay would be aligned to include the point on the vertical portion of the overlay. Points exceeding these vertical lines are not counted. (C) If all points cannot be located within the best-fit bounds, then either data are too scattered to be meaningful or the mechanism does not correspond to a single-hit model and another analysis must be employed.

The Dose-Response Model

The fundamental mathematical relationship between dosage and mortality illustrated by example in the preceding section, has been applied to analysis of data presented in the literature (Goldberg, 1960). The assumptions inherent in the derivation of the dose-response equation and its graphical expression, the linearized mortality grid, are: (1) any one inhaled bacterial cell acts on the host independently of any other inhaled bacterial cell; (2) the effect on the host of any one bacterial cell is independent of the time that this cell reaches the host; and (3) there is a measurable probability that this one cell will reach an environment favorable for its multiplication to the point where disease and subsequent death will occur (Goldberg et al., 1954).

The dose-response model has been verified by experiments in which daily mortality was successfully predicted from daily exposure levels over a period of 30 days. Mortality is a function of dosage and time during prolonged exposure; a given dosage of organisms administered for as long as 6 hours via the respiratory route results in the same mortality observed for an equal dosage administered over a 15-minute period.

To be able to predict the numbers of immune survivors expected after an epidemic, or after a prolonged series of exposures to an infective or lethal agent, the effect of active immunity on the dose-response model was studied. When the dose-response pattern of mice immunized with living avirulent P. pestis was determined, it was found that less than 100% immunization was achieved regardless of the level of the immunizing dose. The role of multiple immunization is, therefore, one of persistance of immune response rather than of degree of protection; quantitative or multiple immunization in excess of a definable minimum will not affect the proportion of the population successfully immunized. An attempt to determine a dose-response model for mice actively immunized with virulent P. pestis-139L was unsuccessful, since death is the most probable result—immunity was therefore an extremely improbable finding.

In mice passively immunized with hyperimmune P. pestis serum, it was found that a ten-fold increase in challenge dose resulted in a twofold increase in mortality until 100% mortality was observed; a relatively large increase in challenge dose is therefore required to produce an increase in mortality. Although passive immunization prolonged the survival time of mice challenged with massive doses of the organism, the final percent mortality was not significantly different from that of non-immunized controls. The fraction of a population protected by passive immunization will vary markedly with challenge dose; in contrast, the fraction of a population actively immunized by surviving a virulent challenge is influenced to only a small extent by the size of the challenge dose.

From the relationship between respiratory infection and lung retention it would be desirable to be able to determine the most effective use of antibiotics in animals exposed to infectious aerosols. The effect on herd mortality of giving varying doses of streptomycin at varying intervals after exposure of mice to aerosols of *K. pneumoniae* was determined. It was found that the optimal dosage-time schedules for maximal survival with the least total quantity of drug could be calculated by using the dose-response model.

Herd Transmission (Cross Infection)

Druett *et al.* (1956) found that there was a positive correlation between the number of deaths in control animals and the number of dead or dying experimentally infected animals with which the controls were in contact. The size of particles infecting experimental animals also influenced the cross-infection rate. Deaths in control animals were approximately four times greater after contact with animals exposed to particles containing only one cell than after contact with animals previously exposed to multi-cellular particles of 12 μ diameter.

Since a gastro-intestinal carrier state has been reported for mice exposed to *K. pneumoniae,* a simulated herd epidemic was set up by Goldberg *et al.* (1954) who held mice in five channels connected in series. They found that lethal respiratory infections in normal mice can be experimentally produced with liquid suspensions of feces from *K. pneumoniae* carriers. A higher incidence of carriers occurred in mice exposed to fecal suspensions than was observed following exposure of mice to the original culture. However, only a fraction of 1% of *K. pneumoniae* organisms remained viable in fecal specimens after 24 hours at an RH of less than 50%, whereas at a higher environmental RH (where survival was far better) fecal material retained its moisture and resisted powdering and subsequent effective aerosolization. They concluded that natural epidemic transmission of this organism among caged mice as the result of contamination with fecal residues is not significant.

Viability and Infectivity

Infectivity (capacity of an organism to invade a host and multiply detectably) has been shown to be separable from apparent death of the microbe. Drying of group *A. streptococci* diminishes infectivity for humans without destroying viability (Perry *et al.*, 1957). Goodlow and Leonard (1961), Hood (1961), and Schlamm (1960), who exposed mice and guinea pigs to aerosols of *P. pestis* and *F. tularensis*, showed that infectivity and viability may become dissociated during aerosol aging. In addition, it was shown that aerosol aging results in a marked decrease in virulence when measured by direct inhalatory challenge, but not when measured by intraperitoneal inoculation of mice with cells collected from the same aerosol (Schlamm, 1960). These observations were confirmed by Sawyer *et al.* (1966) who reported that respiratory infectivity of aging *F. tularensis* aerosols, tested with man and monkeys, decreased more rapidly than the viability of the organisms.

Goldberg, L. J. (1958, unpublished data) observed that the relative humidity at which an aerosol is aged may affect infectivity and viability in different ways. When aerosols of *F. tularensis* are stored in a rotating drum, virulence decreased more slowly at low RH than at high RH. At high RH, persistance of the organism was good, but increased loss of infectivity occurred. In vivo and in vitro assay are two distinct measures of the biological properties of a cell.

REFERENCES

Akers, T. G., Bond, S., and Goldberg, L. J. 1966a. Effect of temperature and relative humidity on survival of airborne columbia SK group viruses. *Appl. Microbiol.*, **14:** 361–364.
Akers, T. G., Bond, S. B., Papke, C., and Leif, W. R. 1966b. Virulence and immunogenicity in mice of airborne encephalomyocarditis viruses and their infectious nucleic acids. *J. Immunol.*, **97:** 379–385.
Andersen, A. 1958. New sampler for the collection, sizing and enumeration of viable airborne particles. *J. Bacteriol.*, **76:** 471–484.
Andersen, A. A., and Andersen, M. R. 1962. A monitor for airborne bacteria. *Appl. Microbiol.*, **10:** 181–184.
Anderson, J. D., and Cox, C. S. 1967. Microbial Survival. In *Airborne Microbes; 17th Symposium of the Society for General Microbiology* pp. 203–226. (P. H. Gregory and J. L. Monteith, Eds.), Cambridge Univ. Press. London.
Anderson, J. D., and Crouch, G. T. 1967. A new principle for the determination of total bacterial numbers in populations recovered from aerosols. *J. Gen. Microbiol.*, **47:** 49–52.
Artenstein, M. S., and Miller, W. S. 1966. Air sampling for respiratory disease agents in army recruits. *Bacteriol. Rev.*, **30:** 571–572.
Bachelor, H. W. 1960. Aerosol samplers. In *Advances in Applied Microbiology Vol. 2* (Wayne W. Umbreit, Ed.), pp. 31–64. Academic Press, New York.

Beard, C. W., and Easterday, B. C. 1965. An aerosol apparatus for the exposure of large and small animals: description and operating characteristics. *Am. J. Vet. Res.*, **26**: 174–182.

Brachman, P. S., Erlich, R., Eichenwald, H. F., Cabelli, V. J., Kethley, T. W., Madin, S. H., Maltman, J. R., Middlebrook, G., Morton, J. D., Silver, I. H., and Wolfe, E. K. 1964. Standard sampler for assay of airborne microorganisms. *Science*, **144**(3624): 1295.

Brachman, P. S., Kaufmann, A. F., and Dalldorf, F. G. 1966. Industrial inhalation anthrax. *Bacteriol. Rev.*, **30**: 646–657.

Brownlow, W. J., and Wessman, G. E. 1960. Nutrition of *Pasteurella pestis* in chemically defined media at temperatures of 36 to 38 C. *J. Bacteriol.*, **79**: 299–304.

Buchanan, L. M., Decker, H. M., Frisque, D. E., Phillips, C. R., and Dahlgren, C. M. 1968. A novel multi-slit high volume sampler. *Bacteriol. Proc.*, **A77**: 13–14.

Cahn, R. D. 1967. Detergents in membrane filters. *Science*, **155**(3759): 195–196.

Constantine, D. G. 1962. Rabies transmission by nonbite route. *Public Health Rept.*, (U.S.) **77**: 287–289. U.S. Govt. Printing Office.

Constantine, D. G. 1967. Rabies transmission by air in bat caves. *Public Health Serv. Publ. No. 1617*, U.S. Govt. Printing Office.

Converse, J. L., and Reed, R. E. 1966. Experimental epidemiology of coccidioidomycosis. *Bacteriol. Rev.*, **30**: 678–694.

Couch, R. B., Cate, T. R., Douglas, R. G., Jr., Gerone, P. J., and Knight, V. 1966. Effect of route of inoculation on experimental respiratory viral disease in volunteers and evidence for airborne transmission. *Bacteriol. Rev.*, **30**: 517–529.

Cown, W. B., Kethley, T. W., and Fincher, E. L. 1957. The critical orifice liquid impinger as a sampler for bacterial aerosols. *Appl. Microbiol.*, **5**: 119–125.

Cox, C. S., and Baldwin, F. 1964. A method for investigating the cause of death of airborne bacteria. *Nature*, **202**(4937): 1135.

Dahlgren, C. M., Decker, H. M., and Harstad, J. B. 1961. A slit sampler for collecting T3 bacteriophage and Venezuelan equine encephalomyelitis virus. I. Studies with T3 bacteriophage. *Appl. Microbiol.*, **9**: 103–105.

Dannenberg, A. M., Jr., and Scott, E. M. 1956. Determination of respiratory LD$_{50}$ from number of primary lesions as illustrated by Melioidosis. *Proc. Soc. Exptl. Biol. Med.*, **92**: 571–575.

Davies, C. N. 1965. The aspiration of heavy airborne particles into a point sink. In *Aerosols, Physical Chemistry and Application* (K. Spurny, Ed.), pp. 131–138. Publishing House of the Czechoslovak Academy of Sciences, Pràgue, Czechoslovakia.

Decker, H. M., Kuehne, R. W., Buchanan, L. M., and Porter, R. 1958. Design and evaluation of a silt-incubator sampler. *Appl. Microbiol.*, **6**: 398–400.

DeOme, K., and Personnel, U.S. Navy Medical Research Unit No. 1, Berkeley, Calif. 1944. The effect of temperature, humidity and glycol vapor on the viability of air-borne bacteria. *Am. J. Hyg.*, **40**: 239–250.

Dimmick, R. L. 1965. Rhythmic response of *Serratia marcescens* to elevated temperature. *J. Bacteriol.*, **89**: 791–798.

Druett, H. A., Henderson, D. W., Packman, L., and Peacock, S. 1953. The influence of particle size on respiratory infection with anthrax spores. *J. Hyg.*, **51**: 359–371.

Druett, H. A., and May, K. R. 1952. A wind tunnel for the study of airborne infections. *J. Hyg.*, **50**: 69–81.

Druett, H. A., Robinson, J. M., Henderson, D. W., Packman, L., and Peacock, S. 1956. Studies on respiratory infection. II. The influence of aerosol particle size on infection of the guinea pig with *Pasteurella pestis*. *J. Hyg.*, **54**: 37–48.

Furcolow, M. L. 1961. Airborne Histoplasmosis. *Bacteriol. Rev.*, **25:** 301–309.
Gerone, P. J., Couch, R. B., Keefer, G. V., Douglas, R. G., Derrenbacher, E. B., and Knight, V. 1966. Assessment of experimental and natural viral aerosols. *Bacteriol. Rev.*, **30:** 576–584.
Gochenour, W. S., Jr. 1966. Discussion. *Bacteriol. Rev.*, **30:** 584–586.
Goldberg, L. J. 1958. Personal communication.
Goldberg, L. J. 1960. A visual approach to bioassay. In *Quantitative Methods in Pharmacology* (H. de Jong, Ed.), pp. 172–205. No. Holland Publ. Co., Amsterdam (Interscience Publ., New York).
Goldberg, L. J. 1968. Application of the Microaerofluorometer to the study of dispersion of a fluorescent aerosol into a selected atmosphere. *J. Appl. Meteorol.*, **7:** 68–72.
Goldberg, L. J., and Leif, W. R. 1950. The use of a radioactive isotope in determining the retention and initial distribution of airborne bacteria in the mouse. *Science*, **112**(2907): 299–300.
Goldberg, L. J., and Shechmeister, I. L. 1951. Studies on the experimental epidemiology of respiratory infections. V. Evaluation of factors related to slit sampling of airborne bacteria. *J. Infect. Diseases*, **88:** 243–247.
Goldberg, L. J., and Watkins, H. M. S. 1965. Preliminary studies with a continuous impinger for collection of bacterial and viral aerosol samples. In *A Symposium on Aerobiology*, 1963 (R. L. Dimmick, Ed.), pp. 211–216. Nav. Biol. Lab., Nav. Supply Center, Oakland, Calif.
Goldberg, L. J., Watkins, H. M. S., Boerke, E. E., and Chatigny, M. S. 1958. The use of a rotating drum for the study of aerosols over extended periods of time. *Am. J. Hyg.*, **68:** 85–93.
Goldberg, L. J., Watkins, H. M. S., Dolmatz, M. S., and Schlamm, N. A. 1954. Studies on the experimental epidemiology of respiratory infection. VI. The relationship between dose of microorganisms and subsequent infection or death of a host. *J. Infect. Diseases*, **94:** 9–21.
Goodlow, R. J., and Leonard, F. A. 1961. Viability and infectivity of microorganisms in experimental airborne infection. *Bacteriol. Rev.*, **25:** 182–187.
Green, L., and Green, G. 1968. Direct method of determining the viability of a freshly generated mixed bacterial aerosol. *Appl. Microbiol.*, **16:** 78–81.
Green, H. L., and Lane, W. R. 1964. *Particulate Clouds: dusts, smokes and mists* (2nd ed.). London; E. and F.N. Spon Ltd.
Griffith, W. R. 1964. A mobile laboratory unit for exposure of animals and human volunteers to bacterial and viral aerosols. *Am. Rev. Respirat. Diseases*, **89:** 240–249.
Guerin, L. F., and Mitchell, C. A. 1964. A method of determining the concentration of airborne virus and sizing droplet nuclei containing the agent. *Can. J. Comp. Med. Vet. Sci.*, **28:** 283–287.
Guyton, A. C. 1947. Measurement of the respiratory volumes of laboratory animals. *Am. J. Physiol.*, **150:** 70–77.
Harper, G. J. 1961. Airborne micro-organisms: survival tests with four viruses. *J. Hyg.*, **59:** 479–486.
Harris, N. D. 1963. The influence of the recovery medium and the incubation temperature on the survival of damaged bacteria. *J. Appl. Bacteriol.*, **26:** 387–397.
Harstad, J. B. 1965. Sampling submicron T1 bacteriophage aresolos. *Appl. Microbiol.*, **13:** 899–908.
Hatch, M. T. 1967. Effect of relative humidity on inactivation and reactivation of *Pasteurella pestis* bacteriophage in aerosols. *Bacteriol. Proc.*, **V97:** 150.

Hatch, M. T., and Dimmick, R. L. 1966. Physiological responses of airborne bacteria to shifts in relative humidity. *Bacteriol. Rev.*, 30: 597–602.

Heckly, R. J., Dimmick, R. L., and Guard, N. 1967. Studies on survival of bacteria: rhythmic response of microorganisms to freeze-drying additives. *Appl. Microbiol.*, 15: 1235–1239.

Henderson, D. W. 1952. An apparatus for the study of airborne infection. *J. Hyg.*, 50: 53–68.

Hood, A. M. 1961. Infectivity of *Pasteurella tularensis* clouds. *J. Hyg.*, 59: 497–504.

Houwink, E. H., and Rolvink, W. 1957. The quantitative assay of bacterial aerosols by electrostatic precipitation. *J. Hyg.*, 55: 544–563.

Jensen, M. M. 1964. Inactivation of airborne viruses by ultraviolet irradiation. *Appl. Microbiol.*, 12: 418–420.

Kaerber, G. 1931. Beitrag zur Kollectiven Behandlung pharmakologischer reiben-versuche. *Arch. Exptl. Pathol. Pharmakol.*, 162: 480–483.

Kethley, T. W., Fincher, E. L., and Cown, W. B. 1957. The effect of sampling method upon the apparent response of airborne bacteria to temperature and relative humidity. *J. Infect. Diseases*, 100: 97–102.

Kethley, T. W., Gordon, M. T., and Orr, C., Jr. 1952. A thermal precipitator for aerobacteriology. *Science*, 116(3014): 368–369.

Kuehne, R. W., and Gochenour, W. S., Jr. 1961. A slit sampler for collecting T₃ bacteriophage and Venezuelan equine encephalomyelitis virus. II. Studies with Venezuelan equine encephalomyelitis virus. *Appl. Microbiol.*, 9: 106–107.

Leif, W. R., and Krueger, A. P. 1950. Studies on the experimental epidemiology of respiratory infections. I. An apparatus for the quantitative study of airborne respiratory pathogens. *J. Infect. Diseases*, 87: 103–116.

LeNoir, M. M., and Camus, J. 1909. Recherches sur la contagion de la tuberculose par l'air. *Compt. Rend. Acad. Sci. Paris*, 148: 309–312.

Lidwell, O. M., and Noble, W. C. 1965. A modification of the Andersen sampler for use in occupied environments. *J. Appl. Bacteriol.*, 28: 280–282.

Lurie, M. B., Abramson, S., and Heppleston, A. G. 1952. On the response of genetically resistant and susceptible rabbits to the quantitative inhalation of human type tubercle bacilli and the nature of resistance to tuberculosis. *J. Exptl. Med.*, 95: 119–134.

Lurie, H. I., and Way, M. 1957. The isolation of dermatophytes from the atmosphere of caves. *Mycologia*, 49: 178–180.

May, K. R. 1964. Calibration of a modified Andersen bacterial aerosol sampler. *Appl. Microbiol.*, 12: 37–43.

May, K. R. 1966. Multistage liquid impinger. *Bacteriol. Rev.*, 30: 559–570.

May, K. R. 1967. Physical aspects of sampling airborne microbes. In *Airborne Microbes—17th Symposium of the Society for General Microbiology* (P. H. Gregory and J. L. Monteith, Eds.), pp. 60–80. Cambridge Univ. Press, London.

Morris, E. J., Darlow, H. M., Peel, J. F. H., and Wright, W. C. 1961. The quantitative assay of mono-dispersed aerosols of bacteria and bacteriophage by electrostatic precipitation. *J. Hyg.*, 59: 487–496.

Noble, W. C. 1967. Sampling airborne microbes—handling the catch. In *Airborne Microbes—17th Symposium for the Society for General Microbiology* (P. H. Gregory and J. L. Monteith, Eds.), pp. 81–107. Cambridge Univ. Press, London.

Orr, C., Jr., Gordon, M. T., and Kordecki, M. 1956. Thermal precipitation for sampling airborne micoorganisms: comparison with other methods. *Appl. Microbiol.*, 4: 116–118.

Perry, W. D., Siegel, A. C., and Ramelkamp, C. H., Jr. 1957. Transmission of group A streptococci. II. The role of contaminated dust. *Am. J. Hyg.*, **66**: 96–101.

Reed, L. J., and Muench, H. 1938. A simple method of estimating fifty percent endpoints. *Am. J. Hyg.*, **27**: 493–497.

Reyniers, J. A. 1943. *Micrurgical and germ free techniques: Their application to experimental biology and medicine.* A symposium. Charles C Thomas, Publisher, Springfield, Ill.

Riley, R. L., and O'Grady, F. 1961. *Airborne Infection: Transmission and Control,* (Chap. 6) pp. 95–125. The Macmillan Company, New York.

Riley, R. L., Mills, C. C., O'Grady, F., Sultan, L. U., Wittstadt, F., and Shivpuri, D. N. 1962. Infectiousness of air from a tuberculosis ward. *Am. Rev. Respirat. Diseases*, **85**: 511–525.

Rosebury, T. 1947. *Experimental airborne infection.* Williams and Wilkins Co., Baltimore, Md.

Sawyer, W. D., Jemski, J. V., Hogge, A. L., Jr., Eigelsbach, H. T., Wolfe, E. K., Dangerfield, H. G., Gochenour, W. S., Jr., and Crozier, D. 1966. Effect of aerosol age on the infectivity of airborne *Pasteurella tularensis* for *Macaca mulatta* and man. *J. Bacteriol.*, **91**: 2180–2184.

Schlamm, N. A. 1960. Detection of viability in aged or injured *Pasteurella tularensis*. *J. Bacteriol.*, **80**: 818–822.

Shipe, E. L., Tyler, M. E., and Chapman, D. N. 1959. Bacterial aerosol samplers. II. Development and evaluation of the Shipe sampler. *Appl. Microbiol.*, **7**: 349–354.

Tenney, S. M., and Remmers, J. E. 1963. Comparative quantitative morphology of the mammalian lung: diffusing area. *Nature*, **197**(4862): 54–56.

Thiéblemont, P., Marble, G., Perrault, G., and Pasquier, Ch. 1965. Technique d'administration d'aérosols radioactifs liquides par voie respiratorie au singe. *Phys. Med. Biol.*, **11**: 307–312.

Thorne, H. V., and Burrows, T. M. 1960. Aerosol sampling methods for the virus of foot and mouth disease and the measurement of virus penetration through aerosol filters. *J. Hyg.*, **58**: 409–417.

Tyler, M. E., and Shipe, E. L. 1959. Bacterial aerosol samplers. I. Development and evaluation of the all-glass impinger. *Appl. Microbiol.*, **7**: 337–348.

Tyler, M. E., Shipe, E. L., and Painter, R. B. 1959. Bacterial aerosol samplers. III. Comparison of biological and physical effects in liquid impinger samplers. *Appl. Microbiol.*, **7**: 355–362.

Webb, S. J. 1965. *Bound Water in Biological Integrity,* p. 53. Charles C. Thomas, Publisher, Springfield, Ill.

Wellock, C. E. 1960. Epidemiology of Q fever in the urban east bay area. *Calif. Health*, **18**(10): 72–76.

Winkler, W. G. 1968. Airborne rabies virus isolation. *Bull. Wildlife Disease Assoc.*, **4**: 37–40.

Wolf, H. W., Skaliy, P., Hall, L. B., Harris, M. M., Decker, H. M., Buchanan, L. M., and Dahlgren, C. M. 1964. Sampling microbial aerosols. *Public Health Monograph No. 60* (Publ. No. 686), U.S. Govt. Printing Office.

Wolfe, E. K., Jr. 1961. Quantitative characterization of aerosols. *Bacteriol. Rev.*, **25**: 194–202.

Wolochow, H. 1958. The membrane filter technique for estimating numbers of viable bacteria: Some observed limitations with certain species. *Appl. Microbiol.*, **6**: 201–206.

Wolochow, H., Chaitgny, M., and Speck, R. S. 1957. Studies on the experimental epidemiology of respiratory infections. VII. Apparatus for the exposure of monkeys to infectious aerosols. *J. Infect. Diseases*, **100**: 48–57.

Won, W. D., and Ross, H. 1966a. Effect of diluent and relative humidity on apparent viability of airborne *Pasteurella pestis*. *Appl. Microbiol.*, **14**: 742–745.

Won, W. D., and Ross, H. 1966b. A freeze-tolerant solid medium for detection and sampling of airborne microorganisms at subzero temperature. *Cryobiol.*, **3**: 88–93.

Won, W. D., and Ross, H. 1967. A modified freeze-tolerant base for solidifying media for sampling microbial aerosols at subzero temperatures. *Bacteriol. Proc.*, **A121**: 21.

5

ATMOSPHERIC IONS AND AEROSOLS

Albert P. Krueger / Sadao Kotaka / Paul C. Andriese

MEDICAL MICROBIOLOGY AND IMMUNOLOGY UNIT, NAVAL BIOLOGICAL
LABORATORY AND THE SCHOOL OF PUBLIC HEALTH, UNIVERSITY OF
CALIFORNIA, BERKELEY

Since gaseous ions are normal constituents of the earth's atmosphere and their numbers may be artificially increased by a variety of means, it is pertinent to ask whether they can affect microbial aerosols. To answer this inquiry, five ancillary questions must first be considered: (1) What are gaseous ions? (2) How do they originate? (3) How are they measured? (4) What evidence exists that they react with aerosols? (5) What biological effects do they have?

THE ESSENTIAL NATURE OF ATMOSPHERIC IONS

Small (Gaseous) Ions

Paired molecular ions are produced when the application of sufficient energy displaces an electron from a molecule of gas, leaving the molecule with a positive charge. The displaced electron (free electron) is surrounded by 2.7×10^{19} molecules/cm³ of air, and under normal conditions it is promptly captured by another molecule to which it imparts a negative charge. The high rate of collision of monatomic or diatomic ions with surrounding nonionized molecules results, in less than 0.02 sec, in the formation of an aggregate composed of some 4 to 12 uncharged

100

molecules, producing what is known as a "small cluster ion." Obviously, such molecular "rafts" are not uniform in composition, but since they possess a unit charge (1.6×10^{-19} coulomb), they are uniformly mobile, averaging 1 to 2 $cm^2/sec/volt$; their effective molecular diameter is < 10 Å and their diffusion coefficient is $> 10^{-2}$ cm^2/sec.

The energy required to produce ions is supplied by:

(a) Minute quantities of radioactive substances such as radium, thorium, etc., present in the soil. Alpha radiation acts only a few centimeters above the ground; beta radiation effectively ionizes air to a height of several meters, while gamma rays are effective to a height of 1 kilometer.

(b) Radioactive gases such as radon and thoron produced by the disintegration of radium and thorium. These gases escape into the air and serve as sources of energy for ionization.

(c) Cosmic rays.

(d) Miscellaneous energy sources such as lightning, the frictional electricity produced by blowing sand or drifting snow, combustion processes and water in motion, e.g., waterfalls.

Near the ground the number of ion pairs produced per cubic centimeter of air per second varies from 10 to 40, depending on the nature of the soil surface. Generally, basic rocks contain less radioactivity than acidic rocks, and sedimentary rocks less than igneous rocks. Approximately 35% of the air ions close to the surface of the earth are produced by radioactive substances in the soil, while some 50% are produced by radioactive gases. Cosmic rays account for only 15% of the total.

In clean air over land, the total number of small air ions approximates 1,200 to 1,500/cm^3; the ratio of small positive ions to small negative ions is about 1.2:1. This disparity has been attributed to the greater diffusibility of negative ions and to the fact that the earth's surface, being negatively charged, attracts positive ions and repels negative ions. At high elevations the total number of small ions increases, whereas the large ion count decreases. For example, careful measurements of small air-ion densities prevailing at the 14,246-foot level on White Mountain, Calif., gave averages of 2,500 positive and 2,200 negative ions/cm^3. Practically no intermediate or large ions were present.

Ion formation over the sea caused by the radioactive substances is negligible, since sea water has 1/1000 the radioactivity of the soil and ion production is ascribable largely to cosmic rays; consequently the rate of formation is only about 2 ion pairs/cm^3/sec. Nevertheless, the concentration of small ions in air over land and over the ocean are essentially the same, because processes responsible for the depletion of small ions operate at a much lower level in the clean air over the ocean. Unhappily, air-ion measurements conducted over the ocean during the past 50 years have shown a steady decrease in small ion content—a probable indication of the widespread distribution of industrial pollutants which tend to capture small ions.

Condensation Nuclei (Large and Intermediate Ions)

Uncharged particles in the atmosphere—the condensation nuclei of Aitken—are focal points for the condensation of water when the air becomes sufficiently saturated with moisture. Condensation nuclei average from 2.5 to 6.0×10^{-2} μ in diameter and contain approximately 10^6 molecules. Their number usually exceeds the number of charged dust particles in the air by a factor of 2 or more orders of magnitude. Considering the earth's atmosphere as a whole, it appears that most condensation nuclei are derived from natural sources such as forest fires, ocean spray, and volcanic eruptions. As population density increases, so does the contribution of human activities to the total numbers of condensation nuclei in the atmosphere; almost all industrial operations, domestic combustion processes, and transportation activities would come under this heading. When such particles acquire a charge by combination with small ions, they are know as large ions or Langevin ions. This tendency of small air ions to react with condensation nuclei to produce Langevin ions binds the reactants and their products in a very close relationship. Since the formation of large ions is a function of the concentration of condensation nuclei, and since the latter are an important element in air pollution, the number of large air ions per cubic centimeter, or the reciprocal of the number of small air ions per cubic centimeter of air, can be used as a measure of air pollution. It is clear that the numbers of condensation nuclei present in the air will vary greatly with location, but some rough averages can be given. Condensation nuclei per cubic centimeter of air occur in the following ranges: (a) sea air, 1×10^3 to 40×10^3; (b) rural air, 10×10^3 to 300×10^3; (c) urban air, 150×10^3 to $4,000 \times 10^3$.

Because of their large size, the average mobility of Langevin ions is much smaller than that of small ions and falls in the range of 5×10^{-4} to 8×10^{-3} cm²/sec/volt. Their number averages 150 to 250/cm³ of air over the sea and about 70 to 90×10^3/cm³ over cities. Besides the well-defined differences in mobility between small and large ions, they are dissimilar in another important property: when the charge is removed from a small ion, the molecular components of the cluster disperse since the charge itself is responsible for holding them together, whereas a large ion can lose its charge but remains otherwise intact.

In addition to small and large ions, it has been recognized that intermediate ions exist with a mobility range greater than that of large ions but smaller than that of small ions, i.e., from 2×10^{-1} to $1 \times 10^{-3}/cm^2/sec/volt$. Except in very special situations, they do not appear to be important constituents of the air.

Ion Generation

The sources of ion formation in nature have been enumerated above. In laboratory and clinical experimentation it is now common practice to employ the energy derived from radioisotopes, such as polonium[210] or tritium, for ion generation. Polonium[210] is an alpha emitter with a half-life of 138 days. Tritium has a half-life of approximately 12.5 years and is commonly used in 50-mc quantities sealed in zirconium deposited on a stainless steel foil. Tritium is a low-energy, 0.0185-mev beta source and the ionization range from the foil is approximately 1 cm. In still air and in the absence of an electrical field, the positive and negative ion pairs produced immediately recombine in the dense ion plasma adjacent to the foil. However, if an electrical potential difference of 300 to 1,000 volts direct current is applied, unipolar ions will be emitted; their charge is dependent on the polarity of the foil. These foils typically produce $> 1 \times 10^9$ ions/sec. More recently, americium[241], a strong alpha emitter with a half-life of 475 years, has been introduced in commercial applications. In Russia, high-voltage electrical fields usually are employed as ion sources in agriculture and in medicine.

Ion Measurement

Two relatively simple procedures are utilized to measure air-ion concentration. In the first, air is drawn at a uniform rate through a duct in which insulated polarizing plates are exposed to the air. As ions are deposited on the plates, the very small current resulting from their deposition is measured with a femto ammeter. The second method utilizes a target collector; essentially this is an electrostatically shielded metal disc connected to an electrometer. With this device, accurate measurements of ion flow from an ion source in an electrical field are readily made in terms of the numbers of ions impinging per second per square centimeter of surface.

Air-Ion Equilibrium

The behavior of air ions in general is governed by their physical properties and by the substances they encounter in their immediate environment. The basic theoretical concepts involved in these two areas are treated in detail by Massey (1950), Loeb (1961), Chalmers (1957), and others, but in this chapter only those aspects are presented which are helpful in relating air ions to aerosols.

In a given volume of air under constant conditions, equilibrium is reached between the number of ions entering the volume + those produced in the volume and the number leaving the volume + the number destroyed in the volume.

The processes involved in attaining ion equilibrium are:

1. Production of small ions by radioactivity, cosmic rays, etc.
2. Production of large ions by reaction between small ions and uncharged condensation nuclei.
3. Small ion recombination. This is usually a negligible factor except in such special conditions as exist, for example, in the air immediately adjacent to the foil of a tritium generator. Here the concentrations of ions are so great that recombination is favored.
4. The reaction between small ions and large ions of the opposite charge.
5. The reaction between large ions of opposite charge. Usually this is negligible.

Rate of Attaining Equilibrium

Equations have been developed for determining the rate at which ion mixtures attain equilibrium. It has been calculated that small ion equilibrium is attained in about 2 sec, whereas it requires 10 to 30 min for establishment of large ion equilibrium. The difference between the reactability of negative and positive ions is considered to be a result of the fact that negative small ions are more mobile than small positive ions. Unfortunately, experimental evidence indicates that in situations frequently encountered in nature, large ion equilibrium is not attained; consequently the equations that define equilibrium relationships have limited practical application. While a complete mathematical treatment of air ion equilibrium is beyond the scope of this book, the interested reader may refer to articles by Chalmers (1957), Whitby, Liu, and Peterson (1965), and Loeb (1961).

THE EFFECT OF SMALL IONS ON THE BEHAVIOR OF AEROSOLS

The effect of the electrostatic charge carried by aerosol particles on the behavior of those particles has been studied for many years. Most aerosols, whether produced by natural or artificial means, are electrically charged, and under some circumstances they may be highly charged. In nature the numbers of ions and of charged aerosol particles are of a low order, i.e., from several hundred to several thousand per cubic centimeter of air, and at these concentrations the charge effects are extremely limited. However, when high concentrations of ions exist, they may materially affect the course of events occurring in a cloud, i.e., the net space charge may be responsible for high charge concentration gradients, strong electrical fields, and resultant precipitation of ions and aerosols.

When no applied electrical field exists, the process of charging of aerosol particles depends on the diffusion of gaseous ions to the particles as a result of their random thermal motion. The ions impart their electrical charge to the surface of the particle and the charged aerosol comes to equilibrium with the ionic atmosphere. An equation has been developed which, on the basis of limited experimental data, predicts the particle charge of micron and larger than micron particles with acceptable accuracy.

When the electric field is of sufficient magnitude, the random thermal motion of the ions becomes negligible and the predominant mechanism responsible for bringing ions into contact with the particles is the flow of ions due to the field itself.

With particles $< 1\ \mu$, their sizes become of the same order of magnitude as the mean free path of the ions. Consequently, when they are charged in the presence of moderate electric field intensities, both the field and diffusion mechanisms function. While equations have been developed for this situation, they have not been very successful in application.

Under normally prevailing conditions the atmosphere contains both negative and positive small ions. They collide frequently with aerosol particles and come to charge equilibrium with the ionic atmosphere.

Behavior of Charged Aerosols

The Electrical Mobility of Charged Aerosol Particles

In an electric field of intensity T (stat volt/cm), an aerosol particle carrying n_c elementary units of electrical charge is subject to an electrical force, F:

(5-1) $$F = n_c q E$$

where n_c = electrical charge of particle, and
$q = 4.8 \times 10^{-10}$ stat coulomb.

The motion of a particle through the medium is subject to a viscous drag force and the final velocity, $\mathbf{V}$, is attained when the electrical and drag forces become equal as predicted by Stokes' law:

$$(5\text{-}2) \qquad V = \frac{n_c q E}{3\pi\eta D_p C}$$

where η = viscosity, poise(dyne-sec/cm^2),
D_p = diameter of particles, cm,
V = velocity, cm/sec, and
C = Cunningham's correction.

That is, the particle speed is proportional to the intensity of the applied electrical field. The electrical mobility of the particle, Z_p, is by definition the proportionality constant:

$$(5\text{-}3) \qquad Z_p = \frac{V}{E}$$

$$(5\text{-}4) \qquad Z_p = \frac{n_c q}{3\pi\eta D_p C}$$

The particle mobility can also be calculated for the diffusion process and for field charging.

The Decay of Unipolar Ions and Uniformly Charged Monodispersed Aerosols

When small ions are not being generated, the numbers of small ions in a given space will decrease as a result of charge repulsion with resultant migration of ions to the boundaries:

$$(5\text{-}5) \qquad \frac{n}{n_0} = \frac{1}{4\pi n_0 q Z_i t + 1}$$

where n = ion concentration, number/cm^3,
n_0 = initial ion concentration,
Z_i = ion mobility, cm^2/stat volt-sec, and
t = time, sec.

This equation is based on the assumption that ion concentration throughout the space is uniform, and recent experimental tests have shown this to be the case. When ion concentrations are plotted as ordinates against time as abscissa (Figure 5-1), it is clear that the decay process in a cloud of unipolar small ions proceeds very rapidly, and that with high initial densities of small ions the decay curve is essentially independent of n_0. This concept of small ion decay rates is confirmed by recent

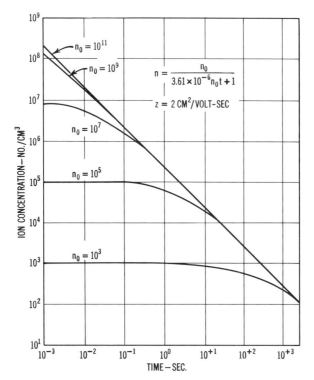

FIGURE 5-1. Theoretical decay curve for small unipolar ions. (Whitby, Liu, and Peterson. 1965. *J. Colloid Sci.*, **20**: 586.) Reprinted with permission of the Academic Press, New York.

experiments demonstrating that in an ordinary room provided with clean air, the ion levels attainable are limited to about 1×10^4 ions/cm³.

In a cloud of unipolarly charged aerosol particles, mutual repulsion will lead to a decrease in concentrations of particles. Such factors as diffusion, interception, and inertial impaction may modify this process. Assuming that decay is due to electrostatic repulsion only, particle parameters may be substituted for ion parameters in equation 5-5.

The decay of unipolarly charged aerosols usually is much faster than that occurring in a neutral aerosol, but it is still several orders of magnitude less than the decay rate of unipolar small ions because of the lower mobility of aerosol particles.

The number of ions and the electric field which will develop when an ion generator functions as a point source, assuming that spherical symmetry prevails about the point source, can be calculated from the following equations:

(5-6)
$$n = \frac{(3Q/2Z_i)^{1/2}(r)^{-3/2}}{4\pi q}$$

(5-7)
$$E = \left(\frac{2Q}{3Z_i r}\right)^{1/2}$$

where Q = ion generation rate, stat amp, and
r = radius, cm.

Evidently the ion concentration decreases as the $\frac{3}{2}$ power of the distance from the source, and the field dereases as $\frac{1}{2}$ power of the distance from the source.

If, on the basis of the preceding discussion, one considers the course of events ensuing when an aerosol decays in the presence of a point source of ions, it would be expected that the particles will become charged very rapidly and that the aerosol will precipitate at a rate which depends on the total charge density of small ions and particles, and on the mobility of the charged particles. Under these circumstances the charge density of the aerosol particles is small relative to that of the ions, the aerosol concentration in the space is uniform, and the particle charging time is much less than the particle decay time:

(5-8)
$$\frac{n_p}{n_{p_0}} = \exp\left[-\frac{q_p C}{3\pi\eta D_p(r_w)^{3/2}}\left(\frac{6Q}{Z_i}\right)^{1/2}_t\right]$$

where q_p = charge on particle, stat coulombs,
r_w = room radius,
η = viscosity, poise(dyne-sec/cm^2),
n_p = particle concentration, number/cm^3, and
n_{p_0} = initial particle concentration, number/cm^3.

This indicates that the particle concentration will decay in an exponential fashion. This equation may be used in predicting the half-life, $t_{1/2}$, of the decaying aerosols:

(5-9)
$$t_{1/2} = 0.691 \frac{3\pi\eta D_p(r_w)^{3/2}}{q_p C}\left(\frac{Z_i}{6Q}\right)^{1/2}$$

Experimental measurements of the decay of aerosols in the presence of a point source have been conducted in a 2,000-ft^3 room provided with clean air, using homogeneous aerosols of methylene blue or uranine and natural atmospheric contaminants. The mass median diameter of the aerosols varied from 0.028 to 3.6 μ, and it was found that the electrical charge developed was somewhat greater for particles exposed to negative ions than to those exposed to positive ions. The values obtained for experimental half-lives of the decaying aerosols agreed quite well with the values predicted by theory. The experimental half-lives ranged from a minimum of 5.5 min for a free needle point source and a particle size of about 3 μ to a maximum of 58 min for a commercially manufactured ion generator and a particle size of 0.26 μ. It is evident that a continuous source of small ions located in a space is capable of precipitating a significant quantity of aerosols.

Coagulation of Electrically Charged Aerosols

Electrical fields pervading an aerosol cloud and surface charges of individual particles may be expected to influence the coagulation rate of aerosols. Studies have been made on unipolar, bipolar, and uncharged particles with the following results:

(a) *Coagulation of a unipolar aerosol:*
Two factors affect the stability of particles carrying a like charge: first, the tendency of particles to repel each other; and second, the development of space charge repulsion. Although coagulation is effectively prevented by even a weak unipolar charge, under most conditions depletion of particles due to space charge is considerably greater than that produced by coagulation.

(b) *Coagulation of a bipolar aerosol:*
If an aerosol is charged to equilibrium in a bipolar ion atmosphere, the observable effect on coagulation is usually negligible. When the charge is very high, coagulation will proceed at a much faster rate than will occur in uncharged aerosols, but the neutralization of particle charge will cause the rate to decrease rapidly.

(c) *Coagulation in an electric field:*
In an electric field aerosol particles develop an induced charge that causes them to behave as dipoles. Only when the strength of the field is very high will the coagulation rate increase significantly.

The Effects of Gaseous Ions on Microorganisms

One would predict on the basis of the evidence presented in the preceding section of this chapter that unipolar small air ions, if present in sufficient number, would produce an increase in the rate of decay of aerosolized bacteria, viruses, or fungi comparable to that produced in aerosols composed of inanimate particles. In addition to this purely physical action, it is necessary to consider the possibility that ions may elicit a biological response quite independent of suspensoid stability.

Actually, the existence of growth inhibition induced by air ions and their bactericidal effects was recognized by A. L. Tchijevski (1933, 1934) and his colleagues in the course of a long series of experiments conducted

during the 1930's at the Central Laboratory for Studies on Aeroionification in Moscow. For example, it was found that 5×10^4 to 5×10^6
negative or positive ions/cm^3 of air retarded the growth of staphylococcal
cultures on plates; negative ions were more effective than positive ions.
In other tests, air ion and electrical field effects on *Micrococcus pyogenes*,
Vibrio cholerae, and *Salmonella typhosa* were studied. In all instances,
colony formation on plates was strongly retarded by large doses of air
ions of either polarity, whereas an electrical field of 300 to 1,000 volts/cm
was itself without effect on the bacteria. Also, various staff members
at the Central Laboratory for Studies on Aeroionification observed that
air in enclosed spaces became sterile as the result of treatment with
air ions.

Similar results have been reported more recently by a number of Russian investigators at the All Union Conference on Aeroionization in Industrial Hygiene which convened during October, 1964, in Leningrad.
When the natural microflora of the air or artificially created bacterial
aerosols were treated with concentrations of 5.2×10^4 to 9.5×10^4 negative ions/cm^3, the number of viable particles was reduced by 85 to 93%.

The earlier Russian work was not available to several Americans who,
within the past 12 years, undertook to determine what action air ions
exerted on bacteria and fungi. The bactericidal properties of air ions
noted by the Russians were independently confirmed for *M. pyogenes*
var. *aureus* (Krueger et al., 1957), for *Escherichia coli* (Kingdon, 1960),
and fungicidal effects were observed for *Neurospora crassa* (Fuerst and
Ball, 1955) and for *Penicillium notatum* (Pratt, 1960).

Of more direct significance to the subject of this text is the recent
work of Phillips, Harris, and Jones (1954) on reactions taking place
between air ions and bacteria in the aerosolized state. Using *Serratia
marcescens* aerosols of $< 5\ \mu$ diameter generated in a 365-liter chamber,
they obtained an exponential decay rate of 23%/min for untreated air,
54%/min with positive ions, and 78%/min for negative ions. The ion
densities averaged 9×10^5/cm^3. They concluded that the increase in
total decay rate occurring with positive ion treatment was due altogether
to the particulate loss, whereas with negative ion treatment the increase
in rate of decay consisted of two components: a particulate loss and
a significant amount of biological decay.

These results are, in general, compatible with data obtained by investigators who have worked with organisms exposed on plates or in small
droplets to high concentration of air ions (Krueger et al., 1957). It
was observed that negative ions were considerably more anti-microbial
than positive ions. The fact that Phillips and his colleagues obtained
only a physical effect of positive air ions on aerosolized *S. marcescens*
could have been a result of the brief periods of exposure employed.

There is strong evidence in the work summarized above on *M. pyogenes* var. *aureus, S. typhosa, V. cholerae, P. notatum* and *N. crassa* that ions of either charge are capable of producing moderate lethal effects, with negative ions being more active than positive ions. Hence, the use of ion sources of whatever nature (e.g., corona discharge, high electric fields, radioactive materials) to charge airborne microbial particles for the purpose of electrostatic collection and subsequent viability assay is of doubtful value.

Electroaerosols

Although aerosolized solutions have been used in inhalation therapy for many years, it was recognized quite early that their instability tended to limit their practical applications. This property led to the development of a different type of aerosol in which droplets receive such a large number of unipolar charges that their mutually repellent forces, opposed by surface tension and internal pressure, break them up into even smaller droplets. In addition to small particle size, the final product displays the following properties: a fine homogeneous dispersion of droplets evaporating slowly because of decreased vapor pressure, manifesting little tendency to form aggregates because of the repellent effect of the unipolar charges, and remaining in dispersion longer than uncharged aerosols. Electroaerosol equipment is so designed that water or saline solutions are aerosolized by air jets in such fashion that the droplets have a mean diameter of 1.35 to 2 μ and receive unipolar electrical charges as they are formed. The extent of charging varies from 4 to 1,600 elementary charges per particle, depending on the type of apparatus employed. Electroaerosol therapy has been successfully used for the treatment of a variety of diseases including asthma, bronchitis, emphysema, migraine, hypertension, etc. Although this form of therapy is well established in Germany, Hungary, Russia, Italy, Spain, Poland, and South Africa, it is only recently that it was introduced into the United States by Doctor Alfred P. Wehner. His work and that of others is described in a special review article prepared by Doctor Wehner for the *American Journal of Physical Medicine.*

REFERENCES

Chalmers, J. A. 1957. *Atmospheric Electricity* (2nd ed.). Pergamon Press, New York.

Dunskii, V. F., and Kitaev, A. V. 1960. Precipitation of a unipolarly charged aerosol in an enclosed space. *Colloid. J.* (USSR), **22**: 167–175.

Fuerst, R., and Ball, R. J. 1955. Biological effects of air ions. Report from the M. D. Anderson Hospital and Tumor Institute, Houston, Texas, pp. 1–12.

Kingdon, K. H. 1960. Interaction of atmospheric ions with biological material. *Phys. in Med. Biol.*, **5**: 1–14.

Kranz, P., and Rich, T. A. 1961. The physics of small air borne ions. *Proceedings of the International Congress on Ionization of the Air,* Article VI, Vol. I. Franklin Institute, Philadelphia, Pa., Oct. 16–17.

Krueger, A. P., Kotaka, S., and Andriese, P. C. 1966. Studies on the biological effects of gasous ions. A review. Special Monograph Series, Vol. 1, Biometeriological Research Center, Leiden, pp. 1–14.

Krueger, A. P., Smith, R. F., and Go, Ing Gan. 1957. The action of air ions on bacteria. I. Protective and lethal effects on suspensions of staphylococci in droplets. *Jour. Gen. Physiol.,* 41: 359–381.

Loeb, L. B. 1961. *Basic Processes of Gaseous Electronics* (2nd ed.). University of California Press, Berkeley and Los Angeles.

Massey, H. S. W. 1950. *Negative Ions,* (2nd ed.). Cambridge University Press. Cambridge.

Phillips, G., Harris, G. J., and Jones, M. W. 1964. The effect of air ions on bacterial aerosols. *Int. J. Biometeor.* 8: 27–37.

Pratt, R., and Barnard, R. W. 1960. Some effects of ionized air on *Penicillium notatum. J. Am. Pharm. Assoc. Sci. Ed.,* 49(10): 643–646.

Spurny, K. (Ed.). 1965. Aerosols in Meteorology and Astronomy. In *Aerosols, Physical Chemistry and Applications.* Proceedings of the First National Conference on Aerosols held at Liblice near Prague, October 8–13, 1962. Publishing House of the Czechoslovak Academy of Sciences, Prague.

Tchijevsky, A. L., 1933–1934. *Transactions of the Central Laboratory for Scientific Research on Ionification* (3 Vol.). Publisher: "The Commune" Voronej, USSR.

Tchijevsky, A. L. 1960. *Air Ionization, Its Role in the National Economy.* Publishing House of the State Planning Commission of the USSR, Moscow.

Wehner, A. P. 1962. Electro-aerosol therapy. (Special Review) *Am. J. Phys. Med.,* 41, part 1. 23–40. Part 2. 67–86.

Wehner, A. P. 1966. Die Entwicklung der Electroaerosologie in den U. S. A. *Zentr. fur Biol. Aerosol Forsch.,* 13: 3–40.

Whitby, K. T., and Liu, B. Y. H. 1965. The electrical behavior of aerosols. August, 1965. Mechanical Engineering Department, University of Minnesota.

Whitby, K. T., Liu, B. Y. H., and Peterson, C. M. 1965. Charging and decay of monodispersed aerosols in the presence of unipolar ion sources. *J. Colloid. Sci.,* 20: 585–601.

NOMENCLATURE

E = electric field intensity (stat volt/cm)

F = force (dynes)

n = ion concentration (number/cm^3); n_0 = initial ion concentration

n_p = particle concentration (number/cm^3); n_{p0} = initial particle concentration

n_c = electrical charge of particle (elementary unit)

q = electron charge = 4.8×10^{-10} stat coulomb

q_p = charge on particle (stat coulombs)

C = Cunningham's correction of Stokes' law

D_p = diameter of particle (cm)

η = viscosity (poise)

r = radius (cm)

V = velocity (cm/sec)

Z_p = electrical mobility of particle (cm^2/stat-volt-sec)

Z_i = ion mobility (cm^2/stat-volt-sec)

t = time (sec)

Q = ion generation rate (stat amperes)

6

APPROACHES TO THE BIOASSAY
OF AIRBORNE POLLUTION

William F. Serat

CALIFORNIA STATE DEPARTMENT OF PUBLIC HEALTH,
BERKELEY AND ST. MARY'S COLLEGE, MORAGA, CALIFORNIA

The total biota of the earth is dependent on the atmosphere which envelops the planet. Changes in the atmosphere profoundly affect those ecological relationships which determine the habitability of our terrain and, consequently, affect the relative distribution of life. From reading man's earliest recordings of history, we have become aware that various pollutants of our atmosphere have always affected living organisms to some extent. Originally, man contributed few contaminates to this seemingly vast supply of our breath of life. With industrialization and the consumption of natural resources for power, coupled with an exponentially increasing transportation system, man's contribution to air contamination has become significant. High concentrations of pollutants in limited areas have produced tragic results on humanity's well-being, but even widespread low pollutant levels over long periods of time cause dramatic changes in the ecology of plants and animals.

Since the ultimate concern with our contaminated environment is the welfare of man and the biological world in general, we are obligated to determine the ways pollution affects us. Aside from medical aspects (e.g., statistical correlation between incidence of disease and observed environmental situations), biological means of determining the extent of pollution offer the most direct possibility of assaying toxicity of airborne contaminants.

Table 6-1 Actual or potential systems for use in the bioassay of atmospheric pollutants

Pollutants	Biological system	Effect	Characteristics, uses or sensitivity of assay system	Reference
Photochemical "smog," products of incomplete combustion given off by automobiles, biologically active components extracted from auto exhaust and ambient air.	Human conjunctival or fetal lung cells in culture.	A linear increase in cell number accompanied by an increase in hydrocarbon concentration until, at higher concentrations, cell number diminished. The relative rate of cell division was determined by the ratio of volatile to residual hydrocarbons with different exhaust extracts.	A sensitive assay procedure for determining the efficiency of filters in removing biologically active hydrocarbons and other substances from automobile exhaust.	Rounds, 1963
Automobile exhaust hydrocarbons irradiated in the long wavelength ultraviolet.	Human conjunctival cells in culture.	A decrease in photodynamic response time, approximately linear with hydrocarbon concentration, was evident with blebbing and death of the cells.	Hydrocarbon concentrations between 5×10^{-10} and 5×10^{-3} mg/ml aqueous suspension were used in the assay.	Rounds, 1965
Dilutions of irradiated auto exhaust fractions, gaseous and particulate.	Cultures of umbilical cord cells.	Increase in growth rate with pollutant concentration followed by a decreased growth rate at high concentrations. Many abnormal nuclei occurred in cells exposed to low concentrations of particulates.	The stimulatory effect was more pronounced at greater dilutions of particulates than of gaseous fractions.	Mueller and Barry, 1962
Peroxyacetyl nitrate	Pinto bean (*Phaseolus vulgaris* L)	Inhibition of photosynthetic pyridine nucleotide reduction and CO_2 fixation dependent on this reduction.	Light is required for photosynthesis inhibition by air pollutants.	Dugger *et al.*, 1965; Darley *et al.*, 1963

Ozone	Lime seedlings	Inhibition of the rate of CO_2 fixation	Inhibition was noted 4 to 5 min after exposure to 0.6 ppm ozone.	Taylor et al., 1961
"City" atmospheres and/or atmospheres polluted artificially with SO_2.	Several species of lichens and, specifically, Parmelia sulcata (Pearson and Skye, 1965)	Incompatible to growth or gross changes in morphology coupled with diminished photosynthetic activities.	The relative abundance of certain species may be used as one index of air pollution.	Skye, 1958; Fenton, 1960; Magdefrau, 1960; Brodo, 1961; Pearson and Skye, 1965
Ozone in low concentration	Escherichia coli	Mutagenic effects attributed to the oxidant. Questions any mutagenic effects: possibly due to in vitro reactions with deoxyribonucleic acid released from lysed cells followed by resorption of the altered polynucleotides.		Davis, 1961 Scott and Lesher, 1963
Photochemical reaction products from butene-1 and NO_2 in air. Products from NO_2 in air.	Escherichia coli	Decreased growth rates.		Estes and Pan, 1965; Estes, 1962.
Products of irradiation of hexene-1 and NO_2 in air	Serratia marcescens Escherichia coli	Inhibition of growth. No response.	Concentrations used, 2 ppm of hexene-1 and 0.5 ppm of NO_2. S. marcescens produced extracellular catalase while E. coli did not.	Jacumin et al., 1964

Table 6-1 (*Continued*)

Pollutants	Biological system	Effect	Characteristics, uses or sensitivity of assay system	Reference
Synthetic pollutant gas streams containing photosynthetic reaction products from *cis*-2-butene and nitric oxide. Ozone, peroxyacetyl nitrate or ambient smog.	*Photobacterium phosphoreum* cells on agar.	Decrease in bioluminescence.	Less than 1 ppm pollutant used; loss in luminescence depended on the time of irradiation of reactants and their concentration ratio. A rapid toxicity assay which may be developed to delineate components in the oxidant fraction of polluted air.	Serat *et al.,* 1965, 1967
Ozone	Luminescent fungi	An initial increase in luminescence is followed by a decrease. Some protection against the effects of the oxidant exhibited by adult cultures of *Armillaria mellea* as compared with young cultures or with *Panus stipticus* cultures may possibly be due to a melanin pigment produced by adult *A. mellea* but not by young cultures or *P. stipticus.*		Berliner, 1963

Material	Test system	Observation	Comments	References
Tobacco smoke, formaldehyde-KCl, or formaldehyde-polystyrene latex particulates in test gas streams	Membrane strip inoculated with *Escherichia coli*	Inhibition of bacterial growth along the area exposed to the aerosols.	A bacteriological analogue test has given qualitative data and to some extent quantitative data which correlates in part with eye irritation. The effect increased with aerosol concentration and corroborates the findings that gaseous pollutants concentrate on particulate surfaces.	Goetz and Tsuneishi, 1958, 1959a, 1959b
Atmospheric pollutants	Mouse	Skin test.	A bioassay for carcinogenicity: Lack of correlation between carcinogenic potency of the pollutants and their concentration of benzo(a) pyrene.	Hueper et al., 1962
Benzene-soluble extracts of organic atmospheric particulates injected in tricaprylin	Neonatal mice	High incidence of pulmonary adenomas and hepatomas, lower incidence of lymphomas, very low incidence of fibrosarcomas.	Pollutants injected at 5.5 mg or less.	Epstein, 1966; Epstein et al., 1966
Photoactivated polynuclear compounds	*Paramecium caudatum,* mice, chick embryo cells	Biological photodynamic response in presence of test compounds, particularly to those showing carcinogenic activity.		Doniach, 1939; Doniach and Mottram, 1937; Lewis, 1935

Table 6-1 (*Continued*)

Pollutants	Biological system	Effect	Characteristics, uses or sensitivity of assay system	Reference
Photoactivated poly- nuclear compounds	*Paramecium caudatum, Tetrahymena pyriformis*	Biological photodynamic response particularly to those compounds showing carcinogenic activity. Motility loss, blebbing, and lysis occur only when exposed to long wavelength ultraviolet light in the presence of the test compound.	A sensitive bioassay technique has been developed in which 0.02 to 0.32 μg of benzo(a)pyrene per ml of aqueous suspension may be determined in a longtime related response, linear over the concentration range used for assay.	Epstein and Burroughs, 1962; Epstein, 1963a; Epstein, 1965
Carcinogenic polycyclic hydrocarbons	*Bacillus megaterium*	Changes in morphology; decreased initial growth rate.		Won and Thomas, 1962
Ozone	Catalase, peroxidase, papain, and urease	Inhibition of enzyme activity with papain showing the greatest sensitivity.	Considerably higher quantities of ozone required for enzyme inhibition in aqueous solution than for production of leaf damage.	Todd, 1958

118

Euglena graeilis	Reduction in photosynthesis quantitatively dependent on ozone concentration. Also, a small decrease of chlorophyll D occurred as measured at 0.8 ppm ozone after 1 hr treatment.	Good quantitation of biological effect with ozone concentration in the range 0.25 to 1.0 ppm makes the assay quite sensitive. Perhaps it may be useful in studying biological effects of many toxicants with high reduction potential.	de Koning and Jegier, 1968
Peroxyacetyl nitrate, ozone	Glucose-6-phosphate dehydrogenase, iso-citric dehydrogenase	Inhibition of enzyme activity; reversible oxidation by peroxyacetyl nitrate and irreversible oxidation of pyridine nucleotides by ozone.	Used high concentrations of pollutants.	Mudd, 1965

Photochemical smog in our environment (nitrogenous gases, oxygen, and organic vapors acted upon by radiation) has implicated certain products of incomplete combustion given off by automobiles. Biologically active components extracted from auto exhaust or ambient air have been assayed in various ways, as shown in Table 6-1.

An extensive literature has been written describing the effects of many pollutants on higher plants. Characteristic symptoms for such contaminants as ozone, oxidants, ozone-olefin reaction products, fluorides, SO_2, ethylene, ammonia, chlorine, peroxyacetyl nitrate, and nitrogen oxides develop on a wide variety of plants. With susceptible species as indicators, these toxicants are injurious at concentrations less than one part per million (ppm), and with fluorides at a few parts per billion. Food crops, ornamentals, tobacco, pine and citrus trees, and weeds are commonly studied in tests for phytotoxicity. However, no common bioassay procedure has been developed from these studies and their review has not been included in this report. Varietal differences among plant species and the history and cultural conditions of plants affect their response to pollutant treatment. Several authors have remarked on the radiant energy required by air pollutants to inhibit photosynthesis (Darley *et al.*, 1963; Dugger *et al.*, 1965). Such an area of investigation offers promise for the development of sensitive assays for some phytotoxicants in the oxidant fraction of polluted atmospheres. The deleterious effect of pollutants on photosynthetic tissue in higher plants is paralleled by their effect on lower plants. Druett and Packman (1968) indicated that pollutants can be assayed by bacteria held on spider webbs.

Since atmospheric contaminants play heavily upon the ecology of plants, and thus indirectly as well as directly upon animals, their effect on microorganisms is equally striking. From certain easily measured characteristics of some bacteria, protozoa, and other single-cell organisms, we might expect fruitful ground for the development of sensitive air pollution bioassay procedures for contaminants present at less than 1 ppm. With few exceptions, the development of such methods has been largely neglected. It is possible that a number of microorganisms possess a characteristic response which can be measured with sufficient ease and rapidity to be practical. Won (1969, personal communication), for example, has obtained preliminary evidence that survival of *Rhizobium* sp. in the aerosol form is affected by the presence of NO_2.

As ozone is the principal oxidant formed in photolysed mixtures containing oxygen, hydrocarbons, and nitrogen oxides, the biological activity of such mixtures generally is a function of ozone concentration. At low concentrations, gaseous pollutants alone may not effect any noticeable biological response. Nevertheless, our atmosphere contains a wide variety of particulate matter that can absorb contaminant gases and probably enhance their apparent toxicity.

With social and economic pressures toward urbanization and its concomitant air pollution, there has been an associated rise in lung-cancer mortality. Among the hydrocarbon contaminants in our atmosphere are carcinogenic polynuclear substances whose concentration in American urban environments remains low—approximately 0.1 to 3.0 g per 100 cubic meters of air for benzo(a)pyrene (Sawicki et al., 1960).

Application of bioassays to individual atmospheric polynuclear pollutants has produced evidence of a significant degree of association between photodynamic and carcinogenic activities (Epstein et al., 1963b; Epstein et al., 1965). However, carcinogenicity of a compound could not be qualified; only the probability of carcinogenicity (Epstein et al., 1964) could be estimated. Indices of potential carcinogenic activity were established, but the presence of specific pollutants could not be ascertained.

Realistic efforts to limit and control pollution in our environment will require more extensive knowledge than presently available of the effects of pollution on living systems. Quantitation of these effects through continuing development of bioassay procedures is a worthwhile pursuit—and we have only begun to examine the vast dumping ground that our atmosphere has become.

REFERENCES

Berliner, M. D. 1963. *Armillaria mellea:* An ozonophilic basidiomycete. *Nature,* **197:** 309–310.

Brodo, I. M. 1961. Transport experiments with corticolous lichens using a new technique. *Ecology,* **42:** 838–841.

Darley, E. F., Dugger, W. M., Jr., Mudd, J. B., Ordin, L., Taylor, O. C., and Stephens, E. R. 1963. Plant damage by pollution derived from automobiles. (A review) *A. M. A. Arch. Environ. Health,* **6:** 761–770.

Davis, I. 1961. Microbiologic studies with ozone. Mutagenesis of ozone for *Escherichia coli.* School of Aerospace Medicine Report, June.

Doniach, I. 1939. A comparison of the photodynamic activity of some carcinogenic with non-carcinogenic compounds. *Brit. J. Exptl. Pathol.,* **20:** 227–235.

Doniach, I., and Mottram, J. C. 1937. Sensitization of the skin of mice to light by carcinogenic agents. *Nature,* **140:** 588.

Druett, H. A., and Packman, L. P. 1968. Sensitive microbiological detector for air pollution. *Nature,* **218:** 699.

Dugger, W. M., Jr., Mudd, J. B., and Koukol, J. 1965. Effect of PAN on certain photosynthetic reactions. *A. M. A. Arch. Environ. Health,* **10:** 195–200.

Epstein, S. S., and Burroughs, M. 1962. Some factors influencing the photodynamic response of *Paramecium caudatum* to 3,4-benzopyrene. *Nature,* **193:** 337–338.

Epstein, S. S., Small, M., Koplan, J., Mantel, N., and Hunter, S. H. 1963a. A photodynamic bioassay of benzo(a)pyrene with *Paramecium caudatum. J. Nat. Cancer Inst.,* **31:** 163–168.

Epstein, S. S., Small, M., Koplan, J., Mantel, N., Falk, H. L., and Sawicki, E. 1963b. Photodynamic bioassay of polycylic air pollutants. *A. M. A. Arch. Environ. Health,* **7**: 531–537.

Epstein, S. S., Small, M., Falk, H. L., and Mantel, N. 1964. On the association between photodynamic and carcinogenic activities in polycylic compounds. *Cancer Res.,* **24**: 855–862.

Epstein, S. S., Small, M., Sawicki, E., and Falk, H. L. 1965. Photodynamic bioassay of polycylic atmospheric pollutants. *J. Air Poll. Contr. Assn.,* **15**: 174–176.

Epstein, S. S. 1965. Photoactivation of polynuclear hydrocarbons. *A. M. A. Arch. Environ. Health,* **10**: 233–239.

Epstein, S. S. 1966. Two sensitive tests for carcinogens in the air. *J. Air Poll. Contro. Assn.,* **16**: 545–456.

Epstein, S. S., Joshi, S., Andrea, J., Mantel, N., Sawicki, E., Stanley, T., and Tabor, E. C. 1966. Carcinogenicity of organic particulate pollutants in urban air after administration of trace quantities to neonatal mice. *Nature,* **212**: 1305–1307.

Estes, F. L. 1962. Analysis of air pollution mixtures: A study of biologically effective components. *Anal. Chem.,* **34**: 998–1001.

Estes, F. L., and Pan, C. H. 1965. Responses of enzyme systems to photochemical reaction products. *A. M. A. Arch. Environ. Health,* **10**: 207–212.

Fenton, A. F. 1960. Lichens as indicators of atmospheric pollution. *Irish Nat. J.,* **13**: 153–159.

Falk, H. L., Kotin, P., and Miller, A. 1960. Aromatic polycylic hydrocarbons in polluted air as indicators of carcinogenic hazard. *Int. J. Air Water Poll.,* **2**: 201–209.

Goetz, A., and Tsuneiski, N. 1958. A bacteriological analogue test for eye irritation by aerosols. *Clean Air Quart.* **2**: 7–10.

Goetz, A., and Tsuneiski, N. 1959a. A bacteriological irritation analogue for aerosols. *Arch. Industr. Health,* **20**: 167–180.

Goetz, A., and Tsuneiski, N. 1959b. Bacteriological test for airborne irritants. *Industr. Engr. Chem.,* **51**: 772–774.

Hueper, W. C., Kotin, P., Tabor, E. C., Payne, W. W., Falk, H., and Sawicki, E. 1962. Carcinogenic bioassays on Air Pollutants. *Arch. Pathol.,* **74**: 89–116.

Jacumin, W. J., Johnston, D. R., and Ripperton, L. A. 1964. Exposure of microorganisms to low concentrations of various pollutants. *Am. Industr. Hyg. Assn. J.,* **25**: 595–600.

Kotin, P., Falk, H. L., and Thomas, M. 1955. Aromatic hydrocarbons III. Presence in particulate phase of diesel engine exhausts and carcinogenicity of exhaust extracts. *Arch. Industr. Health,* **11**: 113–120.

Lewis, M. R., 1935. The photosensitivity of chick embryo cells growing in media containing certain carcinogenic substances. *Am. J. Cancer,* **25**: 305–309.

Magdefrau, K. 1960. Lichen growth and the city climate. *Naturw. Rundschau* (Stuttgart), **13**: 210–214.

Mudd, J. B. 1965. Responses of enzyme systems to air pollutants. *Arch. Environ. Health,* **10**: 201–206.

Mueller, P. K., and Barry, W. H. 1962. Separation of disperse and particulate phases of auto exhaust and their effect on cells *in vitro.* California State Department of Public Health Epidemiology Training Program Report.

Pearson, L., and Skye, E. 1965. Air pollution affects pattern of photosynthesis in *Parmelia sulcata,* a Corticolous Lichen. *Science,* **148**: 1600–1602.

Rounds, D. E. 1963. A biological assay method for mixtures of hydrocarbons using human cells in tissue culture. Presented at the Fifth Conference on Methods in Air Pollution Studies, California State Department of Public Health, January 31 and February 1, 1963, Los Angeles.

Rounds, D. E. 1965. Private communication.

Sawicki, E., Elbert, W., Stanley, T. W., Hauser, T. R., and Fox, F. T. 1960. The detection and determination of polynuclear hydrocarbons in urban airborne particulates. I. The benzopyrene fraction. *Int. J. Air Water Poll.*, **2**: 273–282.

Scott, D. B. M., and Lesher, E. C. 1963. Effects of ozone on survival and permeability of *Escherichia coli*. *J. Bact.*, **85**: 567–576.

Serat, W. F., Budinger, F. E., Jr., and Mueller, P. K. 1965. Evaluation of biological effects of air pollutants by use of luminescent bacteria. *J. Bact.*, **90**: 832–833.

Serat, W. F., Budinger, F. E., Jr., and Mueller, P. K. 1967. Toxicity evaluation of air pollutants by use of luminescent bacteria. *Atmosph. Environ.*, **1**: 21–32.

Skye, E. 1958. Luftfororeningars Inverkan På Busk-och Bladlavfloran Kring Skifferoljeverket I Narkes Kvarntorp. *Svensk Bot. Tidskr.*, **52**: 133–190 (Summary in English).

Taylor, O. C., Dugger, W. M., Thomas, M. D., and Thompson, C. R. 1961. Effect of atmospheric oxidants on apparent photosynthesis in citrus trees. *Plant Physiol.*, **36** (Suppl.): 26–27.

Todd, G. W. 1958. Effect of low concentrations of ozone on the enzymes catalase, peroxidase, papain and urease. *Plant Physiol.*, **11**: 457–463.

Won, W. D., and Thomas, J. F. 1962. Effects of arene-type hydrocarbon air pollutants on *Bacillus megaterium*. *Appl. Microbiol.*, **10**: 217–222.

PART 2

AEROSOL CHAMBERS

7

STIRRED-SETTLING AEROSOLS AND STIRRED-SETTLING AEROSOL CHAMBERS

Robert L. Dimmick

NAVAL BIOLOGICAL LABORATORY, SCHOOL OF PUBLIC HEALTH,
UNIVERSITY OF CALIFORNIA, BERKELEY

INTRODUCTION

Whether the first aerobiologists deliberately constructed chambers in which a desired kind of aerosol could be created, or whether chamber types have led to the invention of aerosol species is perhaps only important to the philosopher. Nevertheless, aerobiologists now find it convenient to speak of static aerosols and dynamic aerosols. Static aerosols are contained in barrels, boxes, flasks, or drums. Dynamic aerosols are contained in hoses, tubes, ducts, or pipes. What we mean is that dynamic aerosols are under the condition of being transported while we study them, whereas static aerosols are confined to a stationary volume. The former are currently useful for short-time studies in the laboratory; that is, "zero" to 15 or 20 min; the latter can be studied from about 2 min to as long as 2 weeks.

In chapter 1 we defined an aerosol as consisting of a mixture of particles and air; that is, the state of matter being in the aerosol form. It is interesting to observe that if we started with a cubic yard of aerosol containing a billion particles, and let those particles settle out until only 10 remained, we would somehow consider the mixture as still being aerosol; we would say, however, that the aerosol has "decayed."

On the other hand, if we started with a cubic yard of air that had only 10 particles in it, we would tend to talk about this as not being an aerosol, because such an aerosol is not very useful in the laboratory. In natural air, < 1 particle per cubic foot would usually be referred to as just air with a "low particle count."

The process of aerosol decay, whether by settling of particles or by dilution (commonly termed "aerosol behavior"), is really what Part II is about. The techniques and equipment used to study aerosols influence the concept of "behavior," so the two are inseparable. When pathogens are employed, the nature of aerosol behavior directly influences the extent of the hazard, so chambers must be designed to accommodate both scientific and safety principles. We shall study the simplest type of aerosol in detail (the static aerosol with stirred settling) because almost all principles of aerosol behavior, both physical and biological, can be illustrated by that study. Particle sizing has been inserted into this chapter because the size parameter of interest is the "aerodynamic size," a measurement related directly to aerosol behavior.

THEORY

The typical stirred-settling chamber is simply a sealed box or barrel with some device to gently agitate the air (Figure 7-1). The air is stirred to provide a homogeneous situation so that any sample is equivalent to any other sample. At first thought one might believe that stirring would add to the precipitation of particles by increasing the average fallout rate. On second thought it would be evident that the total velocity vectors of air streams caused by stirring must be zero. Since no air enters or escapes from the chamber, a volume of air moved in one direction must be compensated by another volume moving in the opposite direction; hence, the vectorial sum remains zero. However, a small volume of rapidly moving air can compensate for a large volume of slowly moving air. Therefore, if true spatial homogeneity of particles of various sizes is to be maintained, there must exist at least one air velocity vector sufficient to waft the largest particle to the highest point in the chamber.

Let us first examine the essential principles of stirred settling and return to operational details later. Consider an enclosed body of air containing particles with exactly the same rate of fall—say, 1 cm/sec— and consider two 1-sec discrete intervals of time, the two repeated sequentially. During the first interval, air is mixed so that particles are distributed evenly in all spaces. During the second interval, assume the air to be motionless. Finally, assume a volume of air immediately above

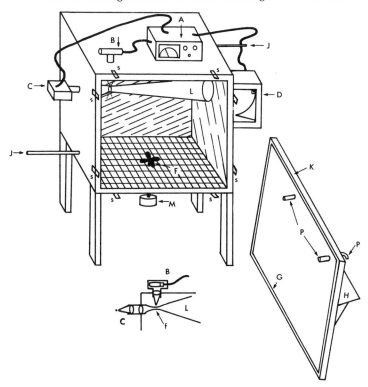

FIGURE 7-1. Suggested design for a stirred-settling chamber. (A) Electronics package: high voltage supply, regulated 6-volt lamp supply, amplifier-readout. (B) Photocell: e.g., RCA 931A. (C) Light source: e.g., microscope lamp. (D) Chart recorder. (F) Fan. (G) Gasket. (H) Shelf for samplers. (J) Air access ports. (K) Removable front panel. (L) Light beam. (M) Fan motor. (P) Sample ports; note downward bend for most direct path to sampler. (f) Focal point. (s) Latches: e.g., suit-case-type. Insert: Practical arangement of light source and photocell.

all horizontal surfaces. The height of this hypothetical volume is the distance a particle of given size and density would fall during the second interval, 1 cm in this example. The number of particles that would fall out of this hypothetical volume during the second interval will always be a fixed percentage of the total number of particles remaining in the chamber, which is a function of the ratio of the hypothetical settling volume to the chamber volume. During the third interval (a repeat of the first) mixing would occur, and during the fourth the number of particles in the hypothetical volume would have been again reduced by the ratio of the settling volume to chamber volume. If the process were made continuous, we would observe a simple rate process, i.e., percent per second, and the equation could be written as:

$$(7\text{-}1) \qquad \frac{N}{N_0} = e^{-kt}$$

where N_0 = total number of particles in the chamber initially,
N = number of particles in the chamber at time t, and
k = a constant, essentially per cent per time.

The height of the hypothetical volume is determined by the rate of fall, so k is a direct function of v, the Stokes' velocity. The larger the area of this hypothetical volume (the area being simply the sum of all horizontal surfaces in the chamber), the greater the number of particles that would settle out, so k is directly related to horizontal area (A). However, the greater the chamber volume (V), the less the effective percentage of the settling area; hence

$$k \cong \frac{A}{V} \cong \frac{\text{width} \times \text{depth}}{\text{width} \times \text{depth} \times \text{height}}$$

or

$$k \cong \frac{1}{H}; \qquad \text{therefore } k = \frac{v}{H}$$

and equation 1 becomes

$$(7\text{-}2) \qquad \frac{N}{N_0} = e^{-vt/H}$$

For chambers with vertical walls, H is simply the height of the chamber; for other chambers, "H" is an "effective height." As an example, the maximum cross-sectional area influences the ratio in a spherical volume, not the surface area of half the sphere. That is,

$$(7\text{-}3) \qquad \text{``}H\text{''} = \frac{V}{A} = \frac{4/3\pi r^3}{\pi r^2} = \frac{4}{3} r$$

where r = radius of sphere, and for a cylinder in a horizontal position,

$$(7\text{-}4) \qquad \text{``}H\text{''} = \frac{V}{A} = \frac{\pi r^2 L}{2rL} = \frac{\pi r}{2}$$

where L = cylinder length and
r = radius of cylinder

whereas for the same chamber in a vertical position

(7-5) $$\text{``}H\text{''} = \frac{V}{A} = \frac{\pi r^2 L}{\pi r^2} = L$$

the same as for all chambers with vertical walls.

All horizontal surface areas such as shelves, brackets, etc., contribute to the "effective area" and must be added to areas used in the above calculations. Usually the correction is of no consequence, but if a chamber is filled with gadgetry of one sort or another, the correction can be important. According to Sinclair (1950), Langstroth and Gillespie (1947), and my own studies as well, the number of particles impacting on the walls and ceiling can be neglected in most calculations.

Essentials of Aerosol Behavior

The plot of $\ln N/N_0$ versus t forms a straight line. The slope of this line, all other factors being constant, is a function of the aerodynamic particle size. There are a number of ways of expressing rates of decay. One unambiguous characteristic that is easily visualized and manipulated, and one commonly used (for example, to express the decay of radioactivity) is the half-life. This is the time required for half of the particles to be removed from the total volume of enclosed air.

The equation, derived from 7-2 is:

(7-6) $$t_{\frac{1}{2}} = 0.693 \frac{H}{\mathbf{v}}$$

where $t_{\frac{1}{2}}$ = time interval for particle numbers to decrease by 50 per cent, sec,

H = effective chamber height, cm, and

$\mathbf{v}$ = Stokes velocity of fall for particle of given size and density, cm/sec.

For a fixed chamber height, variables are particle size and density. It is helpful to have a nomograph that relates half-life to size and density, as shown in Figure 7-2. In the nomograph, the chamber height was assumed to be 100 cm, and Cunningham's correction (see Chapter 1) was not included. To use the nomograph for chambers of other effective heights, use the formula:

(7-7) $$t_{\frac{1}{2}}* = \frac{H*t_{\frac{1}{2}}}{100}$$

where $t_{\frac{1}{2}}*$ = half-life for chamber of effective height $H*$, cm. The nomograph will be used in the following analysis.

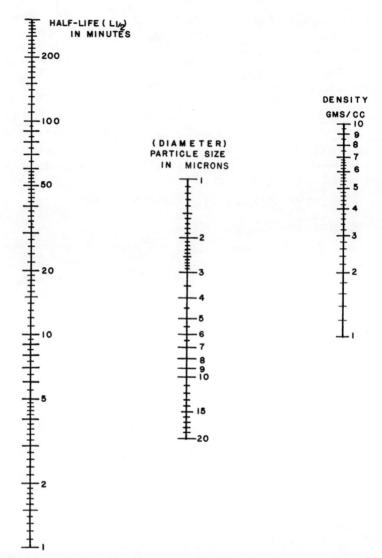

FIGURE 7-2. Nomograph relating density and half-life to particle size for aerosols undergoing stirred settling in a chamber of 100-cm effective height. A straight line drawn from appropriate density to observed half-life intersects at correct particle size. (Dimmick, Hatch, Ng. 1958. *A.M.A. Arch. Ind. Health,* **18: 24.**) Reprinted with permission of the American Medical Association.

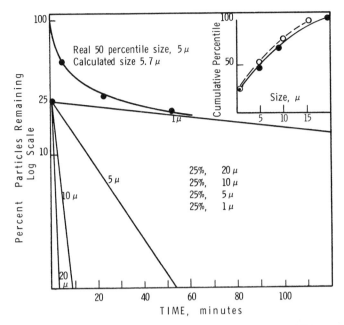

FIGURE 7-3. Construction of a complete decay curve by addition of rates of decay of discrete particle size fractions assumed to be present in equal numbers. Insert is the real cumulative distribution histogram (filled circles) compared with the distribution estimated by subtracting rates from the complete decay (Figure 7-5) shown as open circles.

We can now consider the more realistic case of aerosols with heterogeneous particle sizes. Suppose particles in the aerosol included equal numbers of discrete sizes, for example, 20, 10, 5, and 1 μ. Since the numbers in each size range would decrease independently of the others, we would observe the overall result to be a curved line on semi-log paper. We could construct the total curve, as shown in Figure 7-3, by simple addition of the separate rates. The effect of changing the initial distribution is shown in Figure 7-4. Note that in both instances the final decay rate reaches that of the particles of smallest size. The aerosol becomes more and more homogeneous with respect to particle size as a function of time.

We could reverse this process, with respect to an observed decay, by subtraction. By assuming that the final slope was representative of the amount and decay rate of the smallest size particle, one could subtract this rate and draw a new curve with a new "final decay rate," and repeat this until nothing remained. I have performed this subtraction

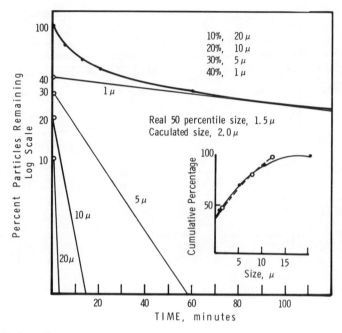

FIGURE 7-4. Construction of a complete decay curve by addition of rates of decay of discrete particle size fractions assumed to be distributed as 10%, 20 μ; 20%, 10 μ; 30%, 5 μ; 40%, 1 μ. Insert is the real cumulative distribution histogram (filled circles) compared with the distribution estimated by subtracting rates from the complete decay (Figure 7-6) shown as open circles.

for the two synthetic curves (Figures 7-3 and 7-4) and the results are in Figure 7-5 and 7-6. The size associated with each slope was determined from the nomograph and the percentage of that size from the abscissa. The estimated cumulative distribution of sizes, compared to the known distribution, is shown as open circles in the inserts of Figures 7-3 and 7-4.

Note that the estimated size by the subtraction process yields apparent fractions not always related to the known fractions. In the examples shown in Figures 7-5 and 7-6, it would have required additional time before the final slope would become a true logarithmic decay; eventually, the slope would have been exactly that of the real value. The selection of a slope with a decay slightly greater than the real one caused the value for the percentage of the smallest size to be over-estimated—and this caused a decrease in the other estimates, as well as a change in apparent size. Nevertheless, the difference between the real and the calculated 50-percentile size was less than 1 μ, and if I had chosen a more correct initial slope I would have been more accurate.

Of course, aerosols are rarely composed of particles in discrete size ranges, so we could not ascertain, by examination of the total curve alone, whether discrete sizes existed or not. Our subtraction process merely told us that the curve seemed to contain certain amounts of hypothetical, discrete sizes. The cumulative distribution of the percentages of these sizes is fairly close to the real distribution. A final example will show this. The curve in Figure 7-7 is made up of the reverse percentages shown in Figure 7-4; again, the real and calculated 50-percentile sizes differ by about 1 μ, and the shapes of the distributions are about the same.

This simplified approach does not consider possible effects of coagulation (Devir, 1965). If coagulation occurs, however, the result would be approximately the same as if large particles had been in the original aerosol, and the answers obtained would remain a useful estimate of the behavior of the whole aerosol. Generally, we presume that the initial concentration is less than 10^9 particles/liter (Chapter 1). If particles were not spherical, as we have been assuming, then analysis by this method would be incorrect. On the other hand, we could say that the

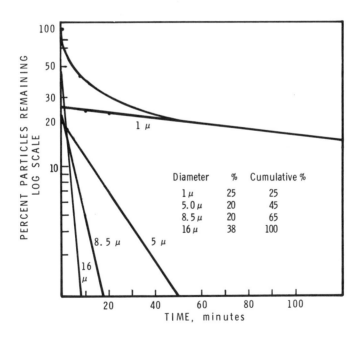

FIGURE 7-5. The complete decay shown in Figure 7-3 with apparent discrete rates and sizes determined by a subtraction process.

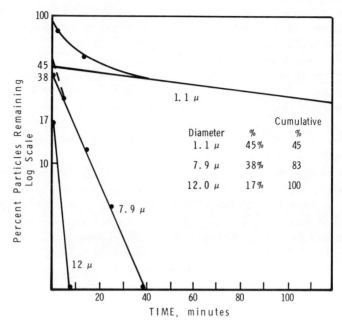

FIGURE 7-6. The complete decay curve shown in Figure 7-4 with apparent discrete rates and sizes determined by a subtraction process.

aerosol exhibited aerodynamic properties (referred to as the Stokes' diameter or settling diameter) equivalent to one made up of sperical particles of an assigned size and of unit density—whether they are nonsperical particles or aggregates of smaller particles—and these properties are, indeed, the ones in which we are interested. It is this aerodynamic property that determines whether a certain particle will be impacted or entrained at given velocities, and is ultimately the property that determines whether or not a given airborne cell can reach an infective position in the host. In summary, when we have described the aerodynamic size distribution of an aerosol, we have described the aerosol behavior.

PARTICLE SIZE

We have talked of particle size as if this parameter were a definite and measurable attribute of particles in an aerosol. The literary confusion about sizing of particles is one unit larger than the number of papers on the subject. Although measurement of particle size is not required to produce (or use) aerosols, size measurement is a necessary activity with respect to characterizing and predicting particle behavior. We will, therefore, turn our attention to particle sizing before discussing other aspects of aerosol behavior.

Actually, we should be interested in the size of particles more for purposes of aerosol control than for record keeping. To infect animals with airborne microbes, or to study relationships of particle size to microbial survival, we would not routinely measure airborne particles solely for the purpose of adjusting our data to account for the observed size, but instead, if the size were not correct, we ought to change our dispersion or exposure procedures to create the desired particle for the experiment. This means that our sizing method should be simple and rapid as well as precise enough for our purposes. In fact, we do not really need to know the "true size" as long as we can measure some property related to the lung-retention efficiency of hygroscopic, aqueous particles. Fortu-

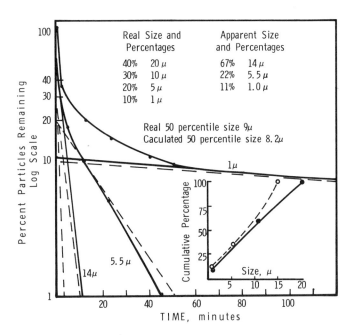

FIGURE 7-7. Construction of a complete decay curve by addition of rates of decay of discrete particle fractions assumed to be distributed as 10%, 1 μ; 20%, 5 μ; 30%, 10 μ; 40%, 20 μ. Apparent rates obtained by a subtraction process are shown as dotted lines. Insert is the real cumulative distribution histogram (filled circles) compared with the distribution estimated by the subtraction process (open circles).

nately, these criteria eliminate about 99% of the generally available methods. In the absence of realistic knowledge, we shall arbitrarily accept an accuracy of ± 10%; this is certainly within the range of accuracy of our knowledge of lung-retention, although newer methods hold promise of decreasing this variance (Sheien et al., 1967).

There is also a question of how to report the size. Probably no aerosol is truly mono-disperse, so size is a distributed parameter—often log-normal, but not always. Kethley et al. (1963) suggest that the surface mean diameter (SMD) is most appropriate. Those interested in infectiousness of aerosols prefer a number, or count mean diameter (CMD). Persons interested in aerosol production and behavior prefer the mass median diameter (MMD), which is the one that will be principally employed in the following discussion. Regardless of which is utilized, it is most inappropriate to refer to an aerosol as having a particle size without reference to both the measured parameter and the distribution, if known. It should be noted that, whereas surface area and volume are directly related to diameter as a function of r^2 and r^3, median measurements are not; Kethley et al. (1963) discuss transformation of these values.

Direct Methods

The diameter measured by means of a microscope with an eyepiece graticle (calibrated by a stage micrometer) is the absolute standard. It is comforting to obtain a sample from an aerosol composed mostly of hydrolyzed and hygroscopic plant or animal debris, plus a scattering of multi-hydrated salts, and observe under the microscope that particles are, indeed visible, and that they possess a unique size usually distributed so that one sees both large and small particles. To measure size by this method requires skill, patience, attention to established routine, and sometimes, I suspect, a fair amount of prescience. Electronic and mechanical methods have been tested (Lang, 1954). I shall not attempt to recapitulate the several excellent papers devoted entirely to particle-size determination by the counting and measurement method (Cadle, 1965; DallaValle, 1948; Ettinger and Posher, 1965; Griffiths and Lampitt, 1947; Lang, 1954; Irdni and Callis, 1963; Orr, 1966; Orr and DallaValle, 1959) but I would like to summarize certain aspects important to our special needs, and to mention briefly some current methodology.

Particles seen with the aid of the microscope are quite often not the same size as they were in the aerosol. For example, if the sample was not taken iso-kinetically, or if preimpaction on sampling lines, etc., had occurred, then the distribution may not be representative of the original population of particles. This is true, of course, for any method of sampling.

Additionally, particles may grow or shrink as a result of changes in relative humidity. I have seen particles of heart infusion broth medium on a slide grow to two or three times their original diameter as the result of a misplaced breath. If the force of impaction is severe, as it must be to collect particles as small as 1 μ, larger particles can be disrupted to form "satellite" particles. More than one particle can "land" in the same space—a situation leading to "coincidence counting." Finally, it has been shown that even experienced technicians tend to make unconscious errors and arrive at answers that often do not agree, even when the same field is recounted.

The oil-immersion lens of most microscopes is made to be used with coverslips of a certain thickness, and the numerical aperature of the lens should match that of the condensing lens. If immersion oil is used without coverslips, then optical aberration can occur. Obviously, water cannot be used under the coverslip as is done when bacteria are being observed. The use of oil can cause distortion, or "sagging" of particles, as well as affect disastrous changes in refractive index; often aqueous particles move together and coalesce in an oily environment.

The best that can be done, if it must be done, is to collect samples on a coverslip, preferably by allowing particles to settle on the glass for a long enough time to collect *all* sizes, to invert the coverslip on a ring or cell, and to observe the particles through the coverslip. Other techniques include the use of a film of deeply stained oil, and opaque material such as carbon films or magnesium oxide (May, 1950), or a recent technique using metal films (Horstman and Wagman, 1967); "holes" produced by the particles are counted and measured. The use of films of chemicals that react with particles to produce stained circles has also been reported. Goldberg (1950) used an amine alkyd compound. None of these is entirely satisfactory for 1- to 10-μ particles, and it must be accepted that if we desire to measure the particle size and distribution of material *as it exists in the aerosol*, then indirect means will have to be employed. By "indirect means" I imply the measurement of some property related in known ways to particle diameter. Although the relationship may be theoretically known, each technique that provides indirect analysis must be shown to be capable of actually measuring particles of known diameter; that is, the instrument must be calibrated.

Indirect Methods

Advantage can be taken of the "impaction parameter." Several graded impaction stages, in series, form what is known as a cascade impactor (May, 1945; Wilcox, 1953). Each stage is built so that the characteristic diameter (see Chapter 1) of that stage is larger than the following stage; the last stage is designed to impact the smallest particle to be considered. The unit can be used in several ways: (a) particles can be collected on coverslips and observed under the microscope; (b) the collected material can be rinsed off and the mass determined by chemical means (a four-stage impactor would provide four points from which to construct a distribution curve); (c) the numbers of viable organisms on the slide of each stage can be assayed by standard microbiological methods; (d) particles passing through each stage can, in turn, be measured in the aerosol form by photoelectric means (Thompson, 1957; Lundgren and Long, 1967), and a weighted distribution curve can then be constructed; and (e) the collecting surface can be nutrient agar in a petri plate, a technique we will discuss later. An eight-stage impactor was constructed and tested by the Naval Biological Laboratory, but it has not been used as widely as the less bulky four-stage units.

Cascade impactors are not easily calibrated, and there is some controversy over interpretation of data from these instruments. Mercer (1962, 1965) points out that the distribution of particles on a given stage will depend on the distribution of sizes in the test aerosol, and he discusses other inherent errors. Two errors of major importance include slippage (the measurement of which usually includes particles that fail to impact, particles re-entrained in the air stream, and fragments resulting from shattering of particles), and wall loss. Mitchell and Pilcher (1959) discuss wall loss and, with respect to their impactor design, consider the instrument to be rapid and accurate with the added benefit that, once a design has been adapted and calibrated, new units built the same way need not be recalibrated. I concur, if one speaks of replicability, but in terms of accuracy with respect to what was in a sampled aerosol, I reserve judgment until more data are available. It is not the general size of aerosol particles that interests the aerobiologist—instead, the size of those particles that contain microbes is the parameter of concern.

An entirely different approach to the use of impaction, and one that has not been sufficiently explored, was utilized by Goldberg (1950). Instead of using a constant airflow rate, he varied the rate and measured the mass slipping past two identical fixed stages. He derived the expression,

(7-8) $$D_{50}V = K$$

where D_{50} = impaction parameter for 50% efficiency, microns,
V = air flow rate, liters per minute, and
K = a constant

for a give particle density and jet dimension. In his instrument, the applicable size range was small (0.5 to 2 μ), but there is no reason why this technique of varied air flow could not be applied to standard cascade impactors, or to Andersen samplers (see below) to increase their versatility.

The "Aerosol Spectrometer" (Goetz et al., 1960) also utilizes inertial parameters to provide the separation of particles according to their aerodynamic diameter. It is ideally suited for particles 2 μ or less, but the essential measurement relies on microscopic examination, and the instrument is rather expensive compared with other instruments for obtaining equally valid data.

Light Scatter

Another suitable indirect method is that of measuring the amount of light scattered from particles (Hodkinson, 1966). Monodisperse aerosols can produce diffraction bands related to particle size, but the method is not very useful for the aerobiologist, although a recent application of computer techniques may yield new utility (Haughey, 1967). Aerosols can be drawn through an illuminated orifice, and a sensitive photomultiplier tube will respond to the "twinkle" of particles as small as 0.1 μ if the source of illumination is intense enough. The amount of light scattered is a function of particle size, so the electrical signal from the photomultiplier tube can be transformed by electronic, pulse-height analyzer circuits to count only those signals within preset limits (Gucker, 1952). The numbers so accumulated can be employed to construct distribution histograms. The channels of the pulse-height circuit can be standardized by use of aerosols of nonhygroscopic particles of known size such as dioctyl phthalate or polystyrene latex.

One encounters certain difficulties with these instruments. For example, the concentration of particles that can be counted per second is low compared with the concentration of airborne particles required in common laboratory experiments, so the aerosol must be diluted with clean air (Katz et al., 1956). This dilution, if the air is not the same temperature and relative humidity as the sample air, can cause changes in the apparent size distribution by (a) selective removal processes, (b) growth or shrinkage of particles, and (c) possible changes in the scattering coefficient.

Although scattered light intensity is related to particle area, it is also related to the nature (scattering coefficient) of the dispersed material, the wavelength of the light, and the angle of scatter; it is actually a very complex phenomenon. The pulse height of a standard particle might not be the same as the pulse height of the desired biological material (Gucker *et al.*, 1965). Particle counters of this type could, however, be calibrated satisfactorily by utilizing stirred settling aerosols as described below.

Altogether, photometric counters with electronically transformed output represent highly sophisticated and rapid means of obtaining useful comparative estimations of particle-size distributions. A number of commercial instruments of this type are available.[1] A miniaturized, airborne system has also been reported (Reist *et al.*, 1967).

Light Scatter Combined with Settling

There is a method of obtaining an indirect measurement of particle size while, in addition, gaining knowledge of the relative concentration of an aerosol. This method utilizes the principle of the photometric method, above, combined with the analysis applied to the behavior of aerosols undergoing stirred settling (Dimmick, 1960). If the light scatter photometer, which is usually aligned so that pulses from single particles can be detected, is made slightly less sensitive so that light detected by the photocell represents a mean scatter value of a number of particles, then the instrument becomes an "aerosol monitor." This can be accomplished in most photometric instruments by drawing the sample through the light path at a rate faster than the ability of a readout meter or counter circuit to respond to light scattered from individual particles, or the virtual image of the light source can be magnified; one obtains an integrated signal (the relative light scatter area). This technique requires that samples be drawn frequently from the aerosol; since samples must be obtained for assay of biological properties, the two operations can be combined.

Operation of a Stirred Setting Chamber with Continuous Monitoring

Particle Size. At the start of an assay, the aerosol is drawn through the photometer until a constant value is registered, then the aerosol is directed into other samplers. For this purpose a chart recorder output is highly valuable because both a permanent record and a visual indication of the change in concentration of the aerosol is provided (Talvitie and Paulus, 1956). In fact, the combination of a photometer head and the recorded output allows one to control the formation, analysis, and dilution of aerosols in a continuous manner, and changes chamber operation from one of guesswork to one of relative precision. Since monitoring

[1] Bausch and Lomb, Inc., Rochester, N.Y.; Climet Instruments, Inc., Sunnyvale, Calif.; Eldorado Electronics, Concord, Calif.; Royco Instruments, Inc., Palo Alto, Calif.; Phoenix Precision Instrument Co., Philadelphia, Pa.

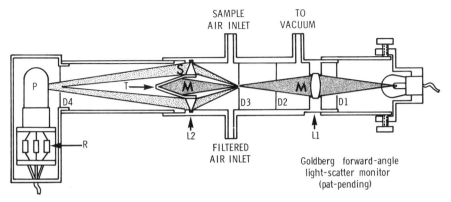

FIGURE 7-8. Simplified diagram of the Naval Biological Laboratory Forward-angle Aerosol Monitor, designed by Leonard J. Goldberg. *D1, D2, D3* and *D4* = Diaphrams; *L1* = Focusing lens (for direct light path, *M*); *L2* = Collimating lens (for scattered path, *S*) with hole containing an inserted, cone-shaped light trap, *T*; *P* = Photomultiplier; *R* = Voltage dividing network (only three resistors shown). This unit, in association with standard electronic readout facilities, has a dynamic range of 7 or more logs, and is ideally suited for filter testing as well as aerosol monitoring.

and analysis are so interdependent, we shall consider them at the same time.

A suitable combination for control purposes would be the photometric head shown in Figure 7-8, or the Gucker (1952) model, in conjunction with, for example, an Eldorado model 201 universal photometer. I have successfully used a battery supply for the photocell, and a vacuum tube voltmeter for readout. The student should consult modern texts on electronics for recent solid-state readout devices, and for proper operation of phototubes. The sensitivity of modern photocells is such that forward-angle scattering is not always needed, so one can often utilize a simplified right-angle scatter arrangement that can be permanently mounted inside most types of stirred settling chambers to provide continouous monitoring as shown in Figure 7-1. A type of "light scatter probe" that utilizes back-angle scatter, and can be inserted into most types of chambers, was described by Dimmick (1961) and utilized by Hayakawa (1962, 1964) for this purpose.

An additional benefit is obtained by either continouous monitoring or intermittent sampling (Dimmick *et al.*, 1958). Data, which would be in terms of the change in relative light scatter area with time, can be plotted on semilog paper to yield a decay curve representative of the behavior of the aerosol in terms of surface area. If the particles were of uniform size, then for every particle that settled, a corresponding fraction of initial surface area available to scatter light would be lost, so the number-decay and surface-decay rates would be identical.

As previously described, we could perform subtraction of slopes and construct a cumulative distribution of surfaces and determine a mean surface diameter. A more meaningful parameter would be the mass median diameter (MMD) if efficiency of dispersion were being studied. Since the relationship between surface and mass is a function of r^2 to r^3, we could obtain relative mass values by simply multiplying the apparent initial light-scatter fraction of each size by the radius, converting to percentages, and drawing the resulting distribution curve. An example of this analysis is shown in Figure 7-9.

Strictly speaking, monochromatic light scattered from individual particles is a complex function of particle size and scatter angle, especially when the size approaches that of the wavelength of light. Above this size, scattered light is a more uniform function of the particle area. Particles near 1 μ in diameter scatter a little more light than their relative area. However, if we consider the situation where the source of illumination is not monochromatic, and where particles scatter light directly from the source as well as from other particles, then in the range of 1 μ and above the area law has proven to be reasonably applicable.

The Cascade Sieve Impactor (Andersen Sampler)

Discussion of the Andersen sampler, probably the most useful sampler the aerobiologist now possesses, has been included at this point in the text to show how stirred settling methodology and impactor methodology are complementary.

The Andersen sampler (Figure 7-10) is composed of a series of graded sieve samplers, each of which is a multiple-jet impactor formed by drilling numerous holes through a plate. A six-stage unit is manufactured by Andersen Samplers and Consulting Service, Provo, Utah (Andersen, 1958), and a useful modification is described by May (1964). The standard unit normally utilizes special glass petri plates to which 27 ml of agar medium are added. Airborne microbes are impacted directly on the agar at a sampling rate of 1 ft³/min. After the plates are incubated, colonies are enumerated with appropriate corrections for coincidence counts (probability of more than 1 particle passing through a given jet), each stage is assigned a collection parameter in terms of a median diameter (MD), a cumulative distribution curve is constructed (sum of colonies equal 100% and each stage assigned a percentile), and the MD of the collected particles is determined from the graph. The total number of colonies divided by the sampling time yields the

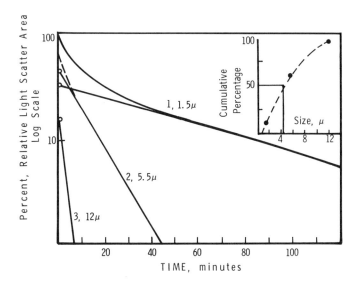

FIGURE 7-9. Size analysis of an aerosol by the light-scatter decay method. The light-scatter decay curve is always assumed to start at 100%, and is traced either by continuous light-scatter recording, or plotted from periodic measurements, until a straight line on semi-log paper is obtained approximately equal in length to the first curved portion. A straight line is drawn through this latter half and extrapolated to zero time. The half-life of this slope, the particle size related to this half-life (from Figure 7-2), and percentage at zero time are recorded. Points on this line are then subtracted from equivalent points on the original curve, and a new curve is drawn. The process is then repeated until only a straight line (12 μ in the figure) remains, each curve being "steeper" than the last. The assumption is made that the total curve was a composite of discrete fractions having approximately the calculated sizes and percentages; the method does not depend on assuming any specific distribution (e.g. log normal). Data are tabulated as shown below, and either the median diameter is determined by plotting as shown in Figures 7-3 or 7-4, or the mass mean diameter is determined arithmetically.

Size, μ (S)	% Light-scatter area, at zero time	Relative mass[a]	Mass % (M)	Cumulative mass %	(S) $\times$ (M)	(S)^2 $\times$ (M)
1.5	30	45	8.5	8.5	13	19
5.5	55	302.5	57.3	65.8	315	1733
12	15	180	34.1	99.9	409	4910
Total		527.5			737	6662

[a] Size times area:

$$\text{Standard deviation} = \sqrt{\frac{(S)^2 \times (M)}{100} - \left(\frac{(S) \times (M)}{100}\right)^2}$$

Standard deviation = 3.46

$$\text{Mass mean diameter} = \frac{(S) \times (M)}{100} = 7.37 \pm 3.46$$

Mass median diameter (graphically) = 4.2 μ

CROSS SECTION OF ANDERSEN SAMPLER

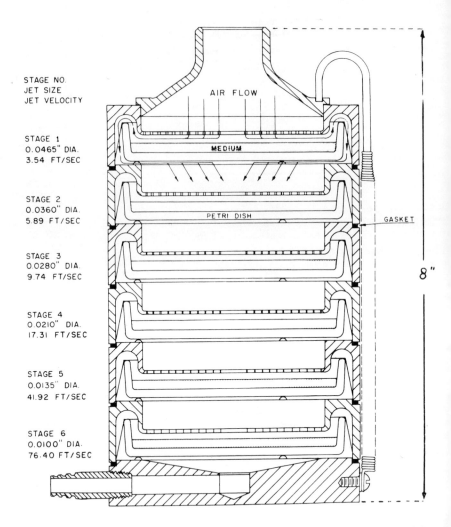

STAGE NO.
JET SIZE
JET VELOCITY

STAGE 1
0.0465" DIA.
3.54 FT/SEC

STAGE 2
0.0360" DIA.
5.89 FT/SEC

STAGE 3
0.0280" DIA.
9.74 FT/SEC

STAGE 4
0.0210" DIA.
17.31 FT/SEC

STAGE 5
0.0135" DIA.
41.92 FT/SEC

STAGE 6
0.0100" DIA.
76.40 FT/SEC

AIR FLOW

MEDIUM

PETRI DISH

GASKET

8"

FIGURE 7-10. Cross section of the Andersen Sampler. Courtesy of Andersen Samplers and Consulting Service, Provo, Utah.

concentration of viable particles (colony-forming units, CFU) per cubic foot of air.

There are some problems. May (1964) points out that the first two stages do not collect properly with respect to larger particles, that the cone-shaped entrance fixture should not be used in field sampling, and that more effective iso-kinetic sampling is obtained if the sampler is tilted into the wind. In the laboratory, the entrance fixture is required unless the sampler is located within a chamber. Slippage of larger particles to lower stages can occur, and is especially bothersome when the sampler is being calibrated. Some slippage is caused by the increase and decrease of air flow when vacuum is applied. Slippage can be decreased by placing a large two-way, quick acting valve between the aerosol source and the sampler. While the vacuum source is being turned on or turned off, the valve directs clean air through the stages, and is turned to direct the aerosol into the sampler only after air flow has equilibrated.

It is reasonable to expect that the same, critical jet-to-surface distances used to calibrate and determine size-parameters for each stage cannot be maintained when petri plates, filled with agar under usual circumstances, replace metal surfaces. Each plate would have to be poured on a precisely level surface, plates would have to be stored under conditions minimizing water loss and shrinkage of agar, and all plates would have to be the same "age." If the laboratory can be equipped to furnish the required numbers of plates this way, well and good. Useful data can be obtained, however, if usual methods of pouring plates are carefully controlled.

A single, detailed example of sampler utility will serve to illustrate results one can expect to find in both laboratory and field studies. We conducted a series of experiments with powdered and atomized aerosols of *Sarcina lutea* (a species that is remarkably stable in the aerosol form) confined in a cubic-meter stirred settling chamber. Several modifications of the Andersen sampler were tested, as well as different kinds of plates and different sized aerosols.

Table 7-1 shows two sets of raw data from samples collected from a single aerosol. Analysis by light-scatter decay indicated that the mass median diameter of all airborn particles during the time of sampling was 1.9 μ A slit sampler and two Andersen samplers, one with glass plates and one with plastic plates, were used to determine "total" decay. Data are divided into three time periods, and average times and values are used to plot decay rates. Figure 7-11 shows the total decay rate as well as particle size (from Figure 7-2) determined by the three different sampling conditions. Rates have been adjusted for sampling volume

Table 7-1 Colony forming units collected from a powdered aerosol[a] of *Sarcina lutea* by a slit sampler and by a standard, 6-stage Andersen sampler.

Aerosol age (min)	Sampling time (sec)	Total colonies						Organisms per liter	SI$_{50}$[b]
		SLIT SAMPLER							
185	60	115						29	
192	60	68						17	
197	60	69						17	
234	75	40						8	
238	75	27						5	
243	75	41						8	
285	120	30						4	
292	120	18						2	
297	120	18						2	

		6-STAGE SAMPLER, PLASTIC PLATES							
stage number →		1	2	3	4	5	6		
186	60	5	11	30	345	200	1	91	3.8
189	60	1	2	19	330	206	0	54	3.8
194	60	3	3	15	311	156	0	38	3.8
230	75	2	2	8	191	137	0	13	3.9
235	75	0	4	10	184	113	0	12	3.9
239	75	0	1	4	180	85	0	10	3.9
281	120	0	1	6	117	135	0	6	4.1
288	120	0	0	7	101	83	0	4	4.0
Half-life, min				47[c]	75	110		mean	3.90; 2.1 μ
Mean diameter, μ[d]				2.6	2.0	1.6			

		6-STAGE SAMPLER, GLASS PLATES							
stage number →		1	2	3	4	5	6		
188	60	3	3	161	355	34	0	86	3.3
190	60	2	4	168	290	25	0	32	3.2
195	60	1	5	143	293	31	0	32	3.2
232	75	0	2	84	214	13	0	13	3.3
237	75	0	0	53	192	17	0	10	3.4
241	75	1	2	52	194	15	0	10	3.4
283	120	0	1	28	155	10	0	5	3.3
290	120	3	2	30	124	15	0	4	3.5
295	120	1	1	21	115	11	0	3	3.4
Half-life, min				37	82	85[c]		mean	3.35; 2.8 μ
Mean diameter, μ[d]				3.0	1.9	1.9			

[a] Temperature, 20°C; Relative humidity 40%.
[b] Fifty-percentile stage index, see text.
[c] Too few colonies to draw accurate slope.
[d] By stirred settling analysis.

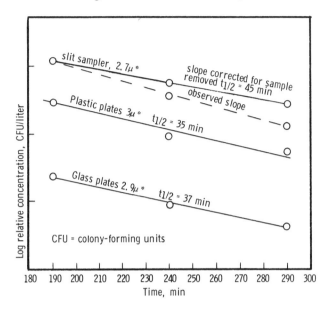

FIGURE 7-11. Total decay rates of an aerosol of *Sarcina lutea* determined by slit and Andersen samplers. Slopes have been corrected for loss caused by the sampling volume (illustrated in top slope) and values were multiplied by a constant to permit separation of slopes in the figure. * From nomograph, Figure 7-2.

removed from the chamber (28.3 liters per sample from a 1,000-liter chamber). These values agree so well that their average (2.9 μ) could be accepted as *the* Stokes' (settling) diameter for those particles containing bacteria. Note the difference between this estimate and the size estimate by the light-scatter method, a situation often observed with powdered aerosols, but seldom noted when bacteria are aerosolized from liquids; particles containing bacteria were a small fraction of the total aerosol mass.

Some of the apparent total decay, however, could have been biological in nature. The Andersen sampler provided additional data to test that suggestion. For example, the number of colonies on each stage of a given sample can be listed as percentages of the total number of colonies collected in that sample. A plot of the cumulative percentage versus stage number permits determination of a 50-percentile stage index (SI_{50}). Figure 7-12 is a plot of data from the initial sample for both the plastic and glass plates; SI_{50} values for all other samples of each type are listed in Table 7-1. The similarity in slopes show that the two conditions (plastic and glass plates) provide estimates of the distribution of particle

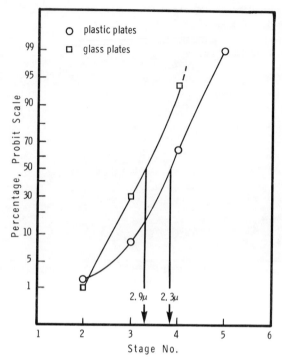

FIGURE 7-12. Cumulative probit plot of distribution of colonies found on each stage of an Andersen sampler (refer to Table 7-1).

sizes that are essentially identical. The complete experiment consisted of 14 aerosols and 53 useful samples (no stage overloaded). Overall replicability, in terms of a 95% confidence interval of the SI_{50}, was ± 3%. Included in this interval, of course, was the ability to produce replicate aerosols.

If the SI_{50} were employed in studies of measurement of lung penetration and retention, then a "true" size would not have to be known; we could simply refer to the SI_{50}. A size parameter can be estimated, however, by referring to Figure 7-13, where a 50-percentile impaction size (data from varied sources) is plotted for each stage; an SI_{50} of 3.3, for glass plates, is equivalent to a particle size diameter of 2.9 μ. For plastic plates, the estimated size (2.3 μ) was smaller than for glass because the same agar volume (27 ml) had been used in each type of plate, hence the slit-to-agar distances were greater with plastic plates (because of their lesser thickness) and particles could penetrate to lower stages than with glass plates.

Since the size determined by total decay rate agreed within reasonable limits to the size determined by the SI_{50}, the bacteria were not dying rapidly under the condition employed; biological and total decay rates were essentially identical. This means we can examine the decay rate per stage to see whether the stages collected particles of about the size that Figure 7-13 predicts. Times and colony numbers were averaged, as was done for data in Figure 7-11. Rates were adjusted for rate of sample-volume removal, the results have been included in Table 7-1, and are plotted in Figure 7-14. Stage 4 (glass), which collected the predominate and most significant number of particles, agreed (within 5%) with the value from May (1964), and all tested stages yielded data within bounds listed by Andersen. The fact that all did not agree with the May values for the three stages where data were available indicates some "slippage" occurred. The fact that the apparent size collected on stage 5, measured by fallout data, was about the same as on stage 4 reflects the extent of the overlap. More importantly, these data, as well as the most recent data of Flesch *et al.*, (1967), suggest that size ranges, established under highly controlled conditions for collection efficiency per stage, may depend upon the distribution of particles in the test aerosol.

The Andersen sampler is an excellent field sampler where vegetative, airborne microbial content is within 10 (a 10-minute sample) to 10,000

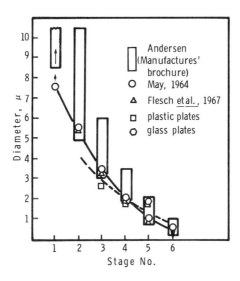

FIGURE 7-13. Relationship between characteristic collection diameter, μ, and stage number for the Andersen Sampler. Data for plastic and glass plates are from Table 7-1. The approximate curve for plastic plates is shown as a broken line.

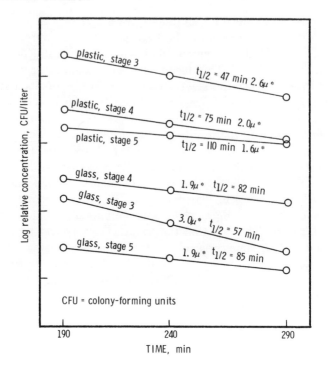

FIGURE 7-14. Total decay rates of an aerosol of *Sarcina lutea* determined by stages 3, 4, and 5 of the Andersen sampler, using either plastic or glass plates. Slopes corrected as in Figure 7-11. * From nomograph, Figure 7-12.

(a 10-second sample) particles/ft³. I have successfully sampled for fungi in the desert (where the relative humidity was 7%) for 30-minute periods. The medium dried severely though not completely, but it was the only sampler of a number tested that yielded useful data. These conditions are undoubtedly too severe for most vegetative microbes, although I found a number of yeast species on the plates. The apparent SI_{50} was not directly related to size because the medium had shrunk, but replicate samples agreed and, on a comparative basis, I was able to show differences in different areas and detect sources of airborne contaminants. This example illustrates the versatility of the sampler.

In this instance I could have set up duplicate conditions in the laboratory and determined a correction factor for the shrinkage. Whether this step is required depends on the experiment. It is my considered opinion that every user of the Andersen sampler (any sampler, for that matter) should have a simple aerosol chamber available, both for the purpose of satisfying himself and his colleagues that he can obtain useful data, and for duplicating field conditions so that appropriate corrections can be made. Again, the required accuracy of such corrections depends on how these data are to be used.

Another important correction, and one that has been overlooked in a number of publications, applies to the estimation of particle size when collected material, such as virus, is removed from the media for assay of viability. Numbers of viable units per stage must be divided by the cube of the characteristic diameter of that stage to obtain an estimate of the relative number of particles. The number of particles per stage is then used to find the SI_{50}.

Miscellaneous Methods

Another indirect method of determining particle size involves the determination of the mass of material that settles out of a restricted columnar volume wherein the aerosol undergoes "tranquil" settling. This method corresponds in many respects to the differential centrifugation of molecules of different sizes and weights used by molecular biologists. As the aerosol settles, the cumulative weight is determined by a sensitive microbalance. An instrument for this purpose, termed a "micromeri-graph," is manufactured by the Sharples Company[2] (Payne, 1948). Difficulties with this instrument are that a minimum of 50 mg of material is required and the lower effective limit for particles of density near 1.1 is greater than 3 μ. Grumprecht and Sliepcevitch (1953) have devised a somewhat similar method using photometric detection. An electrostatic collector and size analyzer is mentioned in Chapter 1.

PRACTICE

Stirring of aerosols need only be vigorous enough to distribute all particles of interest throughout the chamber. A number of very small fans (e.g. 1 inch blades) located at the bottom of a chamber might create sufficient updraft to carry 1-μ particles throughout the volume, but 10-μ particles might not be lifted to the maximum height. In effect, 10-μ particles and above would be confined in a volume smaller than the chamber. If the same fans were placed at the top of the chamber, blowing downward, even less efficient stirring would occur in a vertical direction. A single, larger fan, blowing upward, would provide greatest stirring efficiency.

We could determine from Figure 1-1 that the maximum vertical air

[2] The Sharples Corporation Research Laboratories, Bridgeport, Pa. Bulletin 101.

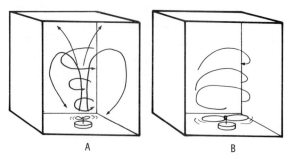

A   B

FIGURE 7-15. Comparison of stirring air in a chamber with (A) a small fan compared with (B) a large fan, assuming equal tip velocities.

velocity, existing by whatever pathways throughout the chamber, would have to be 1 cm/sec for 20-μ particles—which is not very large. A degree of lateral agitation or turbulence must also be provided, so we could guess that if we stirred vigorously enough to provide three to five times this velocity in the vicinity of the top of the chamber, all particles would be uniformly distributed. I have found, empirically, that a 4½-inch fan blade turning at 400 rpm is more than sufficient to distribute 20-μ particles evenly throughout a cubic meter chamber; 300 rpm seemed to be optimal.

The 4½-inch fan blade was selected because it was easily available in a "standard" size from an automotive supply house, was inexpensive, and required little energy to operate. Larger blades require increased power and at equal blade-tip velocities they provide less upward draft vectors, albeit equal overall turbulence. The difference is illustrated in Figure 7-15. Paddlewheels have been used somewhat effectively (Langstroth and Gillespie, 1947), but I have seen chambers where large blades have been used in a downward position!

Three practical factors should be considered. First, rotating fan blades impact particles; the faster the blades move the greater the impaction; large particles are impacted more frequently than smaller ones, so the slowest speed needed to assure homogeneity should be used and we will show how to judge this later. Second, blade area adds to effective floor area, so there are more places onto which particles may fall. This effect is not size-selective. Finally, but not involving impaction, the driving motor should be located outside the chamber because vapors from hot electric motors can influence the biological stability of an aerosol and increase the air temperature, as well as add to the overall particulate content of the chamber. Figure 7-16 shows how a simple fan bearing can be constructed for chambers where noninfectious microbes are to be studied. If the stirred settling chamber must be used for pathogens, the fan should be magnetically coupled to eliminate the possibility of leakage problems sometimes encountered with "sealed" bearings. A stirring motor with a cone-type speed control is ideal for driving the fan.

There is another approach to stirring air in small chambers, those with a volume of 1,000 liters or less. A small shaded-pole motor with a 2-inch fan blade, such as employed to ventilate electronic housings,[4] could be mounted inside the chamber if the fan were operated only for short periods—say, 30 seconds in every 10 minutes, or immediately before a sample is taken. The added heat content would be minimal, hence no harmful amounts of vapor would be introduced. Fans have been used in rotating drums, and I have tested the principle in settling chambers, proving to my own satisfaction that, with particles less than 5 μ, no difference between continuous and intermittent stirring was evident.

The stirred settling chamber (any chamber, in fact) must have at least 3 access ports: air inlet and outlet ports (one of which can serve as a sampling port), and a third to monitor pressure during either the filling or air exchange operation. The simplest and most effective type of pressure monitor for simple chambers is a small toy balloon attached to the chamber by means of a $\frac{1}{4}$- to $\frac{1}{2}$-inch (inside diameter) tubing. The balloon should be changed every day or two of operational time to prevent entrapped particles from being reintroduced into the chamber. This action can occur as a result of minor changes in air pressure, such

[4] For example, Thor-Speedway Tube Fan, Allied Electronics, Chicago, Ill.

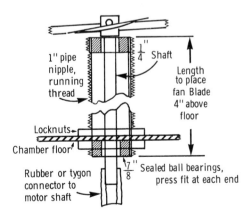

FIGURE 7-16. A simple fan and bearing arrangement for stirred settling chambers.

as opening and closing doors in the room where the chamber is located. Of course, additional ports for any special purpose can be added.

When filling a chamber from an atomizer it is easy enough to balance the pressure by judicious application of a vacuum source. It is expedient that this source be capable of moving at least 30 liters of air per minute at a differential pressure of 15 psi, because this capacity is also required for sampling.

Replacing sample volumes when studying atmospheric effects presents some problems, because the humidity and temperature should be held constant. Ideally, one maintains sources of clean dry and wet air, provides a mixing facility, and uses this source to condition the chamber initially as well as to replace air removed by sampling. One can provide a large balloon attached to the chamber via a valve and absolute filter. The balloon is filled with whatever air is available to fill the chamber initially, and the contained air is released into the chamber to compensate for sampling loss. A simpler, but nonetheless effective, approach is to work at ambient conditions—replace sampled air with room air drawn in through a filter. During a number of months in most climates a variety of meteorological conditions will have occurred and one accumulates data for a range of conditions.

A buffer chamber of about the same volume as the aerosol chamber also permits ambient air to be used with a minimal change of chamber air. For example, assume a 500-liter chamber and a buffer chamber of equal volume. Removal of a 28-liter sample would change the air composition in either chamber by about 6%, but if 28 liters of buffer air, to be replaced by ambient air, were used to replace sample air removed from the aerosol chamber, the change would be about 2%. If, for instance, the chamber and buffer air were set at 80% RH and amibent air was 40% RH, a single 28-liter sample, replaced by buffered air, would change the RH in the chamber to 79.2%. The reverse situation (40% RH inside, 80% RH outside) would cause a change from 40.0 to 40.8% RH—changes not measurable by known techniques.

The chamber should be metallic and should be well grounded electrically. Plywood, cardboard, plastic, or glass are not suitable construction materials because they exchange moisture with the air, and plastic and glass also accumulate electrostatic charges. If light-scatter measurements are to be made within the chamber, the wall should be coated first with a metallic oxide paint and then dull black paint; otherwise the walls may simply be scrubbed to remove oily contaminants. Paint vapors can affect bacteria, hence painted chambers must be thoroughly dried and aerated before they can be used with microbes. We have built double-walled insulated chambers with sheet aluminum and others with stainless steel sheet. Figure 7-1 shows essential details of such chambers.

Chambers for the study of nonpathogens need not be highly complex or expensive, and can be used in student laboratories. It should be possible to construct a chamber from two 30- or 40-gallon galvanized garbage cans, placed with open ends together and sealed with gaskets or with plastic electrical tape covered by heating duct tape. In order to facilitate access for repair or cleaning, do not solder or weld the cans together, but instead use suitcase latches or other reasonable fasteners. Oil drums certainly could be adapted to make stirred settling chambers. If constructed and operated according to the basic principles mentioned here, simple chambers, though not so convenient, should be as useful as more expensive ones.

If pathogens are to be used in the chamber (any of the types mentioned in the book) then construction and utilization of the device are vastly more difficult. The chamber must be capable of withstanding autoclave conditions or steam and formaldehyde treatment; bearings must be leakproof, and all joints must be welded or sealed with pressure gaskets. Figure 7-17 shows such a chamber as an example rather than a recommended design. The simplest way of handling pathogens is to locate chambers in a small, sealed room equipped with an external, vapor-tight, shallow hood with an open back (so that access to sampling ports can be via gloves) and to locate all control functions outside the room. The room can be sterilized by any of the several vapor disinfectants (e.g., beta-propyl lactone) if access is required for maintenance purposes.

Ideally, chambers should be located in temperature-controlled (air conditioned) rooms. They should not be located near heat sources, and they should be shielded from direct sunlight from windows; I have observed as much as 1°C wall-to-wall differential as the result of sunlight falling on a chamber located 10 feet away from a window.

Operation of the stirred settling chamber (and this also applies generally to any chamber) with continuous light scatter monitoring is done as follows: The chamber is purged with air of the desired temperature and RH until the monitor indicates that a minimal "zero" steady state has been attained. If incoming air is not sufficiently filtered, then the "background" level can be surprisingly high. The best way to distinguish between an aerosol background and background caused by stray light or leakage current (a constant value) is to allow the purged and closed chamber to remain undisturbed for 4 to 5 days; the measured background can then be considered to be a "zero" value for stray light, and can be eliminated electronically (for example by "zero" controls on servo recorders), or the value can be subtracted from subsequent readings.

When purging is complete, the chamber is filled with aerosol by whatever means desired. During the operation, the monitor will indicate in-

FIGURE 7-17. View of a stirred-settling chamber designed to be used with pathogens. The U-shaped bypass (upper left) contained a right-angle light-scatter unit with a small fan to provide continuous monitoring. The main fan was located at the chamber bottom, coupled magnetically to the motor to prevent possible leakage through a bearing seal.

creasing values and it is convenient to cease filling when the readout device indicates near full scale. Or, more effectively, the readout device should be adjusted to indicate full scale when the aerosol concentration is at a desired level as previously determined by other means (e.g., biological sampling). The moment filling is stopped becomes "zero" time, calibrated either by starting a timing device, moving the recorder chart drive forward until the pen is on a "set mark," or, prosaically, noting the time on a clock.

If the aerosol is being run for purposes of particle sizing, it is not necessary to continue monitoring until background level is reached, but rather to continue until a nearly true logarithmic decay is observed. The decay rate can appear to be logarithmic fairly early in the aerosol life, and it is best to continue monitoring at least 2 hours past the point where the data seem to indicate the presence of a homogenous aerosol.

To set the fan speed on a new chamber, it is necessary to observe a number of aerosols, all created with the same material in the same way. Start with the slowest speed attainable with whatever drive system is used, observe the decay of an aerosol, as above, then test another aerosol with double the fan speed, etc., until the fastest possible speed is tested. One should observe first a decrease, then an increase in the apparent particle size as the speed of the fan is increased. This will usually not be a result of changes in the slowest decay rate (smallest particles) observed, but rather a change in the rate during the first 10 to 15 minutes. If the speed is too slow, larger particles will settle too rapidly and initial equilibrium will be lengthy—more than 10 seconds. If the speed is too fast, impaction on the blade will remove large particles selectively; a compromise which limits the stirred settling method to sizes between 20 and 0.1 μ diameter is the best that can be attained.

Particles dispersed from the Collison or Wells-type atomizers, using usual bacteriological media with or without bacteria, are really too small and homogeneous to test fan speed effectively. One could use a ten- to twenty-fold concentrations of such media; a solution of sodium chloride just under saturation will produce heterogeneous aerosols from reflux atomizers (Equation 2-1). Do not use the same portion of a test fluid twice—remember that solvent loss in these atomizers is greater than solute loss. Aerosols from pressurized cans ("spray starch" for example) produce a variety of particle sizes and 1- to 2-second "puffs" are reasonably identical.

The stirred settling chamber, with continuous monitoring, can be a precision instrument. It is valuable not only as a particle size analyzer, wherein determinations are directly referrable to aerodynamic diameter, but also for studying effects of changes in media concentration on aerosol formation, for assaying biological behavior, for determining disperser characteristics, and as a source of aerosol for animal exposure. Since the final decay rate of the light-scatter area of initially poly-disperse aerosols is independent of light-scatter intensity as a function of size (the slope is dependent only on change of concentration), the assignment of a size by this method allows the investigator to use the smallest remaining particles in the aerosol as standards for calibrating other instruments. Furthermore, methods other than continuous light-scatter measurement can be used to determine the overall fallout pattern, or decay (Miller *et al.*, 1961). One could determine the mass decay by chemical analysis of impinger samples, as Hayakawa (1964) has done in the case of ammonium chloride aerosols. A radioactive or fluorescent "tracer" could be employed in the same way (Dimmick, 1960). A relative total particle count could be achieved by use of the impactor, or the particles in impinger fluid could be counted by electronic methods (Curby *et al.*, 1963). And, of course, bacterial spores that are deathless in the airborne state are a favorite tool of the aerobiologist (Kethley *et al.*, 1957) for determining physical decay, although data are not available at the time the aerosol is tested.

In the case of aerosols dispersed from Collison and Wells-type atomizers, the situation is greatly simplified compared with the variety of size distributions I have used for illustrative purposes. The distribution of particle sizes is narrow, and median diameters usually range from 1.3 to 4.5 μ when common bacteriological medium is dispersed.

Finally, when using static aerosols to study biological phenomena, one should be aware that the distribution of "ages" of the airborne microbes is set by dispersion time. In small chambers, the time to disperse aerosols of adequate concentration may be as short as 3 seconds if the atomizer fluid has a high viable microbial content. But with virus, for instance, where concentration may be low—or with large chambers—filling time may require 5 minutes or longer. The effect of age-distribution decreases as run-time continues; the age-effect of a 3-minute dispersion time after 10 minutes is large, but after 100 minutes it is negligible. Whether such distribution of ages can affect the validity of data depends on the nature of the experiment.

SUMMARY

In summary, an aerosol is just a collection of particles falling slowly through the air. Most particles (approximately 98%) are removed from air in containers by settling on horizontal surfaces. Some particles are removed by coagulation, and some by impaction on walls, either by inertial or electrostatic forces. When the particle concentration is less than 10^8 per liter, coagulation is negligible. Under usual circumstances, electrostatic precipitation is small. The particles fall independently, according to Stokes' law, and they are transported within an air volume by air currents. When air currents are sufficiently turbulent, particles are distributed uniformly throughout the air in a container—this process is called stirred settling.

Stirred settling can be described mathematically. When we have described, statistically, the number of particles of given size and density we have described the "behavior" of the aerosol. The reverse process allows us to deduce the mean size of particles in the aerosol. Chambers for studying properties of aerosols undergoing stirred settling are easily built, and are also useful for studying particle size, survival of microorganisms, efficiency of dispersers, and for calibrating other particle-sizing instruments.

The stirred settling chamber can embody almost any kind of metallic, sealed chamber with a small fan located in a central position near the bottom of the chamber, and with some method of measuring relative particle content. Various access ports can be attached, the chamber can be filled with an aerosol, and, at constant temperature, particles will settle out at a rate precisely related to the effective height of the chamber, the aerodynamic size, and the density of the contained particles. The stirred settling chamber is ideally suited for studies with nonpathogens, although it could be located in a sealed room and operated through a vented hood arrangement if pathogens are to be used.

REFERENCES

Andersen, A. 1958. New sampler for the collection, sizing and enumeration of viable airborne particles. *J. Bacteriol.,* **76:** 471–484.

Cadle, R. D. 1965. *Particle Size Theory and Industrial Applications.* Reinhold Publishing Corp., New York.

Curby, W. A., Swanton, E. M., and Lind, H. E. 1963. Electrical counting characteristics of several equivolume microorganisms. *J. Gen. Microbiol.,* **32:** 33–41.

DallaValle, J. M. 1948. *Micromeritics, the Technology of Fine Particles.* Pitman Publishing Corp., New York and London.

Dimmick, R. L. 1960. Measurement of the physical decay of aerosols by a light scatter method compared to a radioactive tracer method. *J. Hyg.,* **58:** 373–379.

Dimmick, R. L. 1961. A light-scatter probe for aerosol studies. *Am. Ind. Hyg. Assoc. J.,* **22:** 80–81.

Dimmick, R. L., Hatch, M. T., and Ng. J. 1958. A particle sizing method for aerosols and fine particles. *A. M. A. Arch. Ind. Health,* **18:** 23–29.

162   Aerosol Chambers

Devir, (Weinstock) S. E. 1965. On coagulation of aerosols. II. Size distribution changes in coagulating aerosol. *J. Colloid and Interface Sci.*, **21** (1): 9–23.

Ettinger, H. J., and Posner, S. 1965. Evaluation of particle sizing and aerosol sampling techniques. *Am. Ind. Hgy. Assoc. J.*, **26:** 17–25.

Flisch, J. P., Norris, C. H., and Nugent, A. E. 1967. Calibrating particulate air samplers with monodisperse aerosols: Application to the Andersen Cascade Impactor. *Am. Ind. Hyg. Assoc. J.*, **28:** 507–515.

Goetz, A., Stevenson, J. R., and Preining, O. 1960. The design and performance of the aerosol spectrometer. In *Aerosols, Physical Chemistry and Application*, (Spurny, K., Ed.), Publishing House of the Czechoslovakian Academy of Sciences, Prague, Czechoslovakia.

Goldberg, L. 1950. Studies on experimental epidemiology of respiratory infection. IV. Particle size analyzer applied to the measurement of airborne bacteria. *J. Infect. Dis.*, **87:** 133–141.

Griffiths, H., and Lampitt, L. H. 1947. Symposium on particle size analysis. *Suppl. to Transactions, Inst. Chem. Eng.*, **25.** London.

Gucker, F. T. 1952. Instrumental methods of measuring mass concentration and particulate concentration in aerosols. Chapter **75**, *Proceedings U.S. Technical Conference on Air Pollution*, (McCabe, Louis G., Ed.), McGraw-Hill Book Co., New York.

Gucker, F. T., Rowell, R. L., and Chiv, G. 1965. A study of intensity functions for dielectic spheres calculated from the Mie theory. In *Aerosols, Physical Chemistry and Application* (Spurny, K., Ed.), Publishing House of the Czechoslovak Academy of Sciences, Prague, Czechoslovakia.

Gumprecht, R. O., and Sliepcevitch, C. M. 1953. Measurement of particle sizes in polydispersed systems by means of light transmission measurements combined with differential settling. *J. Phys. Chem.*, **57:** 95–97.

Haughey, F. J. 1967. Computer techniques applied to research in particle size. Aerosol Studies of Rutgers University. *J. Air Pollution Control Assoc.*, **17:** 596–597.

Hayakawa, I. 1962. Studies on coagulation employing ammonium chloride aerosols. *A. P. C. A. Journal*, **12**(6): 265–271.

Hayakawa, I. 1964. The effects of humidity on the coagulatory rate of ammonium chloride aerosols. *A. P. C. A. Journal*, **14**(9): 339–346.

Hodkinson, J. R. 1966. The optical measurement of aerosols. In *Aerosol Science* (C. N. Davies, Ed.), Academic Press, London and New York.

Horstman, S. W., and Wagman, J. 1967. Size analysis of acid aerosols by a metal film technique. *Am. Ind. Hyg. Assoc. J.*, **28:** 523–530.

Irani, R. R., and Callis, C. F. 1963. *Particle Size Measurement Interpretation and Application.* John Wiley and Sons, New York.

Katz, S., Fisher, M. A., and Leiberman, A. 1956. Automatic measurement of aerosol particles. *Soap and Chemical Specialties*, **22** (9): 137–140.

Kethley, T. W., Cown, W. B., and Fincher E. L. 1963. Adequate expression for average particle size of microbiological aerosols. *Appl. Microbiol.* **11:** 188–189.

Kethley, T. W., Orr, Clyde, Fincher, E. L., and DallaValle, J. R. 1957. Airborne microorganisms as analytical tools in aerosol studies. *J. Air Pollution Control Assoc.*, May: 16–20.

Lang, H. R. (Ed.) 1954. Physics of Particle Size Analysis, Supplement #3. *British J. Appl. Phys.*, The Inst. of Physics, London.

Langstroth, G. O., and Gillespie, T. 1947. Coagulation and surface losses in dispersed systems in still and turbulent air. *Canadian J. Res.*, **B25:** 455.

Lundgren, D., and Long, W. 1967. Particle size-distribution data using an inertial-classification-light-scattering device. *J. Air. Pollution Control Assoc.,* **17:** 594.

May, K. R. 1945. The Cascade Impactor: An Instrument for Sampling Coarse Aerosols. *J. Sci. Inst.* **22:** 187–195.

May, K. R. 1950. The measurement of airborne droplets by the magnesium oxide method. *J. Sci. Instr.,* **27:** 128–130.

May, K. R. 1964. Calibration of a modified Andersen bacterial aerosol sampler. *Appl. Microbiol.,* **12:** 37–43.

Mercer, T. T. 1962. On the calibration of cascade impactors. *Ann. Occup. Hyg.,* **6:** 1–14.

Mercer, T. T. 1965. The interpretation of cascade impactor data. *Am. Ind. Hyg. Assoc. J.,* **26:** 236–241.

Miller, W. S., Scherff, R. A., Piepoli, C. R., and Idoine, L. S. 1961. Physical tracers for bacterial aerosols. *Appl. Microbiol.,* **9:** 248–251.

Mitchell, R., and Pilcher, J. M. 1959. Improved cascade impactor for measuring aerosol particle sizes in air pollutants, commercial aerosols, cigarette smoke. *Ind. and Eng. Chem.,* **51:** 1039–1042.

Orr, Clyde, Jr. 1966. *Particulate Technology.* The Macmillan Co., New York.

Orr, Clyde, Jr., and DallaValle, J. M. 1959. *Fine Particle Measurement.* The Macmillan Co., New York.

Payne, R. E. 1948. The measurement of the particle size of sub-sieve particles. Third National Conference of American Instrument Society. Sharples Corporation Bulletin #1244.

Reist, P. C., Burgess, W. A., and First, M. W. 1967. A small airborne particle detector has been developed. It counts and sizes. Current Aerosol Research at the Harvard School of Public Health. *J. Air Pollution Control Assoc.,* **17:** 608.

Sheien, P., Cochran, J. A., Benander, L., and Friend, A. G. 1967. Determination of the respiratory deposition and size of airborne radioactive particles by graded filtration. *J. Air Pollution Control Assoc.,* **17:** 582.

Sinclair, D. 1950. Stability of aerosols and behavior of aerosol particles. Chapter 5, *Handbook of Aerosols.* U.S. Government Printing Office, Washington, D. C.

Talvitie, N. A., and Paulus, H. J. 1956. Recording photometric particle-size analyser. *Rev. Sci. Inst.,* **27:** 763–767.

Thompson, J. K. Determination of aerosol size distributions by jet impactor-light scattering technique. *Anal. Chem.,* **29:** 1847–1850.

Wilcox, J. D. 1953. Design of a new 5-stage cascade impactor. *A. M. A. Arch. Ind. Health and Occ. Med.,* **7:** 376–382.

8

ROTATING DRUM

Robert L. Dimmick / Lydia Wang

NAVAL BIOLOGICAL LABORATORY
SCHOOL OF PUBLIC HEALTH
UNIVERSITY OF CALIFORNIA, BERKELEY

There are a number of reasons why it would be advantageous to have some method of holding aerosols for periods longer than they can be held in typical stirred settling chambers. One reason is the challenge to discover how long the more resistant microbes can survive in the airborne state. Differences in the lifetime of microorganisms as a result of the interaction of environmental and temporal factors should add to our knowledge of possible survival mechanisms. Another more practical reason is that, in natural conditions, winds can bear particles aloft and carry them great distances. Notwithstanding the extent of diffusion that will occur in nature, the questions of whether infectious cells can be carried across miles of territory and whether this process adds to the possibility of continued inoculation of both human and animal populations must be answered. Mere speculation is not sufficient—we would like to test these conditions in the laboratory.

There has been a number of attempts to produce long-lived aerosols, but we are not aware of their having been published. All were efforts to provide a rising column of air to prevent particles from falling; the attempts failed, we presume, because isothermal conditions and laminar air flow could not be maintained, especially when sampling was attempted. The answer provided by Goldberg et al. (1958) was surprisingly simple—rotate the stirred settling chamber about a horizontal axis!

ANALYSIS OF DRUM BEHAVIOR

A comprehensive and effective analysis of the behavior of particles in a rotating drum has not been published. We will develop a simplified and practical approach, based in part on classroom lectures of Leonard Goldberg. Consider a particle suspended in the air inside a cylinder that can be inverted. If we could see the particle, we might wait until it dropped some distance, then quickly invert the cylinder and allow the particle to fall in the opposite direction with respect to the cylinder. If we timed our inverting procedure perfectly, the particle would never fall to either end of the cylinder; instead it would oscillate, always returning to the original position. We would have successfully defeated the effect of gravity.

The same process could be made to happen by substituting a horizontally positioned rotating cylinder for the vertical one and by causing the contained air to rotate with the cylinder. However, instead of oscillating, the particle would attain a circular path, with respect to the cylinder or an observer, and might be suspended perpetually if the speed of rotation were matched to the inertial properties of the particle. We shall refer to a particle moving in a circular path with respect to the cylinder as a particle "system," and conceive the system to have a center of gravity and to have the inertial properties of the given particle.

Although we gained by eliminating the acceleration of gravity, we have added a new motion—that caused by centrifugal force. The system would migrate, or drift, toward the periphery of the drum and would accelerate faster as it approached the periphery. Figure 8-1 shows the vector forces involved. The radius of the particle system would be a function of the inertial properties of the particle and the rotational speed of the drum. Hence, particles would not remain forever suspended in the drum; eventually all would drift to the cylinder wall to be impacted.

In fact, if particles were large enough and the rotational speed of the drum slow enough, the radius of the particle system would be larger than the radius of the drum, and the particle loss would be the result of gravitational forces rather than centrifugal forces.

Now consider an aerosol of homogeneous particle size suspended in the rotating drum, and assume a condition of stirred air as we did for the stirred settling chamber. The situations are similar—a hypothetical volume of air (the "fallout volume") would exist in contact with the wall of the cylinder such that centrifugal force would cause all systems positioned within that volume to migrate to the wall in a unit of time (Figure 8-1). During the next time interval the remaining sys-

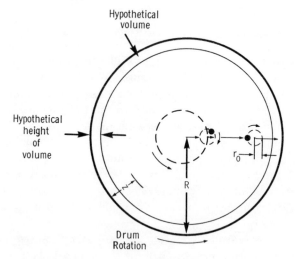

FIGURE 8-1. Migration of particle systems with respect to the rotation of the drum. Small arrows indicate direction of rotation of the systems with respect to the contained air, not to an external point. A stationary observer would see the particle move as shown by the large, dashed circle; it would appear to be off-center by the distance ro.

tems would be uniformly distributed throughout the drum volume, and in the next step particles within the outer volume would again be removed.

The rate of removal of particles would be a percent-per-minute process and, as in stirred settling, we could write the equation

(8-1)
$$\frac{N}{N_0} = e^{-kt}$$

where N = number of particles at time t,
N_0 = original number of particles, and
k = decay constant.

Also, as in stirred settling, k would be related directly to the inertial forces of the particle and inversely to the effective "height" of the chamber. That is,

(8-2)
$$k = \frac{\bar{V}}{\bar{H}}$$

where $\bar{V}$ = particle (or system) velocity in the hypothetical volume, and
$\bar{H}$ = effective chamber height.

$\bar{V}$ can be derived from Stokes' law by substituting $R\omega^2$ for the acceleration of gravity, where ω = radians/sec.

Note that whatever the velocity of the particle may be at various distances from the center, it is the radial velocity at the periphery that determines the height of the fallout volume; so R is effectively the drum radius. We find by substituting in Equation 1-5 that

$$(8\text{-}3) \qquad \bar{V} = \frac{\rho d^2 R \omega^2}{18\eta}$$

The effective height of the chamber ($\bar{H}$) is the ratio of the volume to "horizontal" area, as before. In this case, the area to be considered is the area of the cylinder wall.

We can write,

$$(8\text{-}4) \qquad \bar{H} = \frac{\pi R^2 T \text{ (volume)}}{2\pi R T \text{ (area)}^1} = \frac{R}{2}$$

where R = radius of the drum and
T = thickness or depth of the drum.

Hence,

$$(8\text{-}5) \qquad k = \frac{\bar{V}}{\bar{H}} = \frac{\rho d^2 R \omega^2 / 18\eta}{R/2} = \frac{\rho d^2 \omega^2}{9\eta}$$

This result seems surprising, for the radius of the drum does not appear in the final equation, indicating that the decay rate ought to be constant regardless of drum size. This curious situation becomes obvious if one considers that the larger the drum the greater the "height" of the hypothetical fallout volume. Since drum volume and fallout volume are both a function of R^2, the ratio between the two volumes remains constant. Hence, the fraction of particles removed is identical, regardless of drum size. Also, the thickness of the drum (T) does not influence the ratio. For a given particle size, then, the rotational speed of the drum and minor diffusional forces are the only factors influencing decay rate.

Since the general expression for decay rate in a rotating drum is the same form as that for the stirred settling chamber, we can express the effectiveness of the rotating drum in terms of an equivalent static chamber as follows:

Let $k_1 = k_2$, where k_1 = decay constant for stirred settling chamber

---

[1] Not the horizontal area of equation (4), Chapter 7, but the entire circumferential area.

and k_2 = decay constant for rotating drum. Then,

(8-6)
$$\frac{\rho d^2 g}{18\eta H} = \frac{\rho d^2 \omega^2}{9\eta}, \qquad \text{hence } \frac{g}{H} = 2\omega^2$$

By definition, $2\pi\theta = \omega$ radians/sec, where θ = revolutions/sec; for convenience equation, 8-6 could be written $\frac{g}{H} = 8\pi^2\theta^2$.

Assume a rotational speed of 6 rpm, or 0.1 rps, then

$$H = \frac{g}{8\pi^2\theta^2} = \frac{980}{8\pi^2 \times (10^{-1})^2} = 1{,}250 \text{ cm, or 41 ft.}$$

That is, a drum rotating at 6 rpm would be equivalent, in terms of the ability to contain an aerosol for prolonged periods, to a stirred settling chamber with an effective height of 41 ft. A drum 6 ft in diameter is a convenient size. We can compare the behavior of this drum as a stationary stirred settling chamber with its behavior when it is rotating at 3 rpm, a practical speed.

$$\frac{g}{H} = \frac{980}{\frac{1}{2} \times 3 \text{ ft} \times 30 \text{ cm/ft}} = 6.9 \text{ as a stirred settling chamber}$$

$$\frac{g}{H} = 8\pi^2(0.05)^2 = 0.197 \text{ as a rotating drum at 3 rpm}$$

or approximately 35 times the "effectiveness."

A particle-size nomograph similar to Figure 2, Chapter 7, is not practical for the drum because of the rotational parameter. We have, however, calculated some representative half-lives for a speed of 3 rpm and a particle density of 1.1. They are shown in Table 8-1.

Table 8-1 Half-lives of monodisperse aerosols in a drum rotating at 3 rpm. Particle density assumed to be 1.1.

Particle diameter, μ	Half-life, minutes
1	17,200
2	4,308
3	1,914
5	689
8	269
10	172
12	120

Under ideal conditions, a half-life of 12,000 min has been observed, but we routinely find values between 800 and 1,600 min, depending upon extent of sampling, temperature gradients, mechanical vibrations, and possibly other factors. It is very likely that diffusion of particles to the walls limits our ability to observe the theoretical maximum half-life.

The theoretical half-life of 1-μ particles in a 6-ft *stationary* drum is 488 min.

$$K = \frac{V}{H} = \frac{3.4 \times 10^{-3} \text{ cm/sec}}{144 \text{ cm}} = 2.34 \times 10^{-5};$$

$$t_{\frac{1}{2}} = \frac{0.693}{2.34 \times 10^{-5}} = 488 \text{ min}$$

At the efficiency of 35, derived above, the half-life in the rotating drum would be 17,080 min, which agrees within 1% with the value shown in Table 8-1.

As stated previously, particles in the rotating drum can be said to act as rotating systems with respect to the drum. The rotational speed of these systems is the same as that of the drum. The tangential velocity with respect to air surrounding the particle ($2\pi R\theta$) will be the same as the Stokes' velocity of fall. So:

$$2\pi r_o \theta = \mathbf{v}$$

where r_o = radius of the system
$\mathbf{v}$ = Stokes' velocity of fall for the given particle

and,

(8-7)
$$r_o = \frac{\mathbf{v}}{2\pi\theta}$$

For a 1-μ diameter particle (from Figure 1-1, $v_o = 3.4 \times 10^{-2}$ mm/sec.),

$$r_o = \frac{3.4 \times 10^{-2}}{6.28\,\theta} = \frac{5.24 \times 10^{-3}}{\theta} \text{ mm}$$

If $\theta = 0.1$ rps, then $r_o = 5.24 \times 10^{-2}$ mm, which is quite small. However, for a 10-μ diameter particle,

$$r_o = \frac{3.3}{6.28 \times 0.1} = 5.24 \text{ mm}.$$

a value that is appreciable but still fairly small if we are considering a chamber 6 ft in diameter. Values of r_0 with respect to θ for various particle sizes are shown in Figure 8-2.

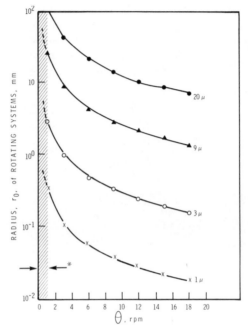

FIGURE 8-2. Radii of rotating particle systems in a drum at various rotational speeds. The shaded area (*) indicates approximate conditions where the rotating drum approaches the behavior of a stirred-settling chamber.

PRACTICAL ASPECTS

Stirring

Well, theory is all right—so long as it does not interfere with common sense. Practicability dictates that there are limits within which these formulae may operate. It is obvious that if we built a drum the size of the radius r_o for a given particle, then those particles would not be held airborne for long periods. If the drum radius were only three or four times r_o, we should discover that R^2 in Equation 8-4 must be modified. As the particle system approached the cylinder wall, precipitation could occur prematurely because of the additional rotation of the particle around its center of gravity. The size of the system, as well as inertial properties, determines the height of the fallout volume, so R^2 in Equation 8-4 should be replaced by $(R - r_o)^2$. As R is increased, the correction becomes less and less, and if $R \geq 20\ r_o$, Equation 8-5 describes the behavior adequately.

At the other end of the scale, if R is very large (say, 50 ft or more), then the particle velocity could become quite large. We mentioned previously that in stirred settling there must be present an air velocity vector sufficient to waft the largest particles to the top of the box. Similarly, stirring in the drum, which was only an assumed condition, must be sufficient to overcome any velocity imparted to particles; otherwise stratification would occur and the aerosol would not be spatially homogeneous. We could make the drum so large that no amount of stirring, which is mandatory for effective sampling, could maintain homogeneity.

The drum described by Goldberg et al. (1958) was equipped with a baffle that served to keep the air rotating, and the slippage that occurred provided satisfactory stirred settling conditions. Newer models were constructed without baffles because it was suggested that the baffle contributed to the impaction area. As a result, we have found it necessary to add a small fan, operated for 30 seconds to 5 minutes at the start of a run, to assure initial homogeneity; after that interval, aerosols appear to remain in a stirred condition.

Although there is no proof, we suspected that in drums without baffles an inner core of air could remain stationary and particles near the bottom might fall out of this volume, where they could be carried upward by air near the cylinder surface to fall back into the stationary volume as they traversed the "top" half of the drum—a condition of intermittent mixing. If the sampling point is located near the drum center, this mechanism would be difficult to distinguish from the theoretical mechanism we have evolved. Intermittent mixing could only occur within distances from the center where radial velocity was less than Stokes' velocity, otherwise particles could not fall into a stationary air mass. A drum at 6 rpm would have to be nearly 80 ft in radius before the radial velocity exceeded Stokes' velocity. Hence, the possibility of intermittent mixing could easily occur in drums of convenient size.

To determine whether an unbaffled drum can rotate the air in the center we must turn to two concepts developed in meterological studies (Sutton, 1953). One is vorticity (Ω), which is a measure of the distribution of air stream velocities between a moving stream and a stationary stream. This distribution is similar to the Maxwell distribution for thermal velocities of molecules. The second is dynamic viscosity (ν), defined as

(8-8) $$\nu = \frac{\eta}{\delta} = 0.15 \text{ at } 20 \text{ C, 1 atm}$$

where η = viscosity of air and
δ = density of air.

Vorticity is defined as:

(8-9)
$$\Omega = \frac{U}{2\sqrt{\pi\nu t}} e^{\frac{-z^2}{4\nu t}}$$

where z = distance over which layers of air are considered (Figure 8-1),
U = velocity of first or driven layer, and
t = time.

This equation can be differentiated with respect to time to yield a root-mean-square value ($\bar{t}$) that tells how long it takes for maximum vorticity (at a point about half initial rotational velocity) to occur at some distance from the moving surface.

Let

$$y = \ln\frac{2\Omega\sqrt{\pi\nu}}{U} = -\left(\ln\sqrt{t} + \frac{z^2}{4\nu t}\right)$$

so

$$\frac{dy}{dt} = -\frac{1}{2t} + \frac{z^2}{4\nu t^2} = 0$$

(8-10)
$$\bar{t} = \frac{z^2}{2\nu} = \frac{z^2}{0.3}$$

where $\bar{t}$ = mean time at which Ω is maximum at distance z.

Air in the drum can be thought of as concentric drums made of thin sheets of air. At some point in time after the drum starts to rotate, some outside layers are rotating with the drum, some near the center are stationary, and between them is a "boundary layer" where velocity gradients occur; this is the "area" of vorticity within which velocities are distributed. Vorticity migrates inward and its thickness increases, until it extends finally from the outside walls to the center. Analysis is complex, but we should be satisfied with mean time values within half an order of magnitude. The problem is analogous to temperature migrating within a metal block.

The boundary, having maximum vorticity at any time $\bar{t}$, can be thought of as separating an outer volume where the mean velocity is that of the drum and an inner volume where the mean velocity is some fraction of that of the drum. If a point half-way between the drum and the center is selected, then mean velocities are distributed so that half are outside and half are inside; from a statistical viewpoint, at time $\bar{t}$ the inner volume has attained one-half the rotational velocity of the outer volume. At time $2\bar{t}$, the inner volume has attained about three-fourths of the rotational

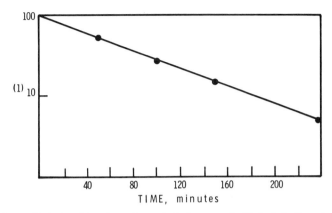

FIGURE 8-3. Relationship of motion of air in an unbaffled, 6-ft diameter rotating drum to time of rotation. (1) Percent of initial average rotational velocity to be acquired by an inner core 30 cm from the drum sides or periphery.

velocity, etc. Relationship between $\bar{t}$ and the rotational velocity at a distance of 30 cm is shown in Figure 8-3. These are maximal $\bar{t}$ values for drums of convenient size because the effect of curvature of the drum periphery has been ignored in this simplified analysis. It is evident, nevertheless, that the time required for air to achieve a reasonable rotational speed is no longer than practicable. (Drums are usually rotated constantly.)

Consideration of the rotating air shell concept reveals two factors:

1. If drum thickness is less than drum diameter, then the distance to be substituted for z in Equation 8-10 to determine $\bar{t}$ would be one-fourth of the thickness—i.e. vertical walls contribute most motion to inner air. Conversely, if thickness is greater than drum diameter, then the distance is one-fourth drum diameter—the periphery contributes most to air motion.

2. Laminar air flow patterns probably exist within an unbaffled drum, and the air is not stirred. Stirring the air with a fan distributes rotational air velocities, and in this sense, probably is as effective as a baffle. To baffle or not to baffle, therefore, is a matter of convenience, but some method of stirring must be incorporated to prevent what we have termed intermittent mixing, just as a liquid to be heated must be stirred to prevent thermal gradients.

The reader should note that this somewhat unsophisticated approach, or any detailed analysis of air patterns in rotating drums, has never been reported as having been tested—despite the fact that drums have

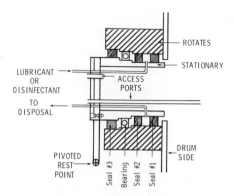

FIGURE 8-4. Cross section of a rotating bearing and seal: Seal #1, 2; Teflon or silicon lip garter, spring loaded. Seal #3; conventional neoprene lip garter, spring loaded (e.g., Garlock Co., Palmyra, N.Y.). Bearing; 9-inch, thin cross section ball bearing (e.g., the Kayden Engineering Corp., Muskeegan, Mich.)

been used by several laboratories for about 10 years. For that matter, no detailed analysis of observed behavior of standard aerosols has been reported; drums have performed so satisfactorily with respect to keeping particles airborne that analysis of behavior is largely an academic matter.

Bearings

Surprisingly enough, bearings for drums 4 ft or larger in diameter sustain considerable wear, despite the fact that the drum rotates slowly. Were it not for the problem of sampling and filling the drum, standard ball bearings could be employed if access ports were placed on the periphery and the drum stopped momentarily for filling and sampling. This has been tested in effect by Leif, W. R. (1958), who placed a light, wooden-framed drum (with transparent sheet plastic walls) on rollers and turned the drum by a belt drive. Although no comparison was made with theoretical capabilities of such operation, aerosols could be maintained for at least 8 hr in a drum about 3 ft in diameter, and with physical decay rates less than if the drum had been used as a stirred settling chamber.

The axle of prototype drums was made hollow and stationary and the drum rotated around it. Access for sampling and filling was by way of tubing that extended through the axle, entered near the center of the drum, and extended about one-third of the distance between axle and periphery. Recently a new design has been tried that is more satis-

FIGURE 8-5. View of a floating bearing for a rotating drum showing sliding hinge attachment to main frame. The round device attached to one of the access pipes is a solenoid, pressure-relief valve.

factory. Figure 8-4 shows essential details of construction. The front plate can be removed easily for modifications and is large enough to accommodate a variety of ports or equipment for sampling (*e.g.* light scatter probe). Of course, the opening of the air inlet tube should be located as far from the sample inlet as practicable to prevent "new" air from streaming into the sampling line.

Properly, both bearings should be hinged, one on a floating pin so buckling or misalignment of drum ends does not cause undue wear. We have operated two drums with a single belt drive on two bearings using this flexible mounting system, which reduces cost per drum considerably (Figure 8-5).

A rotating drum is an ideal laboratory container for aerosol studies. Short-term analysis can be conducted as well as experiments with long-term aerosols. Once the theory of the drum is understood and the behavior properly analysed with aerosols of known particle size, it can be employed in the same manner as the stirred settling chamber to yield particle sizing data as well as biological information.

REFERENCES

Goldberg, L. J., Watkins, H. M. S., Boerke, E. E., and Chatigny, M. A. 1958. The use of a rotating drum for the study of aerosols over extended periods of time. *Am. J. Hyg.*, **68:** 85–93.

Leif, W. R. 1958. Personal communication. The Naval Biological Laboratory, Oakland, California.

Sutton, O. G. 1953. *Micrometeorology*. McGraw-Hill Book Co., New York and London.

NOMENCLATURE

$\bar{}$ = mean value

N = number of particles at time t

N_0 = original number of particles

k = decay constant

V = particle (or system) velocity

H = effective chamber height

θ = rotation speed of drum in rps

R = radius of drum

ρ = density, gm/cm^3

T = thickness or depth of drum

η = viscosity

ω = rotational velocity in radians per sec

g = acceleration of gravity (980 cm/sec^2)

r_o = radius of system

$\mathbf{v}$ = Stokes' velocity of fall for given particle

Ω = vorticity

ν = dynamic viscosity

t = time

z = distance over which layers of air are considered

U = velocity of first layer of air

δ = density of air

9

DYNAMIC AEROSOLS AND DYNAMIC AEROSOL CHAMBERS

R. L. Dimmick / M. T. Hatch

NAVAL BIOLOGICAL LABORATORY
SCHOOL OF PUBLIC HEALTH
UNIVERSITY OF CALIFORNIA, BERKELEY

Dynamic aerosols (as distinguished from static aerosols) exist only while they are being transported through the hoses, tubes, ducts, pipes (and bulges therein) that serve as dynamic aerosol chambers, yet they have a number of characteristics and operational properties sufficiently different from static aerosols to warrant discussion. Since most of these properties are fairly straightforward and are more a matter of common sense than scientific acuity, we shall examine systems in terms of their usefulness to studies of biological behavior not easily accomplished with static chambers.

There are four possible moments when dynamic aerosols are formed in the laboratory:

1. During dispersion, when chambers are being filled.
2. During sampling, when aerosol is being transported from a chamber to a sampler.
3. During sampling, when the air is being accelerated to cause particles to impact.
4. During transport through ducts and chambers, when newly formed aerosol is being examined or utilized.

Dynamic aerosols may be (or have been) used to study:

1. Initial biological decay as related, for example, to culture age. Aerosols can be introduced into small ducts where air flow is rapid and where a sample can be withdrawn during the first to fifteenth second of aerosol time (Webb, 1959; Hayakawa and Poon, 1965; Ferry et al., 1958).
2. Effects of drying rates on survival of airborne microbes (Poon, 1966).
3. Atomizer characteristics that influence particle sizes and numbers (Mercer et al., 1968).
4. Effects of shifts of relative humidity on biological decay (Hatch and Dimmick, 1965).
5. Infectivity of initially formed aerosols (Goldberg et al., 1954).
6. Diffusion principles related to outdoor conditions (Porter et al., 1963).
7. Evaluation of samplers (Goldberg and Shechmeister, 1951).

There are two conditions of air flow that are important and sufficiently different to be considered separately: (1) Fast flow rates (> 2 m/sec); flow is either laminar or turbulent, kinetic forces cause particles to be removed by impaction, and the concept of a critical orifice becomes possible. (2) Slow air flow rates (<1 m/sec); flow is essentially laminar, air volume does not change appreciably, particles are removed by sedimentation and their concentration becomes diffuse, a process sometimes called evanescence. We shall discuss fast-flow briefly; for more general treatment see Thorpe (1967) or Sutton (1953).

FAST AIR FLOW

In Chapter 1 we defined the Reynolds number (dimensionless) which can also be written as

(9-1)
$$\mathrm{Re} = \frac{VD}{\nu}$$

where V = air velocity, cm/sec,
D = duct diameter, cm, and
ν = kinematic viscosity (0.15 cm²/sec).

Flow is laminar until Re becomes about 2 to 3×10^3, when air flow starts to become turbulent; above 4×10^3 it is almost always turbulent. This is a turbulence apparently caused by "packets" of air crossing air streams and adding to or subtracting their kinetic energy from the new stream. In the critical region of 2 to 4×10^3 cm/sec, laminar flow can become turbulent without apparent cause. This turbulence should be distinguished from turbulence caused by the deliberate mixing or stirring of flowing air by means of blades, baffles, or constrictions—they are not the same and we will return to this point later.

As air speed is progressively increased, a point is reached where the velocity approaches the speed of sound, and a new number N_{ma} (the dimensionless Mach number) is defined.

$$(9\text{-}2) \qquad\qquad N_{ma} = \frac{V}{a}$$

where a = velocity of sound (3.44×10^3 cm/sec).

When $N_{ma} = 1$ a critical pressure ratio is reached so that, without usual proof such as the treatise by Thorpe (1967),

$$(9\text{-}3) \qquad\qquad \frac{p}{P} = 0.53$$

where p = pressure (atm) downstream from an orifice, and
P = 1 atm.

If the pressure downstream from an orifice, measured with respect to atmospheric pressure (pressure ratio), is 0.53 or less, air is flowing through the orifice at sonic velocity but the volume flow into the orifice remains constant regardless of fluctuations in the pressure; changing the pressure acts only to change downstream air density. An orifice acting this way is called a critical orifice and is a special case with respect to general behavior of flowing air. It follows that the vacuum system must be capable of moving a sufficient quantity of low-density air to maintain the required pressure drop; many laboratory "house vacuum" systems are not intended to move large quantities of air for this purpose. If this fairly simple way to establish and maintain a constant flow through samplers is to be used, the pressure drop must be monitored to assure proper operation. Calibrated orifices can be purchased or they can be made from thick-walled glass capillary tubing and calibrated with a flow meter (Corn and Bell, 1963).

When an aerosol expands on the downstream side of a critical orifice, many particles are precipitated; this sometimes happens even if the orifice is not critical. Hence, aerosols should not be transported through orifices where an appreciable pressure drop occurs if particle concentration is an important factor in the process. The phenomenon is probably a combination of inertial impaction and thermal precipitation. We are aware of no specific study of the mechanism.

Impingers usually act as their own critical orifices and are calibrated to required flow rates. For instances, the AGI-30 (Wolf et al., 1964) is calibrated for 12.5 liters per minute. For other samplers (Goldberg and Shechmeister, 1951), flow through the collecting orifice or chamber is intended to be less than sonic, so a critical orifice is used *downstream* from the instrument.

We close this section with an admonition that seems unnecessary, although there are reported circumstances where these principles have been ignored. In any series of devices, the principal pressure drop between a low-pressure air source and the atmosphere is always across some limiting orifice that may or may not be critical but that, nevertheless, controls air flow through the entire system. For example, if air is drawn through a system of pipes and chambers at flow rates (0 to 30 liters/min) regulated by a valve connected to a vacuum system, the principle pressure drop is across the valve; pressure in the system is essentially atmospheric. As another example, the pressure in tubing connecting a 1-ft^3/min critical orifice to an operating Andersen sampler is hardly distinguishable from atmospheric pressure.

SLOW AIR FLOW

There are no changes of air density, as a result of pressure or velocity changes when air is slowly drawn through pipes or chambers, that cause significant changes in moisture content of airborne particles, and the velocities are such that inertial forces do not precipitate particles in the 1- to 10-μ range to an extent worthy of consideration.

Except for particle behavior in ducts, air flow through tubes or pipes as small as $\frac{1}{4}$-inch ID is without complication. For example, we have analyzed losses caused by the transportation of aerosol through nominal $\frac{1}{4}$-inch copper tubing, and through bends in tubing, by measuring the number of viable airborne particles of *Sarcina lutea* recovered. In these experiments, *S. lutea* aerosols were generated continuously, and sequential samples were taken through various lengths, or numbers of bends, and the results were compared. Two observations are worthy of note (Table 9-1): (1) Lengths of $\frac{1}{4}$-inch ID tubing as long as 20 ft and with as many as eight right-angle bends did not appreciably alter the number of viable cells collected in an impinger at 12.5 liters/min sampling rate. The count median diameter of these aerosols was about 1.5 μ. (2) As a precautionary note, when using bacteria to determine physical parameters of apparatus, the tested variables should be randomized to allow the effect of possible changes in viable cell output from the atomizer to be included in the analysis. Cell output declined significantly during the experiment (biological loss) when a culture had been stored in the frozen state for 2 months, whereas it increased slightly (deaggrega-

Table 9-1 Effect of length of $\frac{1}{4}$-inch ID tubing, and of number of $\frac{5}{8}$-inch radius, right-angle bends, on the numbers of organism recovered by an AGI-30 impinger, 12.5 liters/min sample rate.

Sample sequence	Experiment 1		Experiment 2		Experiment 3[c]	
	Sample tube length, ft	Mean recovery 10 plates[a]	No. of bends[b]	Mean recovery 5 plates	No. of bends[b]	Mean recovery 5 plates
1	0.5	3.74×10^5	0	2.40×10^5	0	1.90×10^5
2	20.0	3.79×10^5	8	2.58×10^5	8	0.83×10^5
3	1.5	3.97×10^5	2	2.77×10^5	2	0.43×10^5
4	10.0	4.79×10^5	6	2.99×10^5	6	0.24×10^5
5	2.25	5.34×10^5	4	3.35×10^5	4	0.18×10^5
6	5.0	5.22×10^5	0	3.06×10^5	0	0.14×10^5
7			8	3.38×10^5	8	0.08×10^5
8			2	3.47×10^5	2	0.05×10^5
9			6	4.04×10^5	6	0.05×10^5
10			4	3.45×10^5	4	0.03×10^5

[a] Number viable bacteria per liter of aerosol.
[b] Total pipe length 20 ft.
[c] Suspension had been stored, frozen, for two months; Suspension in experiments 1 and 2 were fresh, 24-hr-old cultures.

tion of clumps in atomizer fluid) when a freshly grown culture was used. It would be difficult, if not impossible, to draw valid conclusions by this method unless one were alert to biological influences.

Linear air flow in these sampling lines was 4×10^2 cm/sec, pipe length was 600 cm, pipe diameter was 0.8 cm, the time air was in the pipe was 1.5 sec, and 2-μ particles would have fallen only 1.8×10^{-2} cm in this time—so the results were in agreement with theory. These data indicate that no significant losses are evident with commonly encountered lengths of metallic sampling or transport lines. Where reasonable, larger tubing can be used with increased assurance that particles will not be lost. In preliminary tests with long lengths of rubber tubing of equivalent diameter, we have found particulate loss to be significant.

CHAMBERS

The Henderson Apparatus

The Henderson apparatus (Henderson, 1952) is typical of dynamic aerosol chambers employed to determine both survival capacity and

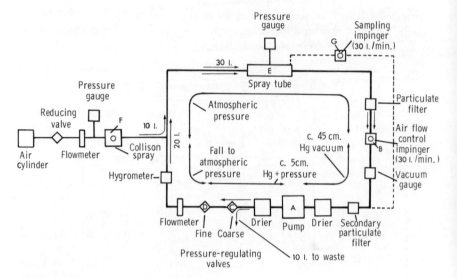

FIGURE 9-1. The Henderson dynamic aerosol apparatus. (Henderson, David W. 1952. An apparatus for the study of airborne infection, *J. Hyg.*, 50(1): 54.) Reproduced with permission of the Cambridge University Press.

infectivity of airborne microbes, although it is more of a concept than a device. Henderson designed and built an apparatus in which animals could be exposed to continuously generated particles rather than to "aging" or static aerosols (Figure 9-1). The concept was not new, but the design was.

Since that time, numerous devices in which animals are exposed to moving aerosols have often been termed "modified" Henderson units (e.g., Bartlema, 1966; Sawyer *et al.*, 1966; Speck and Wolochow, 1957; Wolochow *et al.*, 1957); many are so changed that they resemble the original design only in the mind of the associated author. Use of the term, however, is a convenient and an undoubtedly appropriate way to say, "I used a dynamic system in which aerosol was (*a*) continuously generated, (*b*) led into a mixing chamber of some type, (*c*) moved through a chamber containing animals, (*d*) sampled at some convenient point, and (*e*) disposed of by some more or less efficient burner. Since everyone knows these principles, and since details are not important to this paper, I choose not to completely describe the apparatus." Important details are: (*a*) temperature and humidity of the air at the point of exposure is known and reported, (*b*) samples (reflecting sampling method used) are representative of the true aerosol in terms of both physical and biological content, (*c*) animals are not under undue stress, (*d*) aerosol output is either constant or changes are measured throughout the exposure period, and (e) animals are, indeed, "exposed" to the actual aerosol being measured. The last is important, for it is not sufficient to measure some parameter at a point of aerosol origin and assume it will remain the same when presented to animals after some operation is performed—such as mixing the aerosol with other air or storing it momentarily in an auxiliary chamber.

Modifications (size, shape, number of pipes and valves, etc.) to the "standard" or Henderson-type apparatus are important only if they influence the ability to control or measure the aerosol in such a way that an unusual situation exists. Otherwise the Henderson unit is, with due respect to the British, the "very model" of a well-designed dynamic aerosol instrument.

The Aerosol Transport Apparatus (ATA) and the Dual ATA (DATA)

The ATA is a duct or apparatus through which continuously generated aerosols can be passed. The device also requires continuous sources of air or vacuum, or both. When equipped with sampling ports, the duct provides a device that allows aerosols to be sampled after relatively short intervals of existence. In effect, the tubing conducting an aerosol to or from a static chamber becomes an ATA, and has been used to determine an "initial" decay compared with a "final" decay observed in drums (Webb, 1959). A number of different duct types have been reported (Porter et al., 1963; Jensen, 1965). The essential difference between ducts and Henderson units is that (1) ducts have never been used, as such, to expose animals, and (2) the time during which an aerosol can be measured is extended from about 1 minute to as long as 15 minutes.

The observation time can be extended by using a large duct and a slow flow rate. Leif (1954) used a 100-ft duct 10 inches in diameter to study aerosols during a 15-minute period. He originally termed the apparatus an ATA and found that, without Stairmand baffles to provide turbulence (Stairmand, 1951), sequential samples at a given port were not identical. Porter et al. (1963) used square ducts and reported many measured parameters, but they did not report survival rate studies and their flow rates were quite rapid (40 to 220 cm/sec).

Hatch and Dimmick (1965) placed two ATA's end to end (a 6-inch duct section inserted 2 ft into an 8-inch duct section) and called the apparatus a dual ATA (DATA). The purpose of this instrument was to provide a mechanism whereby the humidity to which airborne cells were initially equilibrated could suddenly be changed without a concomitant change in temperature or pressure, as with adiabatic shifts. Norseth and Mitchell (1963) used dilution techniques in ducts for the purpose

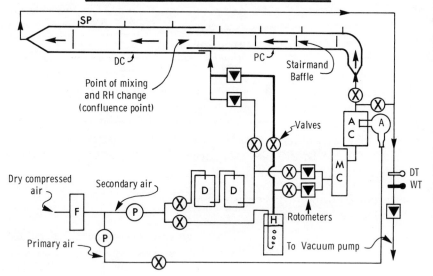

FIGURE 9-2. The dual aerosol transport apparatus (DATA). PC = Primary aerosol duct; DC = Diluted aerosol duct; A = Atomizer; AC = Aerosol mixing chamber; MC = Air mixing chamber; D = Dryer; P = Pressure regulators; F = Filter; SP = Sampling ports; H = Humidifying chamber; DT, DW = wet-dry bulb psychrometer.

of reducing the number of particles to an easily countable concentration, but reported few biological data. Figure 9-2 shows essential details of the DATA apparatus, which we will use as an example of general duct operation.

The Single Duct (ATA)

There are two major problems encountered with ducts: (1) In laminar flow, the enclosed air does not all move at the same speed; (2) since the flow rate is slow, iso-kinetic sampling is almost impossible (Lundgren and Calvert, 1967). Figure 9-3 shows an hypothetical velocity profile of air in a duct.

Poiseuille's solution for velocity distribution (Sutton, 1953) is:

$$(9\text{-}4) \qquad v = \frac{p_1 - p_2}{4\nu L} \left(L^2 - z^2 \right)$$

where v = air velocity at distance z (cm) from center,
$p_1 - p_2$ = pressure drop over length of pipe L, and
ν = dynamic viscosity.

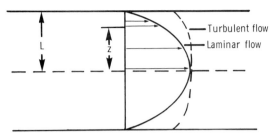

FIGURE 9-3. Symbolic distribution of air velocities in a duct. L, Z, see text.

In ducts 6 to 8 inches in diameter, $p_1 - p_2$ is a very small value and we have not been able to measure it accurately. However, by measuring volume input, a mean velocity can be estimated and substituted for v. By choosing z to be the distance where mean velocity occurs, the pressure drop can be calculated. Thus, the velocity profile at any point in the pipe can be constructed. We have done this for an "80-ft" duct, 6 inches in diameter, assuming a linear flow rate of 4 cm/sec. The cross-sectional area theoretically occupied by particles from a 30-second pulse arriving at the end of a 2,400-cm duct is shown in Figure 9-4.

The volume profile of this short pulse is also shown in the figure. If this pulse were sampled by light scatter measurements, the pulse

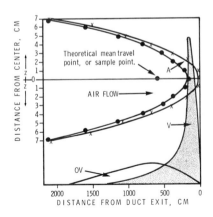

FIGURE 9-4. Theoretical, cross sectional area (A), and volume profile (V), occupied by a 30-second pulse of aerosol, travelling at 4 cm/sec, at the end of an 80-ft duct. This analysis assumed no horizontal mixing. An observed volume profile (OV), where baffles were included, is also shown in the figure. V and OV are in arbitrary, but equivalent, units.

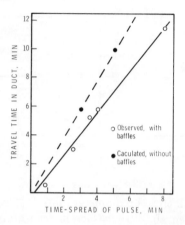

FIGURE 9-5. Diffusion of particles in a duct 6 inches in diameter.

shape would be that of the volume profile which has spread over a distance of about 1,200 cm; at 4 cm/sec velocity the pulse would require 5 minutes to pass the last sampling point.

We have tested a similar duct and will use it as an example. Observed time spreads at different sampling points are shown as a function of mean travel time in Figure 9-5. The observed spread at the 2,400-cm point was 8 minutes, and a typical pulse shape is shown in Figure 9-4. The difference between observed and theoretical pulse duration is attributable to the baffles that, as they stir the aerosol, also decrease the velocity (or increase the distance travelled) of the central core of air, as well as cause pockets of stagnant air that entrain and detain particles. During continuous atomization the number of particles per volume attains an equilibrium, but any sample contains cells with "aerosol ages" distributed as shown by the volume profile. The theoretical profile is relatively sharp, but the observed pulse is wide and flat. The effect of the spatial distribution is to cause samples taken at the final port to contain cells that have left the atomizer as much as 8 minutes apart, whereas those at the first sampling station have less than a 30-second spread.

The mixing caused by baffles brings up a point previously mentioned—that stirring as a result of turbulent flow at a high Reynolds number is different than stirring caused by baffles; baffles produce stagnant pockets. Ideally, turbulent flow would be preferred because (a) average horizontal air velocities are more uniform than in laminar flow, and (b) stirring is directed at right angles to air travel rather than forward or backward. Residence time in the pipe would be too short, however, even if it were practical to move this much air.

Sampling Effects

There are two points of interest that are involved with the volume-sampling relationship. First, the volume of air flowing past a point on our duct is about 700 cm^3/sec, whereas the sampling volume at 12.5 liters/min is 210 cm^3/sec; that is, about one-third of the contents of the pulse is being almost totally removed during sampling. Ideally, the sample should be taken first from the downstream end, then sequentially upstream, with the final sample being the one nearest the atomizer. Otherwise, at least 15 minutes are required before the "disturbance" caused by a sample at the upstream end has been "removed" from the 80-ft duct.

The second point is that the use of a duct would (ideally) allow one to obtain samples representative of rapid biological changes that might be occurring in cells during the atomization period. The apparatus does allow this to be done to some extent. For example, the loss of ability of certain cultures to withstand continuous atomization is shown in Table 9-1, column 7, but only changes that occurred over intervals greater than the time spread of the profile could be detected.

Heckly et al., (1967) have shown that survival capacity oscillates after addition of protective additives to cells that are to be freeze-dried—sometimes within seconds. This effect, if it occurred in aerosols, could never be observed with static chambers, but could be measured if the time spread in ducts could be decreased. The spread could be decreased by removing the baffles and stirring the aerosol at a point immediately downstream from the atomizer; no additional mixing should be required. The spread would be further decreased if a solid core were placed in the duct so that contained air would be shaped as a hollow cylinder. These concepts represent potential uses for ducts that have never been examined.

The Dual Duct (DATA)

Another potential use for ducts that has been tested is to produce sudden shifts in relative humidity without changing temperature. Compared with the wealth of information that can be obtained by the DATA system, discussed in another chapter, and despite the spread of particles in time, the instrument represents a fairly inexpensive investment—$60 worth of stovepipe and $40 worth of copper tubing. Ancillary equipment to produce dry, clean air at 3 ft^3/min could be assembled from vacuum cleaners and small paint-spray compressors, with appropriate filters and drying units, if nonpathogens were employed. As in all other instances, when potentially infectious microbes are employed equipment becomes expensive and complex—an engineering problem covered in a chapter on safety.

Physical Concepts

The DATA concept is relatively simple, and the essential apparatus is shown diagramatically in Figure 9-2. An aerosol is generated at the inlet end of a small duct, swept (by primary makeup air) through the duct at a selected linear flow rate, introduced into a second, larger duct at a confluence point where it is diluted with new secondary makeup air, and swept through the second duct at the same selected linear flow. If primary and secondary air are conditioned alike, the instrument is simply an ATA with an added dilution step; when the two air streams are different, we consider the action to be that of a dual system.

Both physical and biological effects can be studied with the DATA when humidity is shifted. We have used the rate of decrease in relative light-scatter (RLS) area, verified by comparing RLS decay rates with decay of *Bacillus subtilis* spores, to measure physical loss in the apparatus. Loss has varied from as high as 40% per section at relative humidity (RH) conditions above 85% to as little as 10% at 20% RH. Of course, the loss also varied as a function of atomizer action, species of microorganism (size differences), and a variety of subtle factors associated with given experiments. When experimental factors were held constant, physical decay rates varied no more than ±5%. In most instances, biological loss was so much greater than physical decay that the correction was less than the variance of the measured biological parameter.

Physical Measurements

Measured light-scatter areas (RLS, a function of particle size and number) showed that the rate of particle loss in the DATA was exponential, but biological decay could not always be represented by a straight line on semilog paper. It was convenient, therefore, to tabulate all data in terms of the ratio of observed concentration at the inlet of each duct compared with the concentration at the outlet. When the aerosol was diluted, the RLS value at the first sampling port (inlet) of the second duct divided by the RLS value at the last port of the first duct was a measure of apparent dilution (the apparent dilution ratio, ADR). The real ratio, assuming air flow volumes are properly balanced, is a function of the area ratios of the two ducts, or in this case simply $(6^2)/(8^2) = 0.56$. A slight physical loss occurred at the confluence point.

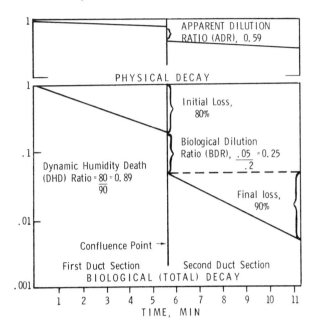

FIGURE 9-6. Example of analysis of behavior of a hypothetical aerosol in a dual aerosol transport apparatus. In this example, a different humidity is assumed to exist in each duct section. The physical loss at the confluence point (ADR) was 0.47/0.80 = 0.59, indicating agreement with the calculated ratio, 0.56. The initial biological loss was 80%. This represents both biological and physical loss, but the latter is so small in this example that correction is unimportant. The final biological loss was 90%, so the effect of changing the airborne microbes from the first condition to the second is the DHD ratio, 0.89. In this example, the effect of the sudden shift at the confluence point is obvious (BDR = 0.25) compared with the ADR, indicating that the rate of moisture exchange on, or in, the microbes did influence their biological nature.

Figure 9-6 graphically shows an example of physical decay and the effect of dilution. When the secondary makeup air had greater moisture content than primary air (a shift up in humidity), the ADR was 0.67 (particles became larger); when no shift was imposed, the ratio was 0.59; the shift-down ADR was 0.51 (particles became smaller). Differences between these values exceeded the 0.1% level of significance ($n = 85$). Although we have not done so, a comprehensive study of changes in particle size as a function of RH, and with different particle content, could be conducted by this method. Small differences in physical decay rates as a result of RH shifts have also been noted.

The rate of physical decay is important, however, if a more comprehensive analysis of the physical behavior of ducts is to be conducted. As an example, do particles contained in hypothetical sections of air, while travelling down the duct, undergo simple stirred-settling action, or do other more complex factors influence apparent physical decay?

As an example, the mean physical decay rate (measured by light-scatter methods) of 24 aerosols in a 6-inch diameter duct was observed to be 0.023 with a 95% confidence interval of ±5% when cells were sprayed from dilute solutions of growth medium. The effective height, "H", of a horizontal, cylindrical chamber is $\pi r/2$; for the 15 cm duct the value becomes 11.8 cm. The half-life of the aerosol can be determined by substitution in Equation 9-6 to yield 30.13 minutes. The equivalent half-life, if the aerosol had been in a 100 cm chamber, can be calculated from Equation 9-7 to provide the value of 250 minutes; by the nomograph in Chapter 1 the equivalent particle size is found to be 1.05 μ.

This apparent particle size is smaller than the real value (1.3 to 1.9 μ) but the error is not large enough to conclude that unusual effects have occurred in ducts. A reasonable theory is that the duct can be considered to be a stirred settling chamber with air in lateral motion. As mentioned for other instruments, the above data should not be considered as fixed, but only as a guide-line to show how analysis for a given situation can be conducted. If reasonable values are not found by such analysis, then the operator can suspect errors in either his technique or equipment structure.

Biological Concepts

The change in ADR as a function of shifts in RH does not represent particle loss. Hence, the real change of numbers of viable airborne microbes as a result of only dilution, i.e., the biological dilution ratio (BDR), should be the real value 0.56, or at best the observed value of 0.59 if no biological changes occurred at the moment of shift. Since different BDR values have been observed, we must discuss biological factors that appear to influence operational characteristics of the DATA.

We found mean BDR values of 0.48 when the RH of aerosols of *Serratia marcescens* was shifted up, 0.56 for no change, and 0.60 when a shift down was imposed. Mean values in this instance were not very helpful because sometimes deviations of replicate experiments were large. An observed BDR is significant for a given experiment within the statistical bounds of ability to sample the number of airborne cells. These bounds are influenced not only by the efficiency and constancy of the sampling system, but also by the biological nature of the microbe. For example, in one test (30 samples) using *S. marcescens* grown on a chemically defined medium and then aerosolized, a larger confidence interval for numbers of cells per liter of aerosol was found ($\pm$ 30%) than when cells were grown on a medium enriched with protein ($\pm$19%). With *Sarcina lutea* the interval was less than $\pm$5%. These estimates should not be considered as standards because measures of apparent deviation under different humidity conditions or, as previously mentioned, different sources of cells, yielded different values for the standard deviation. What can be done for each experiment is draw a line of best fit through points representing measured concentrations at each sampling station, as shown in Figure 9-6, and use extrapolated values. We judge the significance of the BDR with respect to how well the points seem to fit the best line. By this criterion, the mean BDR values previously listed are not significantly different from 0.56.

Sometimes there is no need for statistics to establish that an instant biological change occurred at the confluence point. For instance, when airborne *Pasteurella pestis* was shifted from 87% to 61% RH, the BDR was 0.09 (over one log instantaneous loss); when aerosolized phage particles were subjected to a shift up in RH, the BDR ratio varied (depending on factors discussed in another chapter) from 10 to 10^3 (as much as 3 logs of apparent increase)!

We mentioned, that rather than use decay constants to represent total biological decay we used ratios of percent loss in the first duct section to percent loss in the second, as shown in Figure 9-6. A graphic example of why we do this is shown in Chapter 11. If the two "rates" of total decay were the same, the dynamic humidity death (DHD) ratio would equal 1.0. If the shift in humidity tended to increase death rates, the most frequent finding, then the DHD ratio would be less than 1.0. When airborne *P. pestis* cells were shifted from 39% to 26% RH, the DHD ratio was 3.40, although the BDR was 0.70; i.e., no death occurred as a result of the instantaneous shift, but the rate of death after the shift was smaller than before—additional drying caused fewer cells to die during the second 6-minute period than during the first period when cells were more moist. Evidently, drying, of and by itself, is not the sole cause of death in airborne bacteria.

We list these observations (some unusual and, presently, unexplainable) as guidelines and examples, rather than as quantitatively established data, to show that the biological nature of microorganisms *must* be considered when evaluating behavior of apparatus. The investigator who finds anomalous or inconsistent data should not throw them into the waste basket. Be assured that others have noted similar "impossible" phenomena, and keep in mind that knowledge is always advanced by studies of the uncommon—otherwise research would not be required.

Apparatus using the DATA concept could be made more precise. We mentioned the use of an inner core to reduce the spread of particles. Ducts could be positioned with a downward slope equivalent to the vector direction of gravitational and kinetic motion. Controls could be automated to assure constancy of flow rates and humidity. Duct size could be reduced by employing miniature impingers to reduce sample volume. Air impurities that affect biological systems could be studied by using clean, filtered primary air and ambient secondary air. Only a preliminary exploration of the potential utility of ducts or dual ducts to aerobiology research has been achieved.

REFERENCES

Bartlema, H. C. 1966. Discussion, "Aerosol vaccination with tetanus toxoid." *Bacteriol. Rev.*, 30: 633–635.

Corn, M., and Bell, W. 1963. A technique for construction of predictable low-capacity critical orifices. *Am. Ind. Hygiene Assoc. J.*, 24: 502–504.

Ferry, R. M., and Maple, G. T. 1954. Studies of the loss of viability of stored bacterial aerosols. I. *Micrococcus candidus. J. Infect. Dis.*, 95: 142–159.

Ferry, R. M., Brown, W. F., and Damon, F. B. 1958. Studies on the loss of viability of stored bacterial aerosols. II. Death rates of several nonpathogenic organisms in relation to biological and structural characteristics. *J. Hyg.*, 56: 389–403.

Goldberg, L. J., and Shechmeister, I. L. 1951. Studies on the experimental epidemiology of respiratory infections. V. Evaluation of factors related to slit sampling of airborne bacteria. *J. Infect. Dis.*, 88: 243–247.

Goldberg, L. J., Watkins, H. M. S., Dolmatz, M. S., and Schlamm, N. A. 1954. Studies on the experimental epidemiology of respiratory infections. VI. The relationship between the dose of microorganisms and subsequent infection or death of a host. *J. Infect Dis.*, 94: 9–21.

Hatch, M. T., and Dimmick, R. L. 1965. A study of dynamic aerosols of bacteria subjected to rapid changes in relative humidity. In *A Symposium on Aerobiology* (R. L. Dimmick, Ed.), pp. 256–281. Univ. of Calif., Nav. Biol. Lab., Nav. Supply Center, Oakland, Calif.

Hayakawa, I., and Poon, C. P. 1965. Short storage studies on the effect of temperature and relative humidity on the viability of airborne bacteria. *Am. Ind. Hyg. Assoc. J.*, 26: 150–160.

Heckly, R. J., Dimmick, R. L., and Guard, N. 1967. Studies on survival of bacteria: Rhythmic response of microorganisms to freeze-drying additives. *Appl. Microbiol.*, 15: 1235–1239.

Henderson C. W. 1952. An apparatus for the study of airborne infection. *J. Hyg.*, 50: 53–68.

Jensen, M. M. 1965. Inactivation of virus aerosols by ultraviolet light in a helical baffle chamber. In *A Symposium on Aerobiology* (R. L. Dimmick, Ed.), pp. 219–226. Univ. of Calif., Nav. Biol. Lab., Nav. Supply Center, Oakland, Calif.

Leif, W. R. 1954. Personal communication. The Naval Biological Laboratory, Oakland, Calif.

Lundgren, D., and Calvert, S. 1967. Aerosol sampling with a side port probe. *Am. Indust. Hyg. Assoc. J.*, 28: 208–215.

Mercer, T. T., Tillery, M. I., and Chow, H. Y. 1968. Operating characteristics of some compressed air nebulizers. *Am. Ind. Hyg. Assoc. J.*, 29: 66–78.

Norseth, H. G., and Mitchell, R. I. 1963. The dynamic dilution of aerosols by mixing with clean air. *Ann. N.Y. Acad. Sci.*, 105 (Art. 2): 88–133.

Poon, C. P. C. 1966. Studies on the instantaneous death of airborne *E. coli. Am. J. Epidemiol.*, 84: 1–9.

Porter, F. E., Crider, W. L., Mitchell, R. I., and Margard, W. L. 1963. The dynamic behavior of aerosols. *Ann. N.Y. Acad. Sci.*, 105 (Art. 2): 45–87.

Sawyer, W. D., Dangerfield, H. G., Hogge, A. L., and Crozier, D. 1966. Antibiotic prophylaxis and therapy of airborne tularemia. *Bacteriol. Rev.* 30: 542–548.

Speck, R. S., and Wolochow, H. 1957. Studies on epidemiology of respiratory infections. VIII. Experimental pneumonia plague in *Macacus rhesus. J. Infect. Dis.* 100: 58–69.

Stairmand, C. J. 1951. The sampling of dust-laden gases. *Trans. Inst. Chem. Engrs.* 29: 15–44.

Sutton, O. G. 1953. *Micrometeorology*, pp. 66. McGraw-Hill Book Co., Inc., New York.

Thorpe, J. F. 1967. Fluid flow. Chapter 5. ASHREA. *Handbook of Fundamentals.* Am. Soc. Heating, Refrig. and Air Condit. Engr., Inc., New York.

Webb, S. J. 1959. Factors affecting the viability of airborne bacteria. I. Bacteria aerosolized from distilled water. *Can. J. Microbiol.* 5: 649–669.

Wolf, H. W., Skaliy, P., Hall, L. B., Harris, M. M., and Decker, H. M. 1964. Sampling microbiological aerosols. Public Health Monograph #60, U.S. Dept. of Health, Education and Welfare. U.S. Government Printing Office, Washington, D.C.

Wolochow, H., Chatigny, M., and Speck, R. L. 1957. Studies on the experimental epidemiology of respiratory infections. VII. Apparatus for the exposure of monkeys to infectious aerosols. *J. Infect. Dis.*, 100: 48–57.

NOMENCLATURE

v	velocity of a stream of air	N_{ma}	dimensionless Mach number
V	velocity of an air mass	a	velocity of sound
D	duct diameter	p	pressure
ν	viscosity	P	1 atm
Re	Reynolds number	L	length of pipe

10

CONTAMINATION CONTROL IN AEROBIOLOGY

Mark A. Chatigny / Doris I. Clinger

NAVAL BIOLOGICAL LABORATORY
SCHOOL OF PUBLIC HEALTH
UNIVERSITY OF CALIFORNIA, BERKELEY

INTRODUCTION

Contamination control in aerobiology is concerned with control of air involved in specific experiments, with air surrounding experimental equipment, and with environmental air where other, less directly involved, ancillary activities are conducted. It is assumed that control will mean the containment, transport, disposal, and/or measurement of physical parameters of any manipulated air, including microbial, particulate, and chemical constituents occurring both in normal air or in the experimental environment; the question as to whether these constituents are produced either accidentally or deliberately as a part of the experimental procedure is moot. Interaction between air and equipment will be assumed as part of the environment.

Control of such factors as temperature, pressure, relative humidity, and other variables frequently included in the scope of environmental control, will be discussed only to the degree that they may affect contamination control.

The reader should be provided with some basis for evaluating the need for controls, a rational means of evaluating the extent to which these controls need to be applied, as well as some examples of control methodology to assist in their application. No attempt will be made here to duplicate the voluminous literature on process control systems, but attention will be drawn to parameters deemed most important at the outset of an experimental procedure, and to sources from which further information on specifics of control can be acquired.

Aerobiology work is often conducted with pathogenic, allergenic, or toxic agents. It is essential that the safety of the laboratory worker, his co-workers in the immediate area, and the general public be maintained by control of the environment involved directly with an experiment. Equally important is the requirement that experimental integrity be controlled, both to reduce random events and to minimize bias. Control of microbiological contamination will be used as a model for the more general case. That dangerous microbial aerosols may occur in microbiology laboratories is recognized, and protective measures have been developed. The rationale and methodology of this area of contamination control is applicable to other, less obvious, problem areas.

CONTAMINATION CONTROL RATIONALE

During the planning stage of an experiment, the operator is most concerned with controlling the various parameters to be measured. Shortly thereafter he becomes aware that it may be of equal importance to insure the protection of personnel conducting such experiments. Further consideration usually clarifies the need for control of additional factors required to preserve the integrity of experimental procedures. For example, the problem of intra-cage cross infection in an animal population can significantly alter experimental results.

Laboratory Accidents

Although we could cite instances of fallacious results where experimental errors were associated with lack of controls, such data (with the assistance of statistical mathematics) are usually buried like victims of poor medical practice and do not serve as constant reminders of the real problem. A more specific demonstration of the need for contamination control is found in the records of accidental infection in microbiology laboratories.

It may be stated without fear of contradiction that every infectious microbiological agent which has been studied in the laboratory has, at one time or another, caused infection of operators. In some instances, laboratory infections out-number natural infections and have been the only known human infections.

Sulkin and Pike (1951a, b) published an extensive survey of labora-

Table 10-1   Estimated frequency rates for laboratory infections.

Laboratory	Year	Infections per million man hours
European laboratory	1944–1959	50.0
Fort Detrick	1943–1945	35.0
Canadian T.B. laboratories	1947–1954	19.0
Fort Detrick	1954–1962	9.1[a]
Research laboratories	1930–1950	4.1
NIH	1954–1960	3.4[b]
CDC	1959–1962	1.3
Hospital clinical laboratories	1953	1.0
Public health laboratories	1930–1950	0.4
Clinical laboratories	1930–1950	0.1

[a] Includes non-lost-time infections.
[b] Includes diseases of suspected occupational origin.
(From Phillips, *J. Chem. Education*, **42**: A44, 1965a. Reprinted with permission of the American Chemical Society, Washington, D.C.)

tory infections and listed the wide range of etiologic agents involved. Their data were gathered from the literature, and from letter surveys, so are far from complete.

The surveys were adequate, however, to permit one to compare the degree of hazard of this type of work with other similar occupations, as well as to confirm that aerosol studies with infectious agents can be hazardous. Table 10-1 summarizes some typical estimated frequency rates for laboratory infections.

It is worthwhile to note that the infection rate for a large infectious disease research institute (Army Biological Laboratories, Fort Detrick), in which aerobiology is a significant part of the work, was substantially higher than that seen in clinical laboratories and hospitals, despite the fact that a strong safety program is in effect at that institute. Overall accident rates roughly parallel these statistics and the inference may be drawn that the effect of usual industrial type activities in an infectious disease institute is not as significant as the combined effect of aerobiological activities and the presence of infectious disease agents. Indeed, Wedum (1957) suggests that the over-all hospital laboratory rate of 4.2 accidents per million man hours should be augmented by 2.05 accidents per million man hours, to account for the incidence of infectious disease under these conditions.

Wedum and co-workers at Fort Detrick have been leaders in making definitive studies of the sources of infection in microbiological laboratories. They concluded that aerosols, generated deliberately or accidentally, have caused the overwhelming majority of laboratory infections. In a recent updating of the original survey of Sulkin and Pike (Hanson *et al.*, 1967) a total of 2,700 laboratory infections with 107 deaths was reported. Of these only a limited number was the result of proven laboratory accidents; the remainder was attributed to laboratory operations in which no overt accident occurred. The total number of deaths was approximately 4% of the total cases. This is in contrast to the combined case fatality rate of 1.0% for all disabling injuries in the United States in 1963. (The class of injuries resulting in the highest fatality rate—27%—was motor vehicle accidents.) Most laboratory infections occurred in trained scientific personnel (75%). Animal caretakers, janitors, and dishwashers accounted for the next largest grouping (10%); students not engaged in research, clerical workers, and others accounted for the remainder.

With a few notable exceptions, there appears to be little change in the pattern of laboratory-acquired infection in the years since the 1951 Sulkin and Pike survey. Tuberculosis is still the cause of the majority of cases, probably because of the large number of workers involved. As microbial agents are newly isolated, purified, and concentrated, the list of accidental infections grows longer. One European incident, presumably involving an arbovirus from African monkeys, was responsible for 30 infections (Anders, 1967) and 7 deaths.

Coccidioides immitis is a prolific generator of laboratory infections. Fiese (1958) cites several cases and notes that the only occupations more hazardous (with respect to coccidioidal infection) than agricultural and allied pursuits are those which involve handling *Coccidioides* in the laboratory. It is likely that the cases reported represent only a fraction of actual infections, since it has been estimated that less than 40% of natural infections reach the attention of clinicians.

The accident rates prevously cited do not give the full story of what can happen. There have been occasional dramatic episodes of laboratory epidemics caused by accidents or by procedures that spread aerosols of pathogens throughout several laboratories or entire buildings. Table 10-2 lists some examples of these epidemics.

The Brucellosis incident was caused by leakage from a centrifuge in the Bacteriology building of a State University; airborne contagion was spread throughout the building. Ninety-four individuals were infected; 84 were students, there was 1 fatality. Of 45 people hospitalized, 41 were students, 1 was a laboratory stockroom attendant, 1 a plumber, 1 a stenographer, and 1 was a visiting salesman. Although such epidemics are not common, they are very dramatic when they do occur, and they serve to point out the extreme hazard of widespread infectious aerosols.

Sources of Laboratory Infections

Statistics of laboratory infection are of little interest to the laboratory worker until he becomes directly affected; it is helpful, however, to be aware of relative risks of various operations in the laboratory, not

Table 10-2   Laboratory epidemics of infectious disease.

Disease	Year	Number of persons infected
Psittacosis	1930	11
Brucellosis	1938	94
Q Fever	1940	15
Murine typhus	1942	6
Q Fever	1946	47
Coccidioidomycosis	1950	13
Histoplasmosis	1955	18
Venezuelan encephalitis	1959	24
Tularemia	1961	5

(From Phillips, *J. Chem. Education*, **42**: A44, 1965a. Reprinted with permission of the American Chemical Society, Washington, D.C.)

only for purposes of defining safety problems within the aerobiology laboratory, but also for evaluation of contamination control problems in nearby areas.

Aerosols may be generated accidentally by any of several means. Nearly every laboratory technique produces an aerosol of some magnitude. Our primary concern is with apparatus in which aerosols of particle sizes most effective in causing respiratory infection or irritation are generated and held in relatively high concentration. We must, however, consider ancillary activities incidental to aerobiology as another part of contamination control. Culture preparation, sample collection, microbiological assay, and the holding of animals exposed to infectious aerosols are examples.

Johansson and Ferris (1946) photographed aerosols generated in the process of pipetting liquid samples onto agar plates. They observed that when the last drop of fluid in the tip of the pipette was forcibly expelled, a spray of more than 10,000 droplets was produced. Most of these droplets were in the 1- to 10-μ size range. In recent years, other common laboratory techniques have been examined by workers at the Army Biological Laboratories, and at other institutions, who employed non-pathogens in conjunction with extensive air sampling. They observed that aerosols were produced by pipetting, operating a blendor, dropping cultures from various distances on to different types of materials, inoculating mice, inoculating eggs, and by numerous other activities. Tomlinson (1957) and Reitman and Wedum (1956) have provided data on aerosol generation from simulated trials of various manipulations connected with the processes of lyophilization, centrifugation, animal inoculation and autopsy. Table 10-3 summarizes their findings.

Evidence of aerosol generation from infected animals is sparse, but there are many reports of intra-cage infection of laboratory animals. The potential of airborne transfer of zoonoses from infected animals to man in the laboratory is well recognized. Incidents of infectious hepatitis, cat scratch fever, psittacosis, monkey B-virus, leptospirosis, tuberculosis, and many others have been reported. Wedum (1964) has tabulated some admittedly incomplete data on intra-cage transmission, both from his own work and that of others. Cross-transmissions were probably a result of direct contact, but his data do include many cases for which the excretion of organisms in urine and feces was both the source of intra-cage contamination as well as the source of contaminants that may pervade animal rooms.

Aerosol chambers, obvious sources of aerosols, have actually been less involved as sources of laboratory infections than other devices. It is not clear whether this is because of the relative scarcity of such equipment, or the awareness of a requirement for leak control in construction and installation of these rather special devices. The obvious hazard afforded by various techniques for the inoculation of animals via the respiratory route, and for subsequent maintenance of the animals infected with human pathogens, has led many investigators to design protective equipment for such operations.

While judgment as to the degree of hazard involved in any individual operation may vary, contamination control from the aspect of personnel safety alone appears warranted if infectious, toxic, or allergenic agents are employed. If the need to control parameters inherent in each investigation, and the requirement that experimental measurements be as precise as practicable are added to factors of personnel safety, one needs little further justification for imposition of state-of-the art contamination and environmental controls.

Table 10-3   Examples of aerosols produced from some common bacteriological techniques.

Laboratory operation	Average number of colonies on air sampler plate	
	Minimum	Maximum
Agglutination, slide drop technique	0	0.66
Animal injection (guinea pig) undisinfected inoculation site	15	16
Centrifuge, tube broken, culture stayed in cup	0	20
Centrifuge, tube broken, culture splashed outside of centrifuge	80	1,800
One drop *S. indica* falling 3 in. onto:		
Stainless steel	0.2	4.7
Dry hand towel	0.0	0.35
Towel wet with 5% phenol	0.0	0.05
Insert *hot* loop in culture	0.68	25
Insert *cold* loop in culture	0.0	0.22
Break ampule of lyophilized *S. indica*	1,939	2,040
Streaking, rough agar plate with loop	7	73
Pipetting, inoculate culture	0	2
Using blendor with poor fitting parts	77	1,246
Opening screw-cap bottle	0	45

(From Chatigny, M. A. 1961. In: *Advances in Applied Microbiology* (W. W. Umbreit, Ed.) Vol. 3, p. 141. Reprinted with permission of Academic Press, Inc., New York.)

CONTROL CRITERIA

Where Controls Are Needed

It is apparent that controls may be required in virtually every operation in the microbiology laboratory. We will confine ourselves to discussion of:

1. Aerosol chambers or test vessels.
2. Animal exposure systems.
3. Animal holding spaces or enclosures.
4. Bio-assay areas.
5. Process air supplies.
6. Process air exhausts.

The first four listed areas for control have obvious implications regarding safety problems described previously. The last two relate specifically to experimental integrity and to protection of the general public from waste products.

Types of Control

Not only must one be concerned with problems of overt leakage of particles or chemical vapors in or out of experimental devices, but less obvious parameters, such as temperature, relative humidity, electro-magnetic radiation, electrical charge, ionization of the air environment, pressure, and other factors specific to each experimental procedure, must be considered. The former requirements will be discussed as "leakage" whereas the latter will be termed "purity" controls.

In any rational procedure for application of environmental control measures to the type of operations listed, it is necessary at the outset to determine the degree of control required, as related to:

1. Acceptable accidental dose levels for the operators or the test subjects.
2. The effect of dilution or infiltration on the experiment.
3. Requirements for decontamination and feasibility of its application.
4. Ultimate disposition of leakage products, and their potential effects on other workers in the room or building.

It would be desirable to have a quantitative evaluation of all parameters involved in leakage contamination; such data are not likely to be available, but a record of laboratory accidents can aid in that respect.

For overall evaluation of criteria for leakage control, some of the more obvious questions requiring resolution are:

1. If aerosols of microorganisms are used, is there chance of infection of personnel, plants, or animals?
2. Are there likely to be allergenic responses?
3. Are test organisms species specific; if they are known human pathogens, what is the infective dosage?
4. Are prophylactic or therapeutic measures available?
5. Are there wide variations in susceptibility among hosts because of age, sex, or health?
6. Are organisms sensitive to heat, light, desiccation, or ultraviolet (UV) irradiation? If the particles are radioactive, are there specific regulations regarding permissible isotope levels?
7. What is the half-life, biological or radiological, of the material?
8. What kind of emission exists?
9. Where will the material lodge or be concentrated in a host?
10. What, if any, biological clearance mechanisms exist in the host, and what is that half-life?
11. Is the aerosol corrosive?
12. Is there an explosion hazard?

13. What decontamination procedures are available? Are they effective? Is the decontamination material corrosive to the equipment or hazardous to the operators?

For purity parameters, the list of questions to be answered may be less extensive and more amenable to quantitation:

1. If microorganisms are employed, will they be affected by either chemical contaminants, by stray, UV irradiation, thermal energy, or radioactive energy?
2. Are there organisms or particles in the ambient air similar to those in the experiment?
3. Are the microbes sensitive to temperature, pressure, dessication, or hydration; are they sensitive to rates of change of such parameters?

These questions will usually be included in the scope of aerosol experiments, and the nature of the experiment will serve to define the requirements.

For purity control, criteria may be related to:

1. Sensitivity of the contaminant measuring instrumentation.
2. Dose or sample size in the experiment.
3. Objectives of the experiment with respect to effects of trace contaminants on the experimental parameters.
4. Sensitivity of animal (or human) test subjects to contaminants separately or as synergistic elements.

In aerosol experimentation, a "contaminant" may be defined as any material that interferes in an unpredictable manner with experimental aims. It is apparent that the measuring system employed can limit the extent of control. To define the existance of a contaminant, it is necessary that its effect be measurable. That is, if the accuracy of measurement of the contaminant varies more than parameters inherent in the system, then experimental design cannot provide effective control, and one seems to be encountering "unknown variables."

In summary, development of environmental control criteria is an inherent part of experimental aerobiology. Many questions are involved with specific experimental procedures, and a rational approach will consider all important factors within the planning. With respect to safety, it is necessary to exercise the state-of-the-art in containment and control if one is to assure workers, sponsors, management, and regulatory agencies that the proposed work will be done in a responsible manner. With respect to experiment integrity, it is necessary to ensure that the data collected are valid, significant, and reproducible. The questions listed have been included as a basis for establishing control criteria. Such questions must be supplemented by more quantitative information. Effective control can only be evolved by a thorough analysis of experimental procedures.

CONTAINMENT METHODOLOGY

The most widely used method for contamination control is the containment or barrier system. Its precept is simple: provide an enclosure of minimal volume sealed sufficiently tight to maintain the desired integrity. The containment may include a primary enclosure, a secondary housing, and even a third-level barrier.

The primary barrier is the container into which aerosols are generated. It may be a flask, a drum, a cylindrical tank, a box, or even an extended pipe section. In any case, it must have provisions for (a) generation of an aerosol, (b) entrance of primary (aerosol-forming) and secondary (diluting) air, (c) sampling the air mass within the enclosure, and (d) disposition of the test aerosol; frequently the capability for exposing animals to the aerosol is included.

It is of interest to make an informal estimate of the leak-tightness requirements of a typical aerosol system. Consider the widely used Henderson apparatus (Chapter 9) for exposure of two or more small animals to an infectious aerosol cloud. Assume a challenge aerosol of approximately 10^6 viable particles per liter and assume the interior volume of the apparatus is 5 liters, thus providing a total aerosol cloud at any given moment of 5×10^6 particles. If the device is well constructed of glass or metal, with seals at all joints, and is operated at atmospheric or slightly lower pressure, a relatively small leak rate will prevail.

Sampling, animal transfer operation, and aerosol generation, however, require that the integrity of this system be broken during the course of operation. Estimates of the volume of aerosol escaping during these processes range from 0.01% to 1%. If we assume 0.1% ($1/10^3$) of total volume leakage, there is the possibility of discharging about 5×10^3 microorganisms into the atmosphere surrounding the unit. (It should be noted that such a volume loss would cause a barely perceptible reading—<0.003 inch water—on any manometric pressure detection devices incorporated into this unit.)

Assume that the microorganisms employed were infective in doses of 10 to 100 organisms. Saslaw et al. (1961) report that *Pasteurella tularensis*, for example, has an estimated human infectious dose of 1 to 10 organisms; hence, a severe hazard could exist for personnel in the immediate area. It is reasonable to suggest that a secondary enclosure, with ventilation to remove such particles, should be provided around the unit to protect the worker. This enclosure should provide (a) a barrier between the operator and the aerosol chamber, (b) at least one hundredfold dilution of the leakage, and (c) removal of particles in a manner providing an acceptable margin of safety before disposition of this air to any occupied spaces. The latter is most easily accomplished by the use of an open-or-closed-front safety hood. Hood volume should be approximately 500 liters or more, with a high efficiency exhaust to provide rapid removal of the aerosol through a filter with approximately 99.99% collection efficiency (1 in 10^4 penetration).

If leakage from the hood is about 1%, if perfect mixing occurs, and if the dilution ratio is also about 1% for the open-front safety hood, then aerosol concentration near the hood opening should not exceed 1 organism per 10 liters of air. After dilution of approximately a hundredfold by room air, this hypothetical aerosol system would create an aerosol of 1 organism per 10^3 liters; i.e., approximately 10^3 minutes would be required for man to acquire a 10-organism dosage. If all of the equipment is operating satisfactorily, this is a reasonably safe operation.

In the event of equipment breakage or failure, or miscue by the operator, the primary and/or secondary barriers may become ineffective, and the entire aerosol could be liberated into the room. Hence, both a closed-front hood, which would reduce penetration to 1 in 10^8, and a third level barrier for the protection of others in the general area might be required. The room in which the experiment is housed usually provides this third barrier.

The room can be equipped with a ventilation system to maintain a negative pressure differential and to cause air to flow *into* the contaminated spaces. The exhaust air could be discharged directly out of doors if the exit port is clear of building air intakes. The objective of such a ventilation system would be to ensure rapid, thorough dilution of any accidental aerosol (to minimize the time the aerosol would persist) and to provide safe disposition of waste or accidental discharges.

This example is grossly over-simplified; it gives no consideration to the problem of disposition of the aerosol actually in the Henderson unit, or to the problem of maintaining any animals exposed to the aerosol. However, practical experience has shown that (a) the estimates of numbers of particles escaping are not unreasonable, (b) escaping particles are rapidly mixed with the room air, and (c) the typical protective system briefly described here is effective. This hypothetical, three-barrier system is not far afield from practices in laboratories where close attention is paid to safety, and the illustration should suffice to establish reasons why special equipment and facilities must be available for aerobiology studies with hazardous materials.

Primary Barrier Systems

Examples of aerosol vessels comprising primary barriers range from simple flask systems to vessels as large as required. Most such devices have the common requirement that air be contained in an enclosure that is as leakproof as the state-of-the-art of fabrication and the leak-testing methods available to the investigator can provide. Designs and materials of construction have varied widely, depending on intended application of the device. For work with nonpathogenic organisms, simple glass or metal enclosures are frequently used. Glass equipment has the great advantage of visibility of operation, ready availability, reasonable economy, and a lack of reactivity with many test process materials. A simple glass system, which includes all the necessary attributes for studying aerosols in small volumes, has been described by Griffin *et al.* (1956). This system included a 1-liter flask to hold the aerosol, a nebulizer, humidity measurement facilities, a transfer flask from which aerosol samples could be withdrawn for analysis, a slit sampler, and a vacuum pump. The authors proposed that such a system be used only with nonpathogens (*Serratia marcescens* was noted); however, Piggott and Emmons (1960) utilized a somewhat similar apparatus for work with pathogens. Their device was a 1-liter flask with 10 side-arms, each holding a mouse with its snout positioned centrally. A fungal culture was grown on nutrient medium directly within the same flask. For a challenge experiment, animals were placed in the side arms, and mycelial fragments were blown loose from the growth material by a jet of air from a glass tube suspended in the flask. A similar system, built from metal, has been recently described (Southern *et al.*, 1968). While the hazard engendered by a single liter of noninfectious baterial aerosol must be considered nominal, it is questionable that such devices should ever be used for work with pathogenic agents without the added protection of a secondary barrier system.

In general, glass systems have been limited to volumes less than 10 liters. Rugged structure is required for volumes of 10 to 100 liters. Metal, preferably stainless steel, is the material of choice. Aluminum has been used, but the reactivity of most alloys, particularly in humid atmospheres, limits their usefulness; anodizing offers some improvement.

Plastic materials (acrylics, polyvinyls, polystyrene, and others) have also been employed, but the problem of fragility of large, unsupported plastic enclosures, coupled with the severe problem of electrostatic charge on many types of plastics, has restricted their utility.

Most chambers in the 100- to 1,000-liter volume range have been fabricated of metal, glass, or plastic. With a limited number of exceptions (the million-liter sphere described by Wolfe (1961), for example), larger chambers used most extensively in studies of air pollution effects have been fabricated of conventional building materials including wood, plasterboard, asbestos, cement board, tile, and virtually every other construction material. In many cases, these constructions were chosen to simulate building interior environments.

The selection of materials and methods of construction have, for the most part, been made on the basis of experience and specific requirements. Table 10-4 lists some selected aerosol chamber construction practices showing volumes and areas of application.

With a few exceptions, early workers did not give serious consideration to the effect of chamber construction on the aerosol behavior. Fortunately, the most available and widely used material, glass, had very low reactivity, and the data from much of the early work are quite reliable when considered in context. Table 10-5 lists a wide variety of chambers with some information regarding use, materials, and references.

It is only in recent years, when more exotic materials of construction have become available and more sensitive quantitative means of measuring air contaminants have been put into use, that this problem has become important and obvious. For example, workers concerned with moisture content of the air would not normally be concerned with effects of a metal enclosure, but if they are concerned with parts per million moisture in the air, they rapidly become aware that both metallic and glass surfaces can be hygroscopic at these very low vapor pressures, and considerable attention must be paid to surface finishes to minimize this effect.

DallaValle (1948) cited the effect of surface roughness on condensation phenomena at very low vapor pressures. Dimitriades (1967) observed that many plastic films are extremely porous to various vapors, water vapor among them. Such porosity, not limited to plastic materials, also tends to limit the use of vapor-phase decontaminants in aerosol chambers because of the problem of slow, low level, exudation of the absorbed material. Porosity effects frequently defy theoretical calculations of diffusion rates and of decay of compounds. For example, when beta-propiolactone (BPL) is used to decontaminated a space, the vapor can frequently be detected long after the time calculated for clearance from the space, or even for BPL hydrolysis.

For studies of very dilute aerosols in which the total number of small (0.1 to 10 μ) airborne particulates is of importance, materials of construction may be critical. Many materials will shed small particles when

Table 10-4 Summary of typical materials of construction for aerosol chambers.

Chamber volume, liters	Number of users	Materials of construction	Major uses
<10	10	Stainless steel, brass, Pyrex,[a] copper, plastic chambers Plastic, glass view ports Rubber diaphragms and gaskets	Aerosol decay Animal infectivity
10–100	9	Aluminum, stainless steel, glass, plastic, brass, plywood chambers Plexiglas[b] view ports Vinyl foam, polyurethane, rubber gaskets Canvas, rubber, plastic masks	Aerosol decay Animal infectivity Animal response to air pollutants and radioactive aerosols Human infectivity Respiratory immunization Testing air sampling apparatus
>100–500	7	Tin, aluminum chambers Glass, Lucite,[c] Plexiglas view ports Rubber bellows Rubber gaskets	Animal infectivity Aerosol decay Animal response to radioactive aerosols
>500	14	Steel, aluminum, glass, plastic, angle iron, quarry tile, asbestos cement board, Lucite chambers Pyrex, Plexiglas, glass view ports Neoprene gloves Vinyl foam, polyurethane, neoprene, rubber, soft metal gaskets Rubber masks	Aerosol decay Animal infectivity Animal response to air pollutants Human infectivity Human and animal response to space environments Photochemical smog experiments Bactericidal action of vapors

[a] Pyrex: registered, Corning Glass Works, Corning, N.Y.
[b] Plexiglas: registered, Rohm & Haas Company, Philadelphia, Pa.
[c] Lucite: Registered, E. I. du Pont de Nemours and Company, (Inc.), Wilmington, Dela.

Table 10-5 Characteristics of aerosol chambers used by various workers.

Apparatus type and volume	Major applications	Materials aerosolized	Materials of construction	Author(s)	Remarks
Henderson apparatus 1.3 to 3 liters (aerosol generation) 9 to 40 liters (animal exposure)	Aerosol behavior; animal and human infectivity and virulence;	Bacteria, viruses, fungi (pathogens and non-pathogens), toxins	Brass generation-mixing tube. Brass, aluminum animal exposure boxes. Canvas, plastic, rubber masks and helmets (extensions of tube). Natural rubber diaphragms, gaskets. Stainless steel hood or chamber enclosure.	Henderson, 1952. Wolochow et al., 1957. Roessler and Kautter, 1962. Beard and Easterday, 1965. Griffith, 1964.	Dynamic. Large animal exposure. Mobile unit in el III hood in sealed pressure tight tork
Various animal exposure systems 0.3×10^0 to 3.3×10^2 liters	Animal infectivity, virulence and immunogenicity. Design information to achieve dynamic dilution.	Bacteria, viruses, spores (pathogens and non-pathogens), infectious nucleic acids, dyed smoke	Balloons, autoclave, Pyrex[a] flask, copper tubing, stainless steel chambers. Vertical wind tunnel, Plexiglas,[b] Pyrex view ports. Rubber, neoprene gloves. Tinplate tunnel	Ferry et al., 1958. Akers et al., 1966. Young et al., 1946. Norseth and Mitchell, 1963. Ames and Nungester, 1949. Middlebrook, 1952.	Provides mixed aged aerosols for animal exposure. Varied particle size
Cloud chamber 2.1×10^2 liters	Animal infectivity and virulence	Viruses, animal pathogens	Aluminum chamber, acrylic-plastic window(s). Rubber gaskets, stainless steel mesh cages.	Weiss and Segeler, 1952. Gogolak, 1953.	Dynamic. 108 mice, whole body exposure

Static or stirred settling chambers, flasks, boxes, cylinders, rooms, spheres 1 to 1 × 10⁶ liters	Aerosol decay, animal infectivity and virulence, quantitative human medical studies	Bacteria, viruses, fungi (pathogens and non-pathogens), bactericidal vapors, radioactive tracers	Pyrex flasks, steel vessels, building materials, gas masks	Harper et al., 1958. Griffin et al., 1956. Wolfe, 1961. Piggott and Emmons, 1960. Darlow et al., 1958. Dimmick et al., 1958.	Static or stirred chambers.
Environmental, irradiation chambers, spirometers 2 × 10¹ to 6 × 10⁴ liters	Animal toxicity; animal and human response to hypercapnia and hypoxia, photochemical smog, testing of low-volume air sampling instruments.	Toxic solids, liquids and gases, air pollutants, radioactive aerosols	Stainless steel, polyvinyl, asbestos cement board, aluminum, glass, plastic chambers. Rubber, sponge rubber, neoprene, vinyl foam, polyurethane, plastic foam gaskets. Glass, Plexiglas,[b] Teflon,[c] Pyrex windows.	Urban, 1954. Glauser and Glauser, 1966. Hinners et al., 1966. Schwartz and Silverman, 1965. Fraser et al., 1959. Dimitriades, 1967. Spiegl et al., 1953. Thiéblemont et al., 1965.	Dynamic and static; acute and chronic studies of physiologic responses, toxicological studies.
Dynamic aerosol drum (toroid) 1.6 × 10³ liters	Studies of persistant aerosols	Bacteria (pathogens and non-pathogens)	Stainless steel, neoprene rubber, copper piping.	Goldberg et al., 1958.	Provide aerosol aging to 100+ hours.
80-foot long pipe DATA	Short term effects of RH	Bacteria, viruses	Terneplate, brass fittings, glass atomizer	Hatch and Dimmick, 1965.	Usable for infectivity studies with 6-10 minute aerosol.

[a] Pyrex: registered, Corning Glass Works, Corning, N.Y.
[b] Plexiglas: Registered, Rohm & Haas Company, Philadelphia, Pa.
[c] Teflon: Registered, E. I. du Pont de Nemours and Company, (Inc.), Wilmington, Dela.

209

exposed to elevated temperatures, irradiation (ultraviolet, infrared) or vibration. Methyl methacrylate (Lucite),[1] for example, will shed submicron particles under strong sunlight or thermal radiation.

With few exceptions, natural and synthetic rubber elastomers have been employed for flexible tubing connections and for sealing aerosol chambers. Until recently, the most ubiquitous material used was rubber tubing. Although being replaced by vinyl tubing, it is still widely used. Silver (1946) observed that "rubber" (natural or synthetic, unspecified) absorbed organic vapors very strongly. In one experiment, 2.5 mg chlorovinyl dichlorarsine in a 10-liter, air sample was absorbed into a 12-inch section of ¼-inch tubing. He suggested that all rubber-tubing surfaces be coated with lacquer to correct this source of error.

The availability of new synthetic, chemically resistant elastomers may obviate such requirements for a wide variety of vapors.. When flexible plastic tubing (usually vinyl or polyvinyl polymers) is employed, the leaching of plasticizers may be encountered; in addition, the problem of surface electrostatic charge, either positive or negative, may create other problems, particularly where low level, air pollutants are being studied. Specially compounded and processed plastic tubing is manufactured for "medical" applications. This material is least prone to loss of plasticizer.

Synthetic mastic or plastic compounds are widely used as cements and sealants. Among the most useful of these are the silicone and polysulfide compounds, some of which cure at room temperature into very flexible materials. Polymers that are "self-curing" (i.e., depend on solvent evaporation) may release vapors for extended periods of time.

The most desirable elastomers and sealants for use in aerosol chambers are those that have (a) good resiliency, (b) resistance to chemicals, (c) adequate heat resistance, and (d) low gas or moisture absorption; and that are (e) self-curing without toxic exudates or residues and (f) reasonable in cost. Silicone compounds, particularly those of "food grade" quality (U.S.D.A. standard), meet many of these requirements with the exception of moisture absorption and tear resistance. Actually, with respect to aerosol chambers, there is no "perfect" elastomer. Any sealant or elastomer used for connections, construction, or sealing should be tested carefully for detrimental properties on biological agents before being used in chamber construction.

Materials used for piping systems associated with aerosol chambers should be as similar to those of the primary chamber as practicable. The availability of conventional materials such as copper alloys, steel alloys, and glass in particular, make the use of one or more of these

[1] Registered, E. I. du Pont de Nemours and Company (Inc.), Wilmington, Dela.

materials mandatory. With few exceptions, piping is used to transport aerosols to sampling devices, to expose animals or to transport aerosols to other chambers; contact time within the piping may be relatively short. Unless the aerosol is extremely reactive, contamination will be minimal, so just maintenance of a leak-tight piping system may be the most severe problem.

Properly threaded joints in pipe, and well-cleaned, soldered joints in copper tubing will meet all but the most rigid (mass spectrometer) leak tests; frequently, valves used in such lines will not meet such requirements. For routine, leak-tight, shut-off purposes, only ball or diaphragm (Saunders' patent) type valves[2] are recommended. Conventional "globe," "gate," and "needle" valves[3] are economical and have good flow characteristics, but must be modified with "O" ring seals to prevent leakage around their valve stems. For pipe lines larger than 2-inch IPS,[4] butterfly valves with resilient sealing[5] may be used.

The piping itself should be clean and smooth inside, with all ends reamed to remove burr. The entire aerosol plumbing system should permit thorough flushing with cleaning solutions after assembly. If the work involves human pathogens, the piping should be suitable for sterilizing with steam under pressure. Effective cleaning and biological decontamination, and successful operation thereafter require that there be no unvented high places or undrained low places in the lines.

Piping installed in this manner will also reduce the chance of "carry-over" contamination from one experiment to the next. Tests have shown that there is very little deposition inside smooth pipe lines except at sharp corners where some impaction may occur, and in places where condensation and stagnation may occur. Sudden flow changes can produce secondary aerosols from such sources, or even cause annoying pulsations or oscillations as these "elastic" pockets come into resonance at sensitive air-flow rates.

In small pipe (tubing size 1 inch or less) there is a wide range of pipe, tubing, fittings, level controls, valves, and other devices available from refrigeration parts supply houses; other such equipment is generally found in plumbing supply houses or from compressed air specialty suppliers. In some cases, it may be more economical to improvise some special device or fitting rather than having it custom made. In most

[2] Hills McCanna Company, Carpentersville, Ill.; Grinnell Co., Inc., Providence R.I.
[3] Crane Company, Chicago, Ill.; Jenkins Bros., New York.
[4] Iron Pipe Size often used, IPT (Iron Pipe Thread) or NPT (National Pipe Thread). Stated size relates to *approximate* inside diameter of standard wall thickness pipe.
[5] Conoflow Corporation, Blackwood, N.J.

cases, the needed device is in existence somewhere, and it remains only for a diligent purchasing agent to locate the source.

Secondary Barrier Systems

Secondary barrier containers have received considerably less emphasis than have primary enclosures. Originally, the need for thermal control of the environment surrounding aerobiology experiments generated the requirement for enclosures. In recent years, there has been an increasing awareness of the necessity for barrier containment techniques to control contamination. The definition of various levels of control enables us to establish criteria for each level of the enclosure system.

Secondary barriers may be classified with respect to several parameters; first, by size (they may be used to house a bench-top, or a room-, or a building-size operation); secondly, by intended usage, whether for protection of the product, or of the experiment or personnel; and thirdly, by the degree of containment provided. Enclosures may provide "absolute" barriers as tight as the primary enclosure, or partial barriers permitting controlled leakage.

The principle of minimum volume espoused in design of primary enclosures applies equally to design of secondary barrier containment systems. If the aerosol chamber is of such volume that a bench-top, or standing, cabinet enclosure is impractical, the entire room housing the equipment can be the secondary enclosure. The worker can be provided with a ventilated plastic suit, respirator, or protective clothing as the degree of hazard dictates.

Such a procedure is costly and complicated. If materials to be aerosolized are extremely hazardous in low concentrations, it may be desirable to provide the maximum containment possible against the contingency of failure of the primary enclosure. The secondary barrier system would, in the event of major accident or breakage, be required to function at the penetration level of a combined primary and secondary barrier, i.e., on the order of $1/10^8$ penetration from our previous example. In such case, the aerosol device is usually placed within a gas-tight enclosure just large enough to house the equipment, and the work or manipulation is done by mechanical manipulators, or through armlength gloves. Thus for a small-animal Henderson unit, a bench-top hood would probably provide an optimal enclosure.

The hood enclosure or "cabinet," the single most important safety device in the microbiological laboratory, merits detailed discussion. In form, safety hoods may range from the open-front "fume hood" enclosure to tightly sealed, interconnected hood systems often referred to as cabinets rather than hoods.

FIGURE 10-1. Open-front (Class I) safety hood.

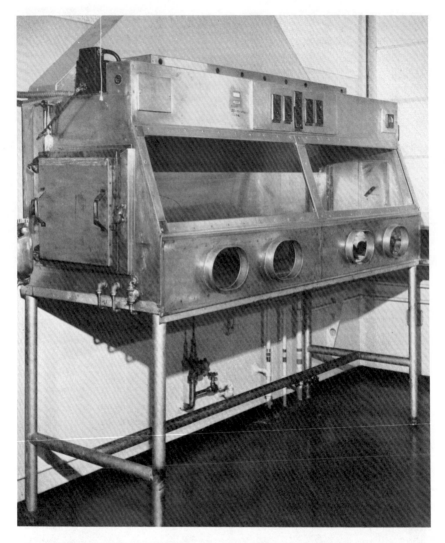

FIGURE 10-2. Open-front safety hood with access ports (Class II).

For convenience, workers in this field have defined three levels of control systems:

Class I   Open-front, ventilated hoods (Figure 10-1).
Class II   Open-front cabinets with restricted entry area and with provision for attachment of gloves. Some uncontrolled leaks are permitted, but the direction of leakage flow is controlled (Figure 10-2).

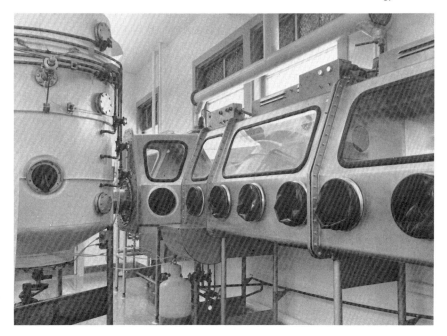

FIGURE 10-3. Closed safety cabinet (Class III) system housing controls for aerosol drums.

Class III   Closed cabinet systems in which all leaks in and out are through controlled openings. Gloves are permanent fixtures and entry and exit are through air locks or a disinfectant bath (Figure 10-3).

Selection of a hood system requires considerable care. For many microbiology operations (considering safety alone), the open-front unit is quite adequate. If aerosols are being generated at elevated pressures, or with strong sprays, it may be desirable to restrict the openings to the Class II level. If protection of the operator is the prime function, the air flow is directed inward. If a "clean" process is being done or specific pathogen free (SPF) animals are being employed, it may be more important that the hood be used with an outward flow of air. In some cases, ventilating air may be moved through the area to be protected, but a neutral pressure condition (with low inward or outward leakage) is maintained. Although a Class III hood system would be (mechanically) the most simple in such cases, there may be instances where it is essential to have the better flexibility of access provided by a Class II system with "neutral" pressure; again, the choice depends on the expected level of contamination. With the safety of personnel as a primary objective, Phillips

(1965b) provided some recommendations for hood use, depending on the agent being aerosolized (Table 10-6).

In this table, columns designated "single cabinets" refer to Class I and Class II systems. The column designated "cabinet system" refers to a Class III system wherein the entire operation, including aerosol generation, animal exposure and holding, and sample collection and processing is carried out within the cabinet system. Animals exposed under such conditions are held in ventilated cage systems maintained with the same integrity of closure as the Class III cabinet system. Frequently, such cabinet systems are constructed in the form of arrays housing incubators, centrifuges, and other equipment. They may be connected directly to other holding spaces wherein animals are maintained in ventilated cages, and are serviced by personnel wearing protective clothing. The Class III system has a significant secondary advantage in that environmental control, for which enclosures were originally required, may be better accomplished by a closed-hood system than by an open-front system requiring air conditioning of the laboratory space for temperature and humidity control.

The design of hood or cabinet systems varies widely. The basic parameters in the design and installation include the work volume, air circulation, materials of construction, and installation procedures. These factors need more detailed examination.

Volume and Work Space

The volume of the hood should be adequate to house the entire test operation. Headspace should be adequate for a full-size vision window, for stacked animal cages or storage of equipment, and for use of pipettes or other hand instruments. Occasionally a two-sided hood may be desirable, particularly if there is need for access to both sides of an animal exposure device. Space for storing and handling air-sampling materials, animals, and atomizers should be enough to obviate the necessity of moving in and out of the enclosure during a test procedure. Commercial single-hood units are offered in a range of sizes with a minimum length of 3 or 4 ft and a maximum of 7 ft; modular units may be connected to form assemblies of greater length. Widths or depths vary from 20 to 36 inches. It is questionable whether a single cabinet smaller than 48 inches long by 26 inches deep by 36 inches high would be useful for any but the very simplest laboratory procedures. A work station for a single individual usually requires approximately 4 ft of hood length. Two workers can conveniently work in a 7-ft space. Some hood designs provide a lift-up front window to permit insertion of large pieces of equipment, but this tends to reduce the protection offered because it

Table 10-6 Correlation of estimation of risk with recommendations for use of protective cabinets.[a]

Disease or agent	Cabinet system aerosol studies	Single cabinets Aerosol studies	Single cabinets Other techniques
Brucellosis	+++	...	+++
Coccidioidomycosis	+++	...	+++
Russian s-s encephalitis	+++	...	+++
Tuberculosis	+++	...	+++
Monkey B virus	+++	...	++
Glanders	++	+++	+++
Melioidosis	++	+++	+++
Rift Valley fever	++	+++	+++
Arboviruses, general	...	+++	++
Encephalitides, various	...	+++	++
Psittacosis	++	+++	++
Rocky Mt. spotted fever	++	+++	++
Q fever	++	+++	++
Typhus	++	+++	++
Tularemia	++	+++	++
Tularemia[b]	...	++	+
Venezuelan encephalitis[b]	...	+++	+
Anthrax	+++	...	±
Botulism[b]	++	+++	±
Histoplasmosis	...	+++	±
Leptospirosis	...	+++	±
Plague	+++	...	±
Poliomyelitis	+++	...	±
Rabies	+++	...	±
Smallpox[b]	+++	...	±
Typhoid	...	+++	0
Adeno, entero viruses	...	++	±
Diphtheria[b]	...	++	0
Fungi, various	...	++	0
Influenza	...	+	±
Meningococcus	...	++	0
Pneumococcus	...	++	0
Streptococcus	...	++	0
Tetanus[b]	...	++	0
Vaccinia[b]	...	++	0
Yellow fever[b]	...	++	0
Salmonellosis	...	+	±
Shigellosis	...	+	±
Infectious hepatitis	...	...	±
Newcastle virus	...	+	0

[a] +++ = mandatory; ++ = strongly advised; + = optional, but in absence of a cabinet a few infections will occur; ± = depending upon technique and supervision; 0 = not required.
[b] For persons receiving live vaccine or toxoid.
(From Phillips, *J. Chem. Education*, **42**: A119, 1965b. Reprinted with permission of the American Chemical Society, Washington, D.C.)

establishes a "leak" immediately in front of the operator. The temptation to use such an entry during a test is a less obvious, though genuine, hazard. An airlock, made only moderately tight to allow some air flushing, can be attached at the side or rear of the hood to permit entry or withdrawal of any but the largest pieces of equipment. An airlock fitted with a drain can be used for washing or decontaminating containers, or for holding samples and other equipment that must be removed during an experiment. If the hood or cabinet is installed with the work surface at "standing" height (usually 36 to 38 inches from the the floor), the operator(s) can work sitting or standing and make maximal use of the depth of the cabinet. It is desirable to have the working floor of the hood depressed 6 to 10 inches below the front opening if the operation to be housed uses equipment more than 6 inches high. While the exact height of the work surface is not critical, working in a hood that is too low can be most irksome. In the case of the Class III system, where glove ports offer great restriction to movement, the height (and often work space) parameters are critical. Doxie and Ullom (1967) have made a study of human-factors design-elements that is extremely valuable, particularly for users of complex multi-hood systems. The user should consider such factors because manufacturers, who cannot predict every possible usage, make most of their equipment to meet the needs of "bench-top" microbiological or chemical activities.

Air Flow

For Class I and Class II cabinets, air inflow or outflow velocity and volume are key factors in effective operation. Air-flow velocity into the hood must overcome or control diffusion of contaminants by local air currents from room ventilating systems or from operator activities in the immedate area of the test process. A person walking at 2 miles per hour creates air currents as strong as 175 lineal feet per minute (lfpm), a velocity that can cause air to be withdrawn from a properly operating Class I hood into the room in sufficiently large quantity to present a hazard under certain conditions. Even the act of withdrawing hands from a hood can create aerosol outflow. It is not usually necessary to overcome any initial velocity imparted to contaminants by accidental spills (or even small aerosol generators) because the air resistance quickly slows all but the largest ($> 25 \mu$) particles to diffusional speeds. Some devices, a high-pressure nozzle for example, do produce heavy sprays with a long trajectory. These devices require a physical (Class III) barrier for effective control.

The current trend in design of open-front hoods, both the restricted area-opening types and full open-area "laminar flow" types, is to provide velocities of 50 to 100 lfpm at the face of the opening. This must be considered minimal, and it is more desirable to have from 125 to 150 lfpm as a design velocity (with clean filters; see later).

Although many cabinet installations are often connected directly to the building ventilation system, an entirely acceptable procedure, the principle of minimizing the volume of potentially contaminated space suggests that each cabinet, or cabinet system, have a fan and filter system exclusively for its own volume; most hoods are so equipped. Centrifugal "squirrel-cage" type fans are most frequently used.

For a closed-front cabinet, or a cabinet that is convertible from open to closed, it is most desirable that the fan static-pressure versus flow-volume characteristics match the requirements of the system. Flow through the conventional open-front hood is a function of fan capacity, air friction losses in the system, and air pressure loss caused by the filter. Fan capacity and, hence, power requirements are governed primarily by the fan diameter and rotational speed. In most hood applications, resistance to flow of all other components in the system is negligible, or minor, compared with that of the air filter.

At the original start-up of an installation, the air filter will be clean and therefore have a relatively low pressure loss, usually 0.75 to 1.0 inch water static pressure (wsp). After some usage, the filter will become dirty and the pressure loss will increase until the filter is changed (usually at ±2.0 inches wsp) with concomitant change in flow volume. Assuming the fan can produce the necessary 2 inches of water suction or pressure to overcome the pressure drop caused by a dirty filter, it is apparent that, if all of the openings at the front of the hood were closed off, pressure differential would rise to at least 2 inches. A high pressure differential tends to destroy "feel" in the rubber gloves and makes operations, under a pressure difference greater than 1 inch, extremely difficult. Clearly, if the Class I hood is to be converted to a Class II unit it is desirable either to permit more air to enter this closed cabinet, or to throttle the blower to decrease internal negative pressure. It should be noted that throttling a fan at flow rates approaching zero is ineffective. Allowing some air to enter the cabinet is feasible, but only in limited quantities because of the problem of drafts and the requirement for filtration of incoming air.

When the main air filter is clean, operation in an open condition will induce a high, or even excessive, flow rate through the hood and will increase power consumption by the blower. Some throttling may be required to provide the desired inlet air velocity and to prevent overloading of the blower motor. This may be done automatically, using various sensing devices and controllers, or may be accomplished manu-

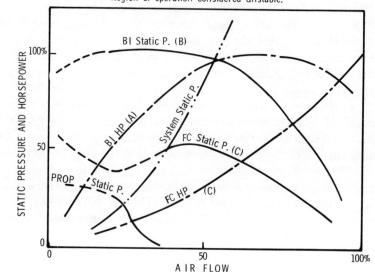

FIGURE 10-4. Approximate static pressure and horsepower characteristics of three types of fans.

ally by periodically adjusting dampers or motor speed. To minimize damper adjustment and reduce problems of over-pressurization in a closed hood, the fan should have flat pressure versus air flow and non-overloading characteristics, as is seen in curve (A) of Figure 10-4 where typical centrifugal-fan pressure, flow, and horsepower curves are illustrated. The backward inclined, BI, centrifugal fan shown in curve (B) most closely approaches the desired static pressure and power demand performance. The forward curved fan, FC, shown in curve (C) is somewhat less effective, and tends to load heavily. Backward inclined fan efficiency tends to be somewhat lower than the FC fan, and the noise level somewhat higher. These deficiencies are more than compensated for by the desirable characteristics of the BI fan.

The influence of the cabinet ventilating system on air currents in the room cannot be neglected. In a small room (10 by 15 ft, for example) the ventilation rate of a properly ventilated, 6-ft Class I hood used with 100% throwaway air [400 to 600 ft³/min (cfm)] may exceed the ventilation rate of the laboratory; in that case, the hood may be used as the total exhaust (or supply) from the room, and be run continuously;

it can operate from a separate source of air, or the hood may be utilized with a large volume of recirculated air. On a continuous basis, making the hood part of a ventilating system can be extremely costly because high efficiency filters, which in this type of installation do not usually have prefilters, will rapidly become loaded with dirt from both the room and outside air. Interconnection of electrical controls of the hood with damper motors in the room exhaust system, to provide a reduction in room exhaust when the hood is in operation, may provide a less costly solution.

Unconditioned outside air has sometimes been used as "make up" for chemical fume hoods, but this is an expensive measure; mechanical complexities outweigh benefit gained. For many applications, serious consideration should be given to recirculation of the major volume of air through these hoods. Recirculated air can pass through particulate filters, carbon filters, or through any other needed air cleaning process. The resultant ventilating air stream will be processed to a degree not usually possible with room ventilating air. A small amount of air can be drawn from or into the cabinet to create the necessary pressure-balance. This mode of operation is particularly applicable when the temperature and humidity control inside the cabinet may be more exacting than for the laboratory.

Cabinets described as "vertical laminar flow" hoods have been used for contamination control in precision small-parts industries and in virology-cell culture laboratories (Coriell et al., 1967) with excellent results. These cabinets use filters over the entire projected vertical area and circulate air from the top, through the filter, the work space, a perforated work surface, and thence to a fan for recirculation. Potential utility of these cabinets in microbiological laboratories is not presently known. There is sufficient data (Coriell et al., 1967; Favero and Berquist, 1968) to indicate that these cabinets may be desirable substitutes for the Class I and II hoods described previously, but they should not be considered acceptable substitutes for Class III units. Any open-front cabinet will have leakage in or out through the work-access space. The laminar flow cabinet will probably have higher leakage rates than the conventional cabinet, but the leakage products will be expelled before they can cause substantial harm to the operator (or the product within the hood). For a majority of applications this is a sufficient, but not "absolute," degree of protection (i.e., penetration less than $1/10^4$). Laminar flow units do not provide the security of a physical barrier that is effective regardless of the availability of electric power or of injudicious actions of the operator.

The particular location of a cabinet may affect the performance of the workers, the operation of the cabinet itself, or the operation of the ventilating system in the laboratory space. The cabinet or hood should be located out of the main stream of traffic, preferrably in a dead-end of the laboratory to prohibit auxilliary activities near the hood front. Such activities, as stated previously, create back drafts that tend to draw material out of the cabinet; they may draw aerosols into the cabinet and onto sensitive operations (e.g., plating procedures). The latter may be particularly true if the cabinet is installed in an animal room; hair, debris, and dander from the animals will rapidly foul air filters, and load all of the sensitive test apparatus with a film of dirt.

Materials of Construction

The design and materials of the cabinet must consider types of operation to be housed and requirements for decontamination or climate controls. Flexible plastic cabinets have been used to provide simple, effective containers for aerosol apparatus and for test animals, as shown in Figure 10-5. Such apparatus is particularly useful for short-term experiments

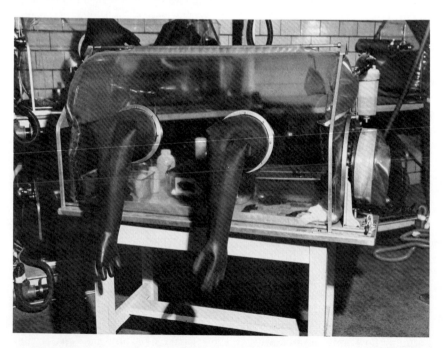

FIGURE 10-5. Flexible plastic isolator. Sterile lock and air outlet trap are on the right side of the isolator. The air filter is on the left and is not visible in the photograph. (Trexler. 1959. *Ann. N.Y. Acad. Sci.*, **78**: 31. 1959, Copyright The New York Academy of Sciences. Reprinted with permission of the New York Academy of Sciences, New York.)

in which flexibility of operation is a primary requirement. However, these polyvinyl barriers have the disadvantage of being easily punctured and torn; if the experiment has a large investment in time and animals, or if there is an extremely hazardous condition, it is mandatory to use more durable construction.

Although many of the early "dry" boxes, used in studies involving radioactive materials, were fabricated of wood, this practice has largely been discontinued; most safety cabinets are now made of metals or reinforced plastics. At present, there is no major difference in the cost of the last two materials, and the more advanced techniques of metal fabrication, coupled with their fire-resistant characteristics, appear to make metal (e.g., stainless steel) a more desirable material. The requirement for the enclosure to withstand various chemicals used in the experiment process, or used for decontamination, is of great importance.

When radioactive or chemical aerosols are studied, it may be necessary to add wetting agents or cleaning compounds to the disinfectants. It may be desirable to dispose of the entire hood or hood interior (strippable coatings) in the event of major spill of radioactive substances; in any event, it must be recognized that there will be the problem of day-to-day cleanup, occasional decontamination of the hood enclosure, and the possible requirement for cleanup after a major spill where heavy contamination extends throughout the entire enclosure. The requirement for a minimal-volume enclosure, and one which will either withstand the cleaning procedures or be economically disposable, is apparent. Although some of the urethane and polyester types of plastics are extremely resistant to chemical treatment and can be fabricated into smooth monolithic enclosures, metal (particularly stainless steel) has considerable advantage in mechanical strength, durability, and in the long run is the most economical.

For experiments in which control of temperature or humidity is of importance, insulated metal or wood cabinets are frequently used. To be most effective, the controlled volume should encompass aerosol originating points, sampling areas, and all of the aerosol containment apparatus, including drums, chambers, and other devices.

Although not properly considered a hood installation, the room in which an aerosol chamber is located may serve as a secondary barrier, and may be associated with Class III hoods, with recirculating airflow throughout the system.

In many instances, it is particularly important that all phases of the aerobiology experiment be under constant temperature conditions, especially in studies with humidity-sensitive microorganisms. For example, at a dry bulb temperature of 70°F, a 1°-F change in wet bulb temperature produces a change of about 2% RH; at 40°F the 1°-F change causes the RH to change about 8%. The use of a room in which all of the aerosol equipment is housed may offer the most desirable way to effect the needed controls.

Figure 10-6, a and b, shows a Class III hood installation connected to a sealed chamber in which a series of rotating drums is installed. The hoods house aerosol generation equipment, sampling devices, and small

FIGURE 10-6a. Class III hood system attached to controlled temperature chamber Figure 10-6b in which 8 aerosol drums are housed. Hoods are at chamber temperature and house aerosol generation and sampling equipment.

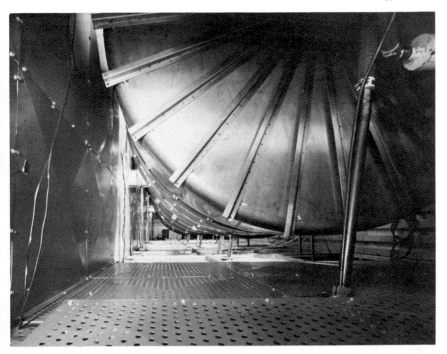

FIGURE 10-6b. View of rotating drums in temperature controlled chamber.

animal exposure equipment. This system provides a very well-controlled isothermal condition for the entire aerobiology experiment with full safety for the operators in the event pathogens are employed. Such a system is extremely costly and requires extensive engineering and maintenance for its successful operation. It is recommended that careful consideration be given to the need to conduct work under such closely controlled environmental conditions.

In summary, the safety hood or cabinet, as a secondary barrier, is an important tool in aerobiology and microbiology laboratories. It can provide effective contamination control at minimal cost, but judicious selection of the type of cabinet to be used, and of the ancillary equipment, is essential. Although design of this equipment is not usually the task of the user, a knowledge of operational parameters is prerequisite to acquisition of a useful and safe apparatus.

Animal-Holding Facilities

The maintenance of animals exposed to infectious agents in aerobiology experiments requires containment facilities paralleling those of the animal-exposure equipment. If the infectious agent must be used in a Class I, II, or III system, then animals so exposed should be afforded the same degree of containment. Jemski and Phillips (1965) have discussed many of the techniques of aerosol challenge of animals and their maintenance after challenge.

The requirement for individual animal isolation and caging usually rests more upon the possibility of cross infection by the microbe being tested than either upon the requirement for protection of the animal from other airborne disease germs, or upon the desire to protect the operator from the exposed animal (Phillips *et al.*, 1956a, b; Kirchheimer *et al.*, 1961; Kruse, 1962). It is suggested that any excess zeal in contamination control should be vented in this area.

The above should not be taken to indicate that there is a lasting problem of widespread contamination from animals exposed to respiratory infection. With few exceptions (e.g., TB, anthrax and coccidioidomycosis), most microbes die rapidly on skin and fur. The sources of contamination are usually the urine and feces of the animal, or exudates from the respiratory tract (Wedum, 1964) that usually provide low levels of contamination over periods coincident with the interval required to assess aerogenic infection.

Effective control is essential if any valid quantitative measure of infectivity of airborne microbes is expected. Figure 10-7 shows simple individual, ventilated cages for holding animals that have been exposed to infectious aerosols, and Figure 10-8 illustrates an open-front, inward air-flow containment system. This rack provides ventilating air directed inward over the top of the cage at a velocity of approximately 150 lfpm and thence to a HEPA filter, a blower, and an exhaust to the outdoor atmosphere. After intranasal challenge in a Class I hood (*Coccidioides immitis* in this example) small groups (8) of animals are held for 72 hr in open-top "shoe-box" cages and are then transferred to conventional animal cage racks. In 10 years of operation, no inter-cage infections, or infection of personnel, have been observed.

Horsfall and Bauer (1940) described an early Class III system where animals exposed to arboviruses were maintained in open cages inside boxed enclosures. These enclosures had input air filters to protect them from contamination by surrounding atmosphere, and high efficiency filtration on the exhaust air ducts leading to a manifold. To service such boxes, personnel must wear protective ventilated plastic suits and service only a single box of cages at one time. Smaller, individually ventilated animal cages may be used if they are connected to an exhaust manifold to circulate air, and if a safety hood for care and cleaning is provided. Small, nonventilated, filter top "shoe-box" cages have been used very

FIGURE 10-7. Individual, plastic, ventilated cages. (Photo courtesy Reginald O. Cook of the National Environmental Health Sciences Center, Research Triangle Park, N.C.)

FIGURE 10-8. Open-front, inward air-flow, animal-cage rack.

successfully for control of airborne viruses causing epidemic infantile diarrhea in mouse breeding colonies (Kraft, 1958). These may be used in conjunction with a Class I hood for feed, water, and transfer of small rodents.

Although there have been no reports of their successful use in studies with human pathogens, these shoe-box cages should be suitable for many types of studies. Germfree and SPF animals have been reared successfully in elaborate stainless steel enclosures (Reyniers, 1959) and in simple, plastic film isolators (Trexler, 1959) ventilated at positive pressure through HEPA filters. The latter technique can provide inexpensive enclosures for the shoe-box cage described above and, in fact, can be adapted to a wide variety of do-it-yourself containers or enclosures if due consideration is given the requirement for reliability and durability. Flexible film enclosures and filter-top cages are particularly suitable for studies involving radioactive materials. The cost is sufficiently low that cages and/or entire enclosures may be disposed of economically.

Many infectious agents may be controlled by application of germicidal ultraviolet (UV) radiation streaming directly over the top of open-top animal cages. Phillips et al. (1957) demonstrated the effectiveness of 250 microwatts per square centimeter (μw/cm^2) of 2537Å irradiation. Guinea pigs are apparently unaffected by such high UV levels, but the eyes of other animals, and of the operators, must be shielded carefully.

In summary, an effective secondary barrier system for animals and ancillary equipment is justified on the basis of operator protection and the need for prevention of cross-infection in the animal population. The general procedures for use of barrier techniques in the laboratory are directly applicable. Animal care should be given attention equal to that afforded the aerosol work.

Tertiary Barrier Systems

A tertiary barrier comprises a facility (e.g., a room or building) in which experimentation is done. The investigator usually has neither responsibility nor authority in matters of facility design; this facility is an important factor, nonetheless, for it must provide a usable working environment for aerobiology experimentation, safe containment or disposal of aerosols accidentally released from the test procedures, and should protect experimental devices from contamination from other sources within the building.

In any aerobiology experiment, whether with pathogens, chemicals, radioactive aerosols, allergens, or other material, possible sequelae to a major accident must be considered. If aerosols are accidentally dispersed throughout the room or building, one must be prepared to cope with any or all of the problems of explosion and fire hazard, massive contamination of rooms, equipment and supplies, exposure of many people (frequently strangers), and a host of lesser problems, many of which can hardly be imagined until after such an incident has occurred.

Since the tertiary barrier may be considered the final line of protection, questions to be considered in barrier selection or design include:

1. What would be the effect of a major accident involving dispersion of aerosols throughout the space?
2. What climate and air quality are needed for the aerobiology apparatus?
3. Will large quantities of animals be housed under controlled conditions?
4. What is the size of primary and secondary barrier systems involved?
5. Are additional, noncompatible activities in progress in the area?
6. What is the effect of a hood or equipment installations on the room or ventilating system?
7. What is the expected traffic flow through the room? Are there other workers in the same space? Is there public access to any part of the room?

Access to the tertiary enclosure should be limited to essential workers, and there must be a mechanism for rapid, safe removal of infectious aerosol.

Isolation of the space is frequently accomplished by allowing access only via an air lock or an adjoining room. Two sequential doorways between any access to a "public" space and the laboratory can be an effective aerosol barrier.

Ventilating air can be used to clear this space, but it should be noted that personnel traversing a lock (or even opening doors) will carry aerosols with them in spite of a turbulent counter flow of air at speeds of 100 lfpm or more. Ultraviolet irradiation is often used in such locks and can be very effective. We will amplify this statement later.

For the most part, laboratories in the United States have forced-air ventilation and heating; those constructed within recent years are usually air conditioned. Although such systems were not meant to control contamination from or into laboratory spaces, air conditioning units have become an important—even an essential—part of contamination-control methodology. In the discussion that follows, we shall assume that the ventilating system can provide:

1. "Once through" ventilating air with clean air supply and disposition of exhaust air clear of occupied spaces after appropriate decontamination treatment, usually through filter banks.

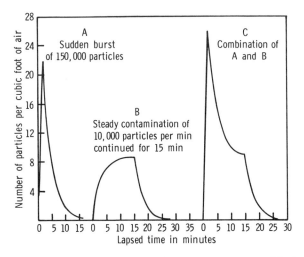

FIGURE 10-9. Dilution ventilation. Room 18 × 20 × 10 feet. Mechanical ventilation, 800 cfm. Bacterial particles 13 μ, settling velocity 1.0 fpm. (Kelthley. 1963. *Proc. Natl. Conf. Institutionally Acquired Infect.*, p. 40. Reprinted with permission of Department of Health, Education and Welfare, Public Health Service, National Communicable Disease Center, Atlanta, Ga.)

2. Pressure-zone or flow-direction control to effect containment or isolation of selected areas, and to eliminate or minimize flow into other spaces in the building.

3. Reasonable control of supply air temperature, humidity, and purity.

We shall also assume that discussion of requirements of ventilation systems is, in general, pertinent to installations in large buildings or structures connected to larger buildings.

Kethley (1963) has computed and demonstrated effects of ventilation rates, stirred settling, and biological decay on the clearance of aerosols from rooms. His data show that whereas relatively high turnover rates are essential for containment within a space, even when relatively large particles are airborne, no ventilation rate that mixes aerosol with ventilating air substantially reduces the dose occupants of the room are exposed to during the initial moments of a burst of aerosol discharge.

Figure 10-9 is a graph of small particle (13 μ) aerosol clearance rates by ventilation, with complete disposal (100% "throwaway") at 13.3 complete air changes per hour. It is apparent that commonly used room ventilating systems, providing from 0 to 15 air changes per hour with good air mixing, reduce the persistence of large "bursts," but are virtually ineffective in reducing a continuous, low-level input to the space. Extremely high ventilation rates would be required to clear the room quickly enough to protect an operator or to reduce an appreciable input to a nonhazardous level. Walter (1966) has suggested that the threshold for "sanitary" air (i.e., air required to minimize cross-infection in a hospital ward) is approximately 1 change per minute, or its proportional equivalent in attentuation or destruction of bacteria. In fact, an increased ventilation rate may actually increase contamination by airborne microbes from outside sources if this air is not cleaned. Additionally, the heating and cooling burden required for tempering outside air would be far in excess of that required to compensate for thermal loads in the occupied space.

"Laminar-flow" room ventilation (Whitfield et al., 1962) is probably the only means by which particles can be removed before they are thoroughly mixed with ventilating air. "Laminar-flow" ventilation, described previously in relation to safety cabinets, is achieved by recirculating room air through HEPA filter banks occupying an entire ceiling or wall of the space. If there is a minimum of intervening objects, the air flows in a "piston" effect to the opposite wall or floor at approximately 100 lfpm, where it is collected in exhaust ducts that occupy a substantial portion of the area. Contamination from sources within a "clean" space is quickly removed and caught by the filter system before it can mix with room air. This operation is in direct contrast to conventional systems that inject air at higher velocities (ca 300 to 600 lfpm) and provide a near-perfect mixing so that contaminants are removed at rates approaching exponential dilution rates. Although "laminar flow" systems may not be laminar in a true aerodynamic sense, the analogy to laminar versus turbulent flow of air is a practical one if the overall degree of mixing of the air is considered equivalent to that encountered in streamlined laminar versus turbulent flow of fluids in pipes.

"Clean room" ventilation, using air downflow from ceiling to floor at rates up to 600 changes/hr, has been effective in reducing biological particulate contamination. This technique is gaining wide acceptance in industrial areas (Federal Standard 209A) and has been tested in hospital surgeries (McDade et al., 1968). With minor rearrangement of components, it can be equally effective in containing hazardous materials. Installation and application of these systems are discussed at length in NASA Contamination Control Handbook, SANDIA Corporation (Sivinski, 1968).

The method most frequently used to decontaminate ventilating air is high-efficiency filtration. If a once-through ventilation system is employed, it is sufficient to provide filters that are 80 to 95% efficient as tested by dioctylphthalate aerosols (DOP test). These filters can clean the air exhausted from a highly contaminated area or clean the supply air to a space requiring an extremely dust-free environment. Recirculation ventilation, as used in clean rooms, permits the use of HEPA filters having efficiencies up to 99.99+%; the burden of outside air dirt is reduced and the life of the filter is extended. This type of ventilation is not well suited to animal rooms, which have associated odor problems. For rooms of normal animals, once-through ventilation, with rough filters (20 to 30% National Bureau of Standards—NBS—efficiency) at each exhaust air duct, is suggested. Rooms holding infected animals should, additionally, be equipped with 80 to 95% (DOP test) filters specified for the laboratory.

In this application, when a continuous low-level contamination output is expected, filter tops for the cages, respiratory protection for the workers or high intensity UV screening over the animal cages may be used (Jemski and Phillips, 1965).

For microbial aerosols, "effective" air exchange rates can be increased by the application of germicidal, UV irradiation. Intensities used in such applications have varied from 0.1 μw/cm^2 (cited by Schechmeister, 1957) to 20 μw/cm^2 (Wolochow et al., 1957). Ultraviolet irradiation can be helpful, but most published data on germicidal effectiveness were collected under special conditions, and results have not been uniformly successful in practice. An effective method is to irradiate air with lamps installed within mechanical ventilating systems where they will be (a) under relatively constant RH and temperature conditions, (b) in a clean air stream, and (c) shielded from direct observation. Wedum et al. (1956) and Phillips et al. (1957) described several other useful applications of UV, including airlock door irradiation, animal cage screening, and in-hood installation; Buttolph and Haynes (1950) and Buttolph et al. (1950) listed application data.

In view of the many technical problems associated with the use of UV, and the finding by Dimmick (1960) that even "naked" cells may be only temporarily "injured," UV is not recommended for routine room-air decontamination application. Ultraviolet energy is probably most useful as a "visible surface" decontaminant. For example, surfaces inside hood systems can be exposed to high intensity UV overnight when the hood is not in use. An intensity level of 20 to 50 μw/cm^2 is feasible, provided the open front is occluded to prevent stray radiation from reaching the workers. Stainless steel surfaces reflect UV fairly well, and there is hazard from reflected radiation. Low ozone, cold cathode lamps offer longer life (albeit at lower output levels) than hot cathode lamps, and produce a minimal deleterious effect on elastomers, plastics, and other ozone-sensitive materials.

Just as the climate or quality of the air in aerosol chambers should be controlled, so should air surrounding an experimental unit be controlled. If animals are used, effective control of air environment to prevent thermal shock or overheating from crowding is an important part of the experiment. Properly applied, conventional air-conditioning controls will provide temperature regulation within ±2°F at a thermostat, and relative humidity control at ±7% RH at a humidistat. Investment in somewhat more sophisticated controls can improve these factors, but the variation can be most directly improved by the addition of higher air recirculation rates within a limited volume area.

Establishment of "pressure zones," or room-to-room gradients within work spaces, is frequently suggested as a means of contamination control, though this type of regulation may not be feasible with conventional ventilating systems. Pressure differences varying from 0.005 to 0.1 inch wsp have been achieved, but even these small differences will rapidly be lost if a door connecting the two spaces is left open. In that event, one is dependent on the flow of air in the desired direction as a control vehicle. Differential pressure, by itself, offers little safeguard against contamination—only air flow that provides both directional control and rapid removal of contaminants is effective. Thus, it is important that a ventilating system be capable of withdrawing (or supplying) large volumes of air through rooms to make open-door flow velocity sufficient to overcome thermal and convective air currents; velocities of 40 to 50 lfpm are suggested. Differential pressure within a building can be effective in minimizing air infiltration from outside wind pressures, and may be used for that purpose.

The interior of spaces used for aerobiology tests should be kept as simple as possible. No excess equipment, storage materials, or animal-holding facilities should be maintained in the space unless they are directly concerned with the experiments on hand. Room finishes should, as in the case of the ventilation system, be chosen with the "worst case" condition in mind. Seamless, monolithic surfaces that are easy to clean and decontaminate are most helpful. The use of vinyl and polyurethane floor finishes with polyurethane and epoxy wall and ceiling finishes is strongly suggested. Where possible, lighting fixtures and other accessories should be watertight and amenable to easy cleaning. These standards would apply to any aerosol work. If work where hazardous microbial pathogens is considered, standards that prevail in a hospital surgery are quite applicable.

Not only must experimental space be considered as a part of a control problem, but one must also consider the exterior of the building, wind currents near buildings, and the consequence of a release of large quantities of aerosol into the environment. Overall control of these factors is not usually within the purview of the facilities engineer, except with respect to input/output filtration.

It may be desirable to limit work during some wind conditions when accidental discharge would be directed into sensitive areas. Simple air-flow tests with smoke clouds or soap bubbles are frequently sufficient to define problem areas. Clarke (1965) has described several types of problems created by air circulation around buildings. He emphasizes that it is not possible to design an installation on the basis of "prevailing" wind directions because the wind does not, in fact, hold to any single direction, and the complex nature of eddy currents around the buildings precludes prediction of exact air-flow direction. Halitsky (1962, 1963; cited by Clarke, 1965) reported studies of air diffusion around buildings indicating that behavior of single small particles may not be directly applicable to re-entry problems; *in situ* testing is the only means by which the hazard of re-entry can be evaluated.

In summary, a tertiary or third-level barrier is an essential element in the event of a major accident, and can provide desirable environmental control for experimentation. Although it is not always within the control of the experimenter, operation of the ventilating system in the experimental space is a critical element. Ventilation can be provided with high circulation rates, high efficiency filtration, and complete air conditioning—particularly in animal spaces where contamination might interfere with an experiment. Standards for such ventilation and air conditioning are not clearly defined, but the application of conventional methods, including increased flow rates, is adequate in most cases. Basic requirements for ventilation and air conditioning environmental control for animal colonies are covered in air conditioning literature (McPhee, 1966).

AIR PURIFICATION SYSTEMS

We should now consider sources of contamination that include:

1. Microorganisms from the surrounding environment.
2. Dusts and pollens from outdoors.
3. Air pollutants in urban areas.
4. The test apparatus itself.
5. Hair, feces, and dander from animals used in the experiment.
6. Decomposition of the air used in the experiment through thermal and UV radiation.
7. Electrical charge on submicronic particles, e.g., large or medium size air ions.

With respect to purity of air supplied to aerobiology apparatus, it is reasonable to conclude that "room" or "normal" air, commonly accepted as a "standard" element, is a will-of-the-wisp definition, valid only to the degree that one does *not* measure contaminants. Restated, the definition of the incoming air purity depends on instrumentation used to evaluate the air, which is in turn dependent on the interests and awareness of the investigator. If one is concerned with effects of ozone in the air, one must first acquire a means of measuring this component, establish some acceptable limits, and then take steps to effect control. If one lacks instrumentation, one may apply some purification process and establish this air as a base-line "standard." Constancy of the quality of such air on a day-to-day basis may be less than ideal.

Sometimes, presence of trace materials may be only of minor consequence. For example, if an apparatus were used to deposit inert particles on a clean test surface, the existence of trace quantities of ozone may be inconsequential. In contrast, if an animal or human host is involved, the physiological implications of trace contaminants may be significant. Workers involved in air pollution studies have a gamut of parameters with which they must be concerned. Their problems may be even further complicated by the presence of evanescent, trace pollutants generated by temperature changes or by solar radiation on gaseous contaminants. Table 10-7 shows the concentration range of the majority of constituents in "normal" atmospheres. It is apparent that there can be wide divergences in air composition dependent on locale. Even these data are conservative because they do not include photo-chemical compounds.

The problem of defining the purity of air discharged from aerosol processes or test subjects is somewhat less severe than the above, but no less important. When working with "pathogens," standards previously applied to leakage from aerosol systems can be applied to discharge; i.e., concentration of materials released should be at some defined acceptable level. When highly infectious human pathogens are used, it is desirable to ensure that contamination is one or more orders of magnitude below the usually acceptable level. In many cases, the same methods used to purify supply air are applicable to discharge air. As an example, the electric air incinerator described by Elsworth *et al.* (1961) (designed to prevent contamination of a liquid culture by the sparging air by providing a 1×10^{12} reduction in count of viable airborne spore formers) would be equally applicable to decontamination of effluent air.

In general, process air treatments may be classified as follows:

A. Particulate removal: positive filters, usually fibrous media:
   1. Passive filters: these are usable for collection of virtually any

Table 10-7   The concentration of materials in the air.

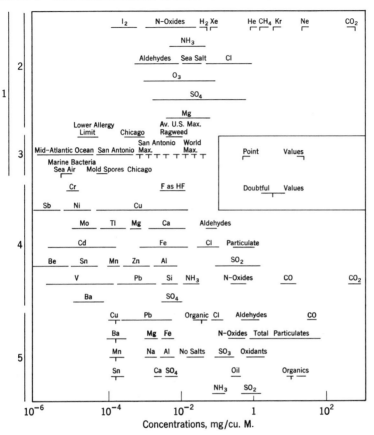

1 = normal air; 2 = chemical matter; 3 = biological matter; 4 = U.S. cities average; 5 = Los Angeles smog, typical and maximum values.
(Courtesy Mine Safety Appliances Company, Catalog Section 10, p. 57, Pittsburgh, Pa.)

particulates including infectious, radioactive, allergenic, and inert particles.

   2. Commercial air cleaners: included are cyclone separators, electrostatic precipitators, wet scrubbers, and a variety of other devices (Perry *et al.*, 1963).

B. Aerosol *destruction* (may not include removal):

   1. Thermal: by application of direct heating.

   2. Radiation: including UV, infrared, alpha, beta, gamma.

   3. Chemical action: by vapor phase decontaminants; beta-propiolactone, formaldehyde, etc.

C. Gas or vapor removal:
  1. Adsorption by broad spectrum adsorbers or specific adsorption chemicals.
  2. Absorption with scrubbers.
D. Ion removal:
  1. Charged plates (electrostatic precipitator).
  2. Injection of oppositely charged ions (neutralization).

Many of these techniques are commonly used in industrial processes. Equipment and operating costs are described in engineering handbooks (Perry *et al.*, 1963) and in other generally available publications (e.g., American Standard Fundamentals Governing the Design and Operation of Local Exhaust Systems, 1960; Whitby *et al.*, 1961; McPhee, 1966; Danielson, 1967). Table 10-8 summarizes information about air-cleaning processes and suggests some applications and limitations. The equipment listed includes much that is considered industrial process machinery, and will not be afforded further discussion. We should, however examine some techniques of particular value in aerobiology and microbiology laboratories.

Particulate Removal

Relatively few of the techniques available for removal of particles from air streams are applicable to most laboratory requirements. Passive air filtration is the most widely used, and in recent years the development of high efficiency, modular sized, HEPA air filters has made possible the use of nearly sterile "clean" benches and rooms. Such modules form effective filter units for small, portable, safety cabinets.

High efficiency. "packaged," air filters with collection efficiencies ranging from 0 to 100% for particles in sizes as small as 50 mμ are available. Filter media consist of glass fibers, asbestos, cotton, synthetic plastics, ceramics, metals, or mixtures of these. Table 10-9 describes some filter types currently available. The efficiencies cited are those described in the footnote to the table, and are for clean filters tested at rated flow capacity.

Within this group, the larger sized HEPA or "absolute" air filter is the type most frequently used in clean benches, clean rooms, and room exhaust systems. This application is discussed in considerable detail in *NASA Contamination Control Handbook* (Sivinski, 1968). The filter medium is accordion-pleated and cemented into a metal, plywood, or composition-board frame. The filter medium and sealing cement composition are selected to meet temperature requirements of the user. Flow velocity may be up to 300 lfpm at the face of the assembly.

Table 10-8 Characteristics of air and gas cleaning devices.

Name of device		Description of device
General class	Specific type	(for each specific type or variation thereof)
Odor Adsorbers	Shallow Bed	Activated charcoal beds in cells or cartridges, molecular sieve.
Air Washers	Spray Chamber	One or two coarse spray banks followed by bent plate eliminators.
	Wet Cell	Wetted glass or synthetic fiber cells followed by bent plate eliminators.
Electro. Precip., Low Voltage	Two Stage, Plate	Ionizing (+) wires followed by collecting (−) plates.
	Two Stage, Filter	Ionizing (+) wires followed by filter (−) cells.
Air Filters, Viscous Coated	Throwaway	Deep bed of coarse glass, vegetable or synthetic fibers in cells.
	Washable	Deep bed of metal wires, screens or ribbons in cells.
Air Filters, Dry Fiber	5–10 micron	Porous mat of 5–10 micron glass or synthetic fibers pleated into cells.
	2–5 micron	Porous mat of 2–5 micron glass or synthetic fibers pleated into cells.
Absolute Filters	Paper	Porous paper of <1-micron glass, ceramic or other fibers pleated into cells.
Industrial Filters	Cloth Bag	Bags made of natural or synthetic fiber fabrics.
Electro. Precip., High Voltage	Single Stage, Plate	Ionizing (−) wires between parallel collecting (+) plates.
Dry Inertial Collectors	Settling Chamber	Straight horizontal chamber—some with shelves.
	Cyclone	Chamber with provisions for spiral flow.
	Impingement	Alternate stages of nozzles and baffles.
Scrubbers	Cyclone	Cyclone collector with coarse radial sprays.
	Impingement	Impingement collector with wetted baffles.
	Fog	Cyclone collector with fine tangential sprays.
	Multi-Dynamic	Power driven normal and reverse flow fan stages with coarse sprays.
	Venturi	Venturi with coarse sprays at throat.
	Submerged Nozzle	Nozzle partially submerged in water.
Incinerators	Direct	Combustion chamber with supplemental fuel firing.
	Catalytic	Combustion chamber with catalyst plus supplemental fuel.
Gas Absorbers	Spray Tower	Vertical-up airflow chamber with downward sprays.
	Packed Column	Tower with counter-currently wetted Rashig rings, Berl saddles, etc.
	Fiber Cell	One or more stages of co-currently wetted fiber cells.
Gas Adsorbers	Deep Bed	Activated charcoal beds in regenerative-recovery equipment, molecular sieve, activated alumina, silica gel.

Table 10-8 (Continued)

Name of device — General class	Optimal size particle (microns)	Limits of gas temperature (°F)	Usual face velocity (fpm)	Usual face velocity (through)	Usual air resist. (WG)	Usual efficiency (% by Wt)
Odor Adsorbers (1)	(Molecular)	0–100	50–120	bed	<0.3	<95
Air Washers (1,2)	>20	40–700	300–500	chamber	<0.4	<25
Electro. Precip., Low Voltage (2)	>5	40–700	200–350	cells	<0.7	>25
	<1	0–250	275–500	plates	<0.3	>90
Air Filters, Viscous Coated (2)	<1	0–180	200–300	cells	<0.2	>50
	<5	0–250	300–500	cells	<0.1	<25
Air Filters, Dry Fiber (2)	<5	0–180	300–500	cells	<0.1	<25
	<3	0–180	5–25	mat	<0.3	>50
	<0.5	0–180	5–25	mat	<0.5	<95
Absolute Filters (5)	<1	0–1,800	4–6	paper	<1	>99.95
Industrial Filters (3)	>0.3	0–180ᵃ	1–30	fabric	<1	<99
Electro. Precip., High Voltage (3)	<2	0–700	180–600	plates	>4	<95
Dry Inertial Collectors (3)	>50	0–700	300–600	chamber	<1	<50
Scrubbers (3,4)	>10	0–700	2,000–4,000	inlet	<0.1	<80
	>10	0–700	3,000–6,000	nozzles	<2	<80
	>10	40–700	2,000–4,000	inlet	<4	<80
	<5	40–700	3,000–6,000	nozzles	>2	<80
	<2	40–700	3,000–4,000	inlet	>2	<80
	<1	40–700	2,000–3,000	inlet	>2 (up to 4″ developed)	<99
	<2	40–700	12,000–24,000	throat	>10	<99
	<2	40–700	2,000–4,000	nozzles	>2	<99
Incinerators (4,5)	any	2,000	2,000–4,000	chamber	<1	<90
	(Molecular)	1,000	500–1,000	chamber	<1	<95
Gas Absorbers (4)	(Molecular)	40–100	500–1,000	tower	<10	<95
	(Molecular)	40–100	300–800	bed	<4	<95
	(Molecular)	40–100	200–300	cells	<10	<95
Gas Adsorbers (4)	(Molecular)	0–100	20–120	bed	<10	<100

ᵃ 500°F if glass fiber.

Characteristics of air- and gas-cleaning devices:

Removable contaminants

(1) Malodors, gases

(4) Gases, vapors, malodors

Table 10-9 Short listing of high-efficiency air filters.

Rated air flow SCFM	Min. filter efficiency, %		Typ. outside dimensions	Mfr.[a]	Maximum initial resistance	Filter media	Separators and adhesives	Enclosure
	DOP[b]	NBS[b]						
1,000 or more	99.97+	…	24″ × 24″ × 12″	1,2	0.9″ H_2O at 1,000 cfm	glass, glass-asbestos, asbestos-cellulose, ceramic or ceramic-asbestos fiber(s) paper, with or without organic binder	Kraft paper, aluminum, plastic, glass or asbestos separators or without separators; rubber, plastic or silastic adhesives or sealed with packed filter fiber	plywood, steel or mineral board box or can; Nat. rubber, neoprene or glass-fiber gasket
	95	100		3,4	0.4″ H_2O at 1,000 cfm			
	65	90		6,7	0.6″ H_2O at 2,000 cfm			
	45	80			0.5″ H_2O at 2,000 cfm			
100 to 1,000	99.97+	…	10″ dia × 10″ lg to 24″ × 24″ × 6″	1,2	0.9″ H_2O at rated flow			
	95	100		3,4	0.4″ H_2O at rated flow			
	65	90		6,7	0.4″ H_2O at 500 cfm			
	45	80			0.3″ H_2O at 500 cfm			
10 to 100	99.999	…	6″ dia × 5″ lg to 8″ × 8″ × 6″	1,2	>1″ wsp at rated flow			
	99.97	…		3,4				
	95	100		6,7				
1 to 10	99.95+	…	3″ × 3″ × 3½″ to 4″ dia × 9″ lg	7				
	95 to 98+	…	3½″ dia × 7″ lg	7	0.1+ psi	as above plus carbon	pleated, metal ends, epoxy cement	light or heavy wall metal or plastic canister
	98+	…	ca. 1½″ × 1½″ × 2½″	2,6	0.1 to 0.2 psi	glass-fiber epoxy		
				7	ca. 1.0″ wsp	glass-fiber		
<1	99.95+	…	½″ to 10″ dia disk	5	13.5 psi at 1 cfm per 1 cm² area	glass-fiber with or without binder	none	machined metal housing, disk or can shaped
			max. 5″ dia disk or max 3″ dia × 15″ lg hollow candle	8	1 psi at 0.26 cfm per 1 ft² area and 1/8″ thickness	microporous porcelain		

[a] Manufacturers (incomplete list):
1. American Air Filter Company, Inc., Louisville, Ky.
2. Cambridge Filter Corp., Syracuse, N.Y.
3. Farr Company, Los Angeles, Calif.
4. Flanders Filters, Inc., Riverhead, N.Y.
5. Gelman Instrument Company, Ann Arbor, Mich.
6. Mine Safety Appliances Company, Pittsburgh, Pa.
7. Pall Corporation, Glen Cove, N.Y.
8. Selas Corporation of America, Philadelphia, Pa.

[b] See Table 10-10.

Three ventilating filter efficiency test procedures are currently recognized as "standard." These are:

1. *Weight Test:* This test compares the weights of a synthetic test dust collected before and after the filter. It is useful for low-efficiency filters.
2. *NBS* (National Bureau of Standards) *Discoloration Test:* This test compares the degree of discoloration produced by passage of a test dust through a sample filter with that caused by passage of "clean" air through similar media. Particle size range is not considered.
3. *DOP Test:* A controlled temperature-generated dioctylphthalate (DOP) smoke is introduced in an air stream and the concentration before and after the test filter is measured by light-scatter techniques. Particle size of this aerosol is predominantly in the 0.2- to 1.0-μ range.

The differing factors in the test methods do not permit direct comparison. Table 10-10 gives approximate relationships, showing the contrast in ratings.

HEPA filters in ventilating systems should be installed in fireproof housings and frames, and fire-retardant or fire resistant media and sealants should be used. If high humidity or water-saturated conditions are expected, then moisture-resistant glass-fiber material should be used. Discussion of the specific application with competent manufacturers'

Table 10-10 Comparison of test-method efficiencies

	Percent efficiency	
Weight	Discoloration	DOP
. . .	. . .	99.95+
. . .	90–95	80–85
99	80–85	50–60
75	8–12	2–5

. . . Test not usable.
(From Sivinski, H. D., (Ed.). 1968. *NASA Contamination Control Handbook.* SANDIA Laboratory Report No. SC-M-68-370, SANDIA Corp., Albuquerque, N.M. Reprinted by permission of the editor.)

representatives is essential for proper application of these high-efficiency filters. A prefilter should be used to reduce the burden of ordinary large-particle dirt. The prefilter should have a rating of not less than 35% by the NBS test and should be installed to permit easy removal.

HEPA filters are shipped from the manufacturers' plants after overall penetration tests have been performed. These tests employ temperature controlled DOP smoke and evaluation by light-scatter methods. Results of each test are inscribed on each filter unit. Two points should be borne in mind:

1. The filter assembly is a relatively fragile device; shipping damage to the medium occurs with disturbing frequency. Usually, this damage is at the junction of the pleats and the cemented edge and is revealed only by persistent inspection. Frequently, the rupture damage creates only minor change in overall filter efficiency, and very careful testing is required to locate the damage. Testing, other than visual inspection, requires that the filter be challenged with a DOP aerosol generated from a Collison atomizer or the like. The downstream side is scanned with a probe leading to a penetrometer, such as the forward-angle monitor described in Chapter 7.

   Readings at various points are compared with values obtained upstream from the filter; the ratio is percent penetration. If this ratio is 10% greater than the rated efficiency (i.e., a filter rated at +99.99 reads 99.90) or if a tenfold rise in the average background reading occurs at some point, then the filter can be considered to be "leaky." Small leaks occurring in accessible places may be sealed with a mastic (e.g., Dow Corning[6] #892 silicone sealant). Leaks in inaccessible places may be tolerated if the *overall* efficiency of the filter is acceptable and if the penetration of a few large particles is not detrimental.

2. The scanning probe test is more sensitive than the standard test made by the manufacturer, and it is entirely possible that the filter will show many leaks. Premium filters, which have been tested by scanning probe, are available at moderate increase in cost. If the application warrants use of a 99.97+% efficiency filter, it also warrants accurate testing as well as full consideration of its installed effectiveness.

The most frequent source of failure of an installed filter is in the ductwork frame, or mount, into which the filter is installed. Frequently

[6] Dow Corning Corporation, Midland, Mich.

this frame is not leaktight; the filter matting edge may be uneven, scored, or cracked in the process of installation, or the filter may not seat properly. Frames should be inspected very carefully and made as tight as possible before the filter is installed. If the surface and seals appear at all imperfect, or if the gasket supplied with the filter is defective, a nonhardening mastic material (3-M EC-1279)[7] may be used to fill voids and seal rough edges of the filter frame. If the temperature of filter application is too high ($>$ 65°C) to allow a mastic material to be used, then self-curing silicone rubber may be employed. The rather "permanent" nature of such an installation is warranted by both the high cost of testing and the expected long life of filters once they are successfully installed. In any case, firm, uniform pulldown of the filter into the frame is essential. *After* installation, each HEPA filter should be afforded a full testing procedure, using techniques described later in this chapter. With adequate prefiltration, the HEPA filter will operate without replacement in all but the dirtiest environments for at least two years. In dusty areas, prefilters must be changed frequently; sometimes high-efficiency filters (e.g., 50 to 60% efficiency, DOP test) must be employed as prefilters.

Some commercially available safety hoods have an exhaust filter assembly of fine glass-fiber material approximately $\frac{1}{2}$ to 1 inch thick.[8] This medium has penetration characteristics similar to the thin medium described above. It is furnished in sheet form to be assembled into a filter by the user, and has pressure loss (on an equal flow per unit area basis) almost triple that of the thin medium. It is most useful in making up filters for special applications, or for short-term use when commercial standard HEPA units are not available; it also can be used for pipeline filters and for packing and sealing HEPA filter housings.

Recent developments in commercial filter material fabrication have permitted fibrous filters to be made with efficiencies (for 0.3-μ particles) ranging up to 99.9999+% with moderate pressure losses. Use of filters with this efficiency is not usually warranted because they are expensive and difficult to test. It would be better to install two HEPA filters in tandem; the resultant sequential filtration will have combined efficiencies approaching 99.9999%. Since leakage measured by DOP test is usually the result of small holes in or around the absolute filter rather than the slippage of minute particles (less than 0.2 μ in diameter). the second filter adds directly to the capability of the first (Breslin, 1967).

[7] 3-M Company, St. Paul, Minnesota.
[8] American Air Filters Company, Inc. (50-FG), Louisville, Ky.; Owens-Corning Fiberglas Corporation (PF-105), Toledo, O.

Open-cell, foamed-plastic filter material, while not usually considered in the HEPA classification, is available with a range of particle-size, efficiency ratings.[9] In part, this material can be useful as a cutoff filter for screening out particles larger than selected sizes which can range from 1.0 to 30 μ, depending on the pore size selected; performance of polyurethane foam filters has been evaluated by Roesler (1966), and a broad screening of filter material performance has been reported by Whitby et al. (1961).

Filtration of air in pipelines, particularly those of aerosol process systems, can usually be done with higher pressure losses than in ventilating systems. Accordingly, packed long-glass fiber or merino wool has been used extensively in shop-made filter assemblies to provide penetration values as low as $1/10^9$ (Henderson, 1952). Glass fiber is readily available, may be heat sterilized, and may be packed to the depth necessary for the efficiency of collection desired. Thin media HEPA filters in canisters, as well as a wide variety of other filtration media with various particle collection efficiencies, are available that permit in-line installation with pipeline canisters of steel, glass, plastic, or stainless steel. These offer substantially less pressure loss than hand-packed filter beds. In selecting canister-type filters, care should be taken to insure that the medium is suitable for the expected moisture conditions. Binding material used in the medium and cements used in the assembly of a filter must be tested for nontoxiciy. If the extent of the pressure drop in the air line is not important, ceramic,[10] sintered metal,[11] or Millipore[12] filters may be employed. The first two types may be cleaned by flaming and ashing, or by treating with acids.

All high-efficiency particulate filters exhibit a nearly linear relationship between flow rate and pressure differential. In a piping system any of these filters, used with a manometer or differential pressure gauge, can provide a convenient flow-measuring system. If two filters are mounted in series, the second filter can measure flow without incurring a substantial dirt burden. Flow may be measured by using the original manufacturer's flow versus pressure drop curves, or filters can be calibrated with other flow measuring equipment.

With minor exceptions, most commercial air cleaners, such as cyclone separators, electrostatic precipitators, wet scrubbers, and other devices are more useful in fixed building ventilating systems or industrial processes than in laboratory studies in aerobiology.

[9] Scott Paper Co., Foam Division, Chester, Pa.
[10] Selas Corporation of America, Philadelphia, Pa.
[11] Pall Corporation, Glen Cove, N.Y.
[12] Millipore Filter Corporation, Bedford, Mass.

Microbial Particles

In some cases, particularly where the extreme hazard of concentrated pathogenic microbial or viral aerosols exists, it may be essential to insure rapid destruction of all effluent viable particles with a substantial degree of certainty (i.e., $P < 10^{-9}$) and to insure that the decontamination device will not be overloaded or penetrated by peak loads. Sequential filtration with selected HEPA filters, carefully tested *in situ,* can provide this capability, and use of such filters should be given first consideration.

If there is a requirement for periodic sterilization of the system as well as microbial burden exceeding the filter capabilities, direct heating of air ($> 250°C$) may be desirable. Electrical heaters or gas or oil burners usually are employed. Decker *et al.* (1954) have described a 75 cfm electric air incinerator for which they report efficiencies of "100%." Elsworth *et al.* (1955, 1961) have described another electric air heater, and they derived a rationale for design of such equipment. Chatigny *et al.* (1968) and Barbeito *et al.* (1968) have described large volume (up to 2,000 cfm) air incinerators. There is general agreement that the temperature/time (T/t) relationship for destruction of *Bacillus subtilis* var. *niger* (BG) spores is an adequate measure of the effectiveness of the air incinerator for microbial contaminants. Bourdillon *et al.* (1948) suggested a T/t relationship of 225°C for 0.4 second or longer. Decker *et al.* (1954) used T/t values ranging from 218°C with 24-second retention to 302°C with 3-second retention to achieve 99.999% reduction. They suggest use of 329°C for 3 seconds for exhaust from a Class I hood. Elsworth *et al.* (1955, 1961) suggested use of a linear Arrhenius extrapolation based on an oxidation reaction rate for protein of 12,000 cal/mole to predict reduction by a factor of 10^{13} for BG, reporting T/t values of 240°C for 6 seconds to 320°C for 2 seconds. Our own work (Chatigny *et al.,* 1968) suggests that the difficulties in making accurate time-temperature measurements, coupled with the variations in hardy spore resistance, make it essential to use a substantial margin of safety in applying these T/t relationships. We have had good success with a burner that provides 6+ seconds of retention at ca. 300°C for 8-log reduction. Our data indicate that one can safely design equipment using 3× the times shown by Elsworth *et al.* (1961) to compensate for poor thermal contact, inaccurate temperature measurement, and other factors.

Although the direct heating of microbe-laden air to temperatures below 250°C is not true "dry heat" sterilization, it is sufficiently similar at the elevated temperatures noted to permit use of published kill rates. If rapid on-stream decontamination of air and surfaces of aerosol vessels is not essential, the application of 60 to 95°C over extended periods of time may be sufficient to destroy most vegetative microorganisms. Higher temperatures (e.g., up to 125°C for 24 hours) may be required for destruction of hardy spore-formers.

This procedure can also be useful for decontaminating surfaces and equipment. The medium on or in which the microorganisms are lodged can cause wide variation in kill rates. Angelotti *et al.* (1968) have demonstrated D values (90% kill) for BG spores at 125°C as follows:

On stainless steel strips	8.3 min
On paper strips	102.0 min
Between mated steel washers	47.5 min
In Lucite rods	3.1 hours
In Epoxy rods	5.3 hours

With the exception of the paper strips, the Z_D value [90% change in D as $f(T)$] for dry heat ranges from 21 to 32°C, in comparison with the Z_D value of 9 to 11°C for wet heat in the temperature range 104 to 132°C. It is suggested that in the temperature range 75 to 150°C, the estimate of a 1-D value increase per 25°C be used for air and surface contamination and that 32°C be used for contamination occurring within mated parts.

Despite a decreased energy input, radiant (infrared) heating of suspended particles has not been employed, because it is difficult to transmit long wavelength radiant energy to submicron particles without heating the airstream.

Short wavelength (2537 Å) UV illumination, on the other hand, is used extensively, although it is somewhat out of favor for decontamination of room ventilating air because of wide variations in effectiveness. Its action is influenced by temperature, humidity, and varied sizes of microbe-bearing particles, and there is difficulty in monitoring and maintenance of the UV sources. Without regard to the beneficial surface decontamination effect of UV, it is probably most useful in supplementing high-efficiency air filters, and tends to be most effective against small-size microbial particles in clean, humidity-controlled atmospheres.

Atomic or nuclear radiation has not been used for decontamination of microbe-laden air and, with minor exceptions, its use is unfeasible at this time. The use of Gamma irradiation of solid materials for sterilization purposes, particularly packaged hospital supplies, is commercially feasible. The limited work in the area of microbial decontamination shows that radiation is effective in production of free radicals which may be involved in the death mechanism of the cells. It is anticipated that there will be exploratory development of air sterilization processes in the near future.

Chemicals and Vapors

Chemical processes for inactivating airborne microbial contaminants include vapor-phase materials, air scrubbers, and other vapor-particle contact devices. Many vapor-phase disinfectants have been suggested, but formaldehyde, ethylene oxide, and beta-propiolactone are three disinfectants currently used. Their primary application is in terminal decontamination of spaces or equipment.

In some instances, a vapor-phase decontaminant may become a contaminant. The vapor should be used with extreme care, since either the disinfectant itself, the products of its decay, or residual products may be toxic to the aerosol or to the operators. Gaseous decontaminants should not be employed in aerosol vessels in which microbial aerosols will be contained unless the vessel can be cleaned thoroughly. Discharge of these vapors into the open air after use must be done with full consideration of reentry problems discussed earlier.

The removal of gas or vapor contaminants from process airstreams is most frequently done by adsorption onto surfaces of solids, absorption in liquids or solids, or condensation by freezing or chilling below the boiling point. The latter, in the form of liquid nitrogen or dry ice-methyl cellosolve traps, is effective for purging small volumes of air, but not for the larger volumes.

Early research in adsorption was directed toward problems of defining adsorption of solids from solutions. In recent years, attention has been directed to the question of selective adsorption of gases. Fundamentals of adsorption processes have been described by Mantel (1945). The total amount of gas adsorbed by any material is related primarily to the total surface of the adsorbent. Other factors effective in adsorption are concentration of the gas, physical and chemical characteristics of the adsorbent, and pressure and temperature of the process.

In practice, adsorbents usually are installed in deep bed canisters, and adsorption proceeds in a "wave" front through the entire bed. For some combinations (water being adsorbed onto silica gel, for example), the heat of adsorption may be quite large, and the adsorption front may be followed by a significant temperature rise at the active front. Cooling the bed during adsorption can increase its capacity.

Adsorbents most frequently used include:

Activated Carbon

It has been shown that the material from which carbon is prepared has a demonstrable effect on the ability of the matrix to absorb various gases. Carbon prepared from logwood, for example, has approximately twice the capacity for adsorption as carbon from rosewood, whereas coconut shell carbon is about twice as efficient as logwood. Primary carbon is not nearly as efficient as "activated" carbon.

Activation is done by selective oxidation of various hydrocarbons upon the surface of the charcoal materials by heating the char to a specific temperature for a given length of time. Activated charcoals of a wide variety of materials, of which coconut shell is most popular, are available commercially. Activated carbon accelerates reactions to which various gases are subject. For particular adsorption characteristics, carbon may be treated with salts of chromium, copper, or other materials. Impregnation with these salts enhances this catalytic power, thus improving breakdown of some toxic gases that cannot be eliminated by adsorption alone (Hassler, 1963).

Industrial gas-mask adsorbents contain additional ingredients or treatments, giving effective protection from such gases as sulphur dioxide, hydrogen sulphide, and chlorine. A different treatment is required to provide improved capacity for ammonia, and still another is effective for mercury vapor. The use of such mask canisters can be very helpful in purifying small quantities of process air. The user is cautioned to select the mask canister or carbon fill with his specific requirements in mind.

For protection against toxic gases, activated carbon is supplied in granular form. With many vapors normally encountered, a good grade carbon will adsorb from 20 to 50% of its own weight before full service life is exceeded, at which time the carbon is replaced. Regeneration of the carbon is not recommended. Table 10-11 lists some typical adsorption factors for use of activated charcoal with no added catalysts.

Other substances shown to be strongly adsorbed on charcoal include acetic acid, benzene, ethyl alcohol, carbon tetrachloride, methyl alcohol, chloroform, acetone, carbon disulphide, diethyl ether, chlorinated hydrocarbons, nitrous oxide, hydrochloric acid, and acid aldehyde. Substances that are less strongly adsorbed, or for which the total adsorption capacity is relatively lower, include ammonia, carbon dioxide, carbon monoxide, formaldehyde, formic acid, and sulfur dioxide; at depressed temperatures adsorption may be improved.

Silica Gel

Silica gel is prepared by coagulating a colloidal solution of silica acid. It is not a true gel, but is a hard, highly porous, glassy form of silicone dioxide. Hydrogen sulphide and sulphur dioxide have been successfully sorbed on silica gel, although it is principally used to extract water vapor. It is an excellent air drying agent because it is a strong water

Table 10-11   Adsorption of gases at 0°C on charcoal

Gas	Boiling point °C	Critical temperature °C	Gas adsorbed cc/g charcoal	
			Wood, 10 cm pressure	Coconut, 76 cm pressure
Helium	−268	−267	...	2
Hydrogen	−253	−241	0.3	4
Nitrogen	−196	−149	1.5	15
Carbon monoxide	−190	−136	...	21
Oxygen	−183	−119	2	18
Argon	−186	−117	...	12
Carbon dioxide	− 78	+131	20	...
Ammonia	− 33.5	+130	50	...

(From Hassler, J. W., 1963. *Activated Carbon*, p. 22. Reprinted with permission of Chemical Publishing Company, Inc., New York.)

adsorbent, is cheap, may be regenerated readily by heating, and is widely available. Application data for silica gels and activated alumina are provided in the *Chemical Engineers' Handbook* (Perry *et al.*, 1963).

Activated Alumina

Activated alumina is a granular material consisting mostly of highly porous aluminum oxide in the trihydrated form. It has limited use as an adsorbent for specific gases, is an excellent water adsorber, and is frequently used in air dryers because it may be readily regenerated by heating. Its cost is somewhat higher than silica gel and it tends to "dust," particularly if overheated in regeneration. It is a robust material and may be reused many times.

Molecular Sieve

This is a synthetic sodium, or calcium aluminum silicate, of very high porosity. Pores within the granules are of uniform diameter. Molecules above a specific size are not adsorbed. Its affinity for water vapor is extremely strong, and it is frequently used in water traps in high vacuum systems.

Adsorbents may be used successfully on carbon dioxide, hydrogen sulfide, acetylene, ammonia and sulphur dioxide. They show promise for adsorption of compounds of low molecular weight. With the exception of carbon, these adsorbents may be regenerated by the application of heat or by purging the adsorbent with a gas less dense than the adsorbed vapor.

Under normal operating conditions, the unused capacity of any adsorption bed at a given time cannot be measured. Tests for determining residual capacity of carbon beds have been suggested, including measurement of penetration rates of SO_2 or halogenated hydrocarbons, but have not received wide acceptance. The efficiency of adsorption tends to be reasonably constant until the capacity of the bed is exhausted, after which penetration will occur. Penetration may also occur if a contaminant gas is driven off by a gas with a higher adsorption tendency.

Absorption air cleaning is a process by which the gaseous contaminant in air is brought into close contact with a chemical (usually in liquid form) with which it will react to form a nongaseous substance. The most important factor effecting gas absorption is contact with the absorbing solution for a period of time sufficient to allow the reaction to take place. Thus, the absorption process efficiency is dependent on the following factors:

1. Size of air bubbles contacting the solution.
2. Contact time.
3. Concentration of the absorbing solution.

The speed of the chemical reaction (including temperature effects) and the rate of flow of air through the absorbing solution are corollaries to these physical parameters. Absorption-type scrubbers are frequently used for humidity control in large ventilating systems, particularly those requiring recirculation air. Lithium hydroxide and triethylene glycol solutions are two typical absorption agents used. With respect to aerobiology, absorption processes can be most useful in providing a means of removing or detecting a specific contaminant.

Ions

With the exception of those arising from controlled condensation of vapors, most dusts are charged to some extent (DallaValle et al., 1954). Whitby et al. (1960, 1965) suggest the use of air-ion generation to provide in-air neutralization of the charge. The initial distribution of charge may be symmetrical with respect to polarity, if the aerosol is produced by atomization from a nonpolar solution or by a process by which dry particles are in mutual contact before being airborne; subsequently, an unbalance of charge may occur because of the selective loss of particles of one polarity by having contacted materials of the vessel or charged particles already in the air stream (Mercer, 1964). In urban areas, for example, air ions (usually of positive charge and of the small and medium size/charge ratio) may range from 500 to $500,000/cm^3$. Such incoming contaminants may be removed by charged plates located in the air stream or by grounded metal-fiber filters. It should be noted, however, that small-particle charges may be modified by almost any action (e.g., heating, compression, expansion, or other treatment of the airstream).

MODEL INSTALLATION

The foregoing discussion of air purification techniques, although brief, is applicable to the control of contamination of the air involved both in the supply to and the exhaust from aerosol processes. Each of the techniques described has been used, at one time or another, in nearly every laboratory. Combinations of the various treatments, in conjunction with containment procedures and temperature-humidity controls, have been used in assembly of sophisticated aerosol systems.

Figure 10-10 is a schematic, simplified diagram of the process system for an aerosol chamber installation (Figure 10-6) at the Naval Biological Laboratory. The installation includes an incoming air purification sys-

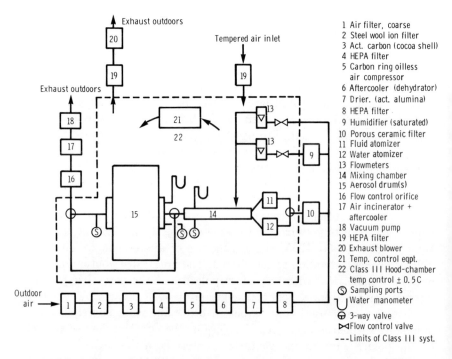

1 Air filter, coarse
2 Steel wool ion filter
3 Act. carbon (cocoa shell)
4 HEPA filter
5 Carbon ring oilless
  air compressor
6 Aftercooler (dehydrator)
7 Drier. (act. alumina)
8 HEPA filter
9 Humidifier (saturated)
10 Porous ceramic filter
11 Fluid atomizer
12 Water atomizer
13 Flowmeters
14 Mixing chamber
15 Aerosol drum(s)
16 Flow control orifice
17 Air incinerator +
  aftercooler
18 Vacuum pump
19 HEPA filter
20 Exhaust blower
21 Temp. control eqpt.
22 Class III Hood-chamber
  temp control ± 0.5 C
Ⓢ Sampling ports
Water manometer
⊕ 3-way valve
⋈ Flow control valve
---Limits of Class III syst.

FIGURE 10-10. Simplified schematic diagram of a model drum facility.

tem, temperature and humidity controls, safety containment, and a contaminated air disposal system that provides a sterile discharge to the atmosphere. A brief description of the features of this system will be of assistance in demonstrating the practical application of many of the principles discussed previously.

One objective of this program is to standardize aerosol generation, holding, and sampling techniques. Great care has been taken in all aspects of the installation, particularly in the assembly of this system to ensure that all materials used in its construction were defined to the maximal extent practicable. Steel piping is used for air supply piping up to the point where the air is humidified and where aerosols are generated. From this point, all piping and aerosol vessels are made of stainless steel. Process air-disposal piping is made of steel. The use of organic materials is kept to a minimum. The carbon filter is installed with metal trays; all HEPA filters have glass-fiber media without binders or waterproofing, and end seals are made of the filter media packed into place. All elastomers are food grade, silicone rubber. Packing glands on valves and equipment are made of Teflon[13] or carbon. The air compressor operates oil-free (carbon pistons and rings). After assembly, all parts of the piping and aerosol vessels were washed with a heated organic solvent, a water miscible solvent, then a neutral detergent, and finally rinsed for several hours with demineralized water. Gas chromatographic analysis of samples of air emanating from this supply system shows penetration only of some light hydrocarbons (possibly methane).

Similar standards of construction and cleaning were employed for the aerosol drums, control piping, and associated components. Rotating seals used on the axle of the drum are stainless steel with carbon and ceramic wear faces. The drums and the aerosol mixing chambers are fabricated of stainless steel polished to a fine brush finish. The drum, mixing chamber, aerosol generator(s), and samplers are enclosed in a Class III hood and chamber to provide temperature control within the range of +40 to +120°F, with a point-to-point variation not exceeding 1°F. Control of the humidity of the process air is achieved by mixing controlled temperature dry air and saturated air.

The air disposal system was not cleaned to the standard of the drums and process air system, but was demonstrated to be leak-tight by a halogen detector system. All process air is exhausted through disinfectant-sealed vacuum pumps and then through an air incinerator that has been tested to allow less than $1/10^8$ penetration of viable BG spores. The entire air disposal system manifold is operated at —17 inches mer-

---

[13] *Registered, E. I. du Pont de Nemours and Company (Inc.), Wilmington, Dela.

cury to provide sampling vacuum. Full emergency electric power backup to normal power supply is provided for operation of all hood blowers, pumps, and other equipment. Although only one aerosol drum is diagrammed in Figure 10-10, the process air supply system and the contaminated air disposal system shown are capable of providing clean, dry air and air disposal facilities for this multi-drum installation and several other major aerosol installations connected to the system. Each of the systems may be sterilized by steam or dry heat, and the product of vapors, condensate liquids, or any other potentially contaminated material may be dumped into the waste disposal system.

It is not suggested that such a system is a requirement for all aerobiology studies. It is a demonstration of the lengths to which one may go to establish a baseline standard for study of survival of microbial aerosols over extended periods. Studies with aerosol equipment using this air system have shown the need for improved control of contamination in the process supply air. The exact degree and area of control must obviously be generated by the requirements of the experiment. The reader, especially the engineer, is cordially referred to Chapters 11 and 14 for biological principles substantiating these stringent requirements.

TEST AND EVALUATION PROCEDURES

Many parameters discussed previously can be evaluated best by experimentation methods. Chemical content of the air environment is a case in point. Other factors may require special test procedures, either because their influence or presence would not be measured during usual experiments, or because the consequences of malfunction may be too great to risk; e.g., when working with highly virulent pathogens. If one considers only potential hazards of aerosol work, good management practices require the operator to know he has exercised the "state of the art" in contamination control for safety of his own group, and of others in areas that may be affected by his work. A program for test and evaluation of contamination control measures, essential to meet safety requirements, can also provide valuable calibration of equipment and training in equipment operation.

The major elements requiring test are the aerosol chambers, the barrier enclosures, and the air quality control systems.

Chambers

Evaluating leakage of aerosol chambers may be simplest if the chamber is a pressure vessel designed in accordance with ASME[14] specifica-

[14] American Society of Mechanical Engineers, code for unfired pressure vessels.

tions, or other pressure vessel codes. Such vessels are hydrostatically tested at pressures exceeding normal use ratings, and any leakage is usually limited to access ports, piping attachments or other openings. Chambers, hoods, animal enclosures, and other devices not designed as pressure vessels present additional problems in leak detection and correction.

The first and simplest test to perform is the static pressure test, which may be done with a water manometer. The chamber may be made positive or negative, and if temperature is held constant, leakage in fractional volumes may be expressed by the rate of change of the water column [a change of 1 inch wsp is approximately 0.033 pound/inch^2 (psi)] divided by atmospheric pressure, 15 psi. If the expected concentration of aerosol in the vessel is known, the quantity escaping or entering may be calculated. If a multibarrier system is used, it may suffice to accept some leakage at this primary level, and the fractions previously observed ($1/10^4$ to $1/10^6$) are readily achievable.

Locating leaks may be done in sequence of increasing sensitivity with increasingly complex equipment. Soap bubbles are the least sophisticated and are first choice. The halide torch detector[15] is inexpensive, has good sensitivity, and uses a burner with small butane or propane gas cylinders. It is used in conjunction with halogenated hydrocarbon refrigerants (Freon[16]) such as R-12, R-11, or other gases. Both the halide torch and the refrigerant gases are available at refrigeration parts supply houses at low cost (less than $10). The electronic halide detector,[17] also sold at these outlets, is more expensive ($100 to $200) but is ten- to a thousandfold more sensitive and has no associated fire hazard. The refrigerant may be added to the chamber in a concentration of 10 oz/30 ft^3 for the halide torch and 1 oz/30 ft^3 for the electronic detector to locate leaks. The chamber can then be pressurized to a few inches of water or to two or three times the pressure at which it will be used. Other leak detection possibilities include generation of fluorescent powder aerosols and subsequent scanning of surfaces with a strong UV illumination; a dye, or chemical or biological particulates can be used in a similar manner. As a last choice, and with greatest sensitivity, a mass spectrometer, leak detector can be utilized.

Any leak location technique requires that test material be nonflammable, readily removed, nontoxic, of small particle size (or a gas), and that the instrumentation be capable of rapid response at parts per mil-

[15] Turner Corporation, Model LP777, Sycamore, Ill.
[16] Registered, E. I. duPont de Nemours and Company (Inc.), Wilmington, Dela.
[17] General Electric, Instrument Department, Type H with R-12 refrigerant vehicle, West Lynn. Mass.

lion, or hundred million, levels. The reader is urged to make use of instrumentation and materials available to him.

These same techniques are applicable to evaluation of secondary barrier systems. To test hoods it will be necessary to limit pressures to ±2 or 3 inches wsp and to block air ducts and access openings with tape and vinyl film. If secondary enclosures are insulated, it is desirable to remove insulation near suspected leak areas. Incidentally, some insulations, closed-cell foamed plastics for example, contain residual refrigerant-type gases that can interfere strongly with the tests previously described.

Barriers; Rooms and Buildings

Detection and location of leaks in and out of rooms or buildings, although inherently less demanding with respect to accuracy, may be more difficult to accomplish. "Background" may interfere with airborne-particle detection, and test gas dilution may demand very high detector sensitivity. Visual observation of smokes or soap bubbles is an excellent rough technique in common use by ventilating engineers. It may be refined by use of electronic smoke detectors or air samplers (bubbler or filter type). Some of the most sensitive measures include use of biological aerosols and samplers, or fluorescent particle aerosols with Rotorod[18] or other samplers. The microaerofluorometer described by Goldberg (1968) has been used successfully for this purpose. It has sensitivity adequate to trace very dilute clouds over great distances and is an "instantaneous" detector. Biological aerosols can be equally sensitive but have the inherent delay of the sample processing. More importantly, if a sensitive microbe is used, estimates of leaks can be grossly underestimated; if a hardy species (spore) is used, one may be left with a decontamination problem. It should be kept in mind that the third-level barrier system may encompass very large space volumes and airflow rates. Testing this part of the system is essentially testing the effect of a major accident or failure of the system, and the tests should be extended to include areas affected only in these "normally unexpected" conditions.

Air Systems

Air filtration devices, which play an important role in both the protective barrier enclosure and the process air quality control systems, also require testing, although commercially acquired filter-canister assemblies and low efficiency ventilation filters can be used directly.

High-efficiency particulate filters, in ventilating systems particularly,

---

[18] Metronics Associates, Inc., Palo Alto, Calif.

should be evaluated *in situ*. Although HEPA filters are individually tested by the manufacturer, in actual use it is the exception rather than the rule to have them perform according to specifications. Reasons for these failures, possible test methods, and measures for correction are described at length in *NASA Contamination Control Handbook* (Sivinski, 1968). In effect, it is difficult to deliver these filters on the site in perfect condition and to make an installation that is as effective as the filter medium. Usually, these rather delicate units are handled by workers who are not accustomed to taking the care required and are installed with equal lack of caution. As a result, damage to the filters and poor installation practice are sources of the vast majority of system failures.

Manufacturers use a small-particle aerosol of DOP to test these filters. Penetration is measured by light-scattering photometers. This test procedure can be repeated in the laboratory using a Collison-type atomizer or a multijet submerged nozzle for aerosol generation. Other workers have employed aerosols of methylene blue dye (Whitby *et al.*, 1961), sodium chloride (Dorman *et al.*, 1965), bacteria (Decker *et al.*, 1957), viruses (Harstad *et al.*, 1967), fluorescent particles (Sullivan *et al.*, 1967), condensation nuclei (Davis and Clifton, 1966), or radioactive materials (Breslin, 1965, cited by Breslin, 1967). Each of these challenge aerosols has its own requirements for assay, but the light-scatter test is the only one of the group providing instantaneous readout. Harstad *et al.* (1967) have shown that filter-efficiency tests using bacteria, viruses, or smaller particles, frequently indicate higher collection efficiencies than the DOP test. There is sufficient evidence to indicate that HEPA filters are at least as efficient in collecting these particles as they are for the small (0.1 to 1.0 μ) spherical particles of the DOP smoke. Thus, ratings by the manufacturer tend to be conservative, and an application of HEPA filters to filtration of biological particles may be done with the assurance of success.

The light-scatter detector (penetrometer, see Figure 7-8) is useful for overall efficiency testing of filters and for probing filters and frames for leakage. Efficiency may be determined directly by measuring the ratio of downstream versus challenge aerosol concentrations, and leak points may be located by probing for a tenfold increase in penetration at any single point.

During the testing process, the aerosol challenge time-concentration should be minimal to prevent excess loading of filters. Although there is no indication that continued heavy challenge will produce breakthrough, it is desirable to minimize fiber wetting with DOP, oil, or aqueous materials.

The use of simulants or tracer materials is somewhat less practical in evaluation of chemical contamination. Test and evaluation of equipment is best done with instrumentation used in the experimental process. Some on-stream monitoring, RH measurements for example, can be useful, but performance of air purification systems employing carbon, silica gel, or activated alumina is difficult to predict on the basis of continuous monitoring. There are some techniques requiring measurement of penetration of light chemicals, or chemicals with a low adsorption affinity, but these methods have not been standardized nor has their sensitivity been established. Developing overall "capacity" of any component on an empirical basis is the simplest and most reliable technique.

As noted previously, monitoring of the effectiveness of air-cleaning devices requires instrumentation with broad span and high sensitivity. One area in which this can be of critical importance is that of monitoring humidity of an air supply. If an adsorption system is used for an air drier, it will be necessary to measure moisture content on the order of parts per hundred million of air (e.g., dewpoints ca. —50°C). If a water-spray, air saturator is used for humidification, then humidities near 99% must be measured. The problems of measurement over such a broad span are severe, and the instrumentation required may be very costly. Measurement of other chemical constituents not having such wide variance requires sensitivity that will permit reliable detection of less than 10% of the minimum quantity of the contaminant that is of interest. Instrumentation should be selected for sensitivity approaching $\frac{1}{100}$ of this value to ensure accuracy in "background" measurements.

Requirements for high sensitivity, broad span, and reliable operation tend to emphasize the need for careful standardization or calibration of air-quality measuring equipment. If on-stream chemical-electronic equipment is used, the operator must conduct parallel tests under rigorous conditions to confirm both the operation of his instrumentation and the continuing effective operation of the air purification system.

CONCLUSIONS

In this chapter we have described numerous contamination control problems and some methods for their resolution; we have related these to operational safety and experimental requirements. In general, the emphasis has been on physical problems and equipment. Aerobiology studies, by their very nature, require close control of the air environment. The experimental procedures tend to be somewhat more complicated, and involve more mechanical equipment, than most other microbiology laboratory techniques. Frequently, it is necessary to utilize many diverse talents in developing and operating an aerobiology laboratory.

All of these requirements place a heavy demand on the aerobiology laboratory worker. In the absence of a competent safety or industrial hygiene group, he may be required to demonstrate that his work creates no hazard to the members of his staff and to other personnel in the area. He will be required to direct the activities of mechanics, electronics technicians, and other supporting personnel, or may develop his own skills in broad areas to a degree that will ensure the reliable operation of the aerobiology apparatus. He will also, of course, be responsible for the quality of the work and the publications ensuing therefrom.

While effective contamination control requires certain minimal equipment, it is even more dependent on workers who are fully aware of, and have an interest in, contamination control procedures. As yet, there is no recognized discipline of contamination control. Such competence must be related to training in safety in the microbiology laboratory, techniques in careful operating procedures, and in measurement techniques.

REFERENCES

Akers, T. G., Bond, S. B., Papke, C., and Leif, W. R. 1966. Virulence and immunogenicity in mice of airborne encephalomyocarditis viruses and their infectious nucleic acids. *J. Immunol.*, **97**: 379–385.

American Standard Fundamentals Governing the Design and Operation of Local Exhaust Systems. 1960. American Standards Association superceded 1966 by United States of America Standards Institute, New York.

Ames, A. M., and Nungester, W. J. 1949. The initial distribution of air-borne bacteria in the host. *J. Infect. Dis.*, **84**: 56–63.

Anders, W. 1967. Obscure disease related to African monkeys. *Morbidity and Mortality Weekly Rep.*, **16**(37): 316.

Angelotti, R., Maryanski, J. H., Butler, T. F., Peele, J. T., and Campbell, J. E. 1968. Influence of spore moisture content on the dry-heat resistance of *Bacillus subtilis* var. *niger. Appl. Microbiol.*, **16**: 735–745.

Barbeito, M. S., Taylor, L. A., and Seiders, R. W. 1968. Microbiological evaluation of a large-volume air incinerator. *Appl. Microbiol.*, **16**: 490–495.

Beard, C. W., and Easterday, B. C. 1965. An aerosol apparatus for the exposure of large and small animals: Description and operating characteristics. *Am. J. Vet. Res.*, **26**: 174–182.

Bourdillon, R. B., Lidwell, O. M., and Raymond, W. F. 1948. Studies in air hygiene. *Med. Res. Council (London) Spec. Rep. Ser.*, **262**: 190–203.

Breslin, A. J. 1967. Validity tests of in-place air filters. *Health Physics*, **13**: 93.

Buttolph, L. J., and Haynes, H. 1950. Ultraviolet air sanitation. Pamphlet LD-11, General Electric Co., Cleveland, O.

Buttolph, L. J., Haynes, H., and Matelsky, I. 1950. Ultraviolet product sanitation. Pamphlet LD-14, General Electric Co., Cleveland, O.

Chatigny, M. A. 1961. Protection against infection in the microbiological laboratory: Devices and procedures. In: *Advances in Applied Microbiology* (W. W. Umbreit, Ed.), Vol. 3, p. 141. Academic Press, New York.

Chatigny, M. A., Sarshad, A. A., and Pike, G. F. 1968. Unpublished data.

Clarke, J. H. 1965. The design and location of building inlets and outlets to minimize wind effect and building re-entry of exhaust fumes. *Am. Indust. Hyg. Assoc. J.*, **26:** 242–248.

Coriell, L. L., McGarrity, G. J., and Horneff, J. 1967. Medical applications of dust-free rooms. I. Elimination of airborne bacteria in a research laboratory. *Am. J. Public Health*, **57:** 1824–1836.

DallaValle, J. M. 1948. *Micromeritics*. Pitman Publishing Corp., New York.

DallaValle, J. M., Orr, C. Jr., and Hinkle, B. L. 1954. The aggregation of aerosols. *Brit. J. Appl. Phys. Suppl. 3*, pp. S198–S206.

Danielson, J. A. 1967. (Ed.) *Air Pollution Engineering Manual*. Public Health Service Pub-999-AP-40, Cincinnati, O.

Darlow, H. M., Powell, E. O., Bale, W. R., and Morris, E. J. 1958. Observations on the bactericidal action of hexyl resorcinol aerosols. *J. Hyg.*, **56:** 108–124.

Davis, R. E., and Clifton, J. J. 1966. A new method for in situ testing high efficiency air filters using condensation nuclei as the test aerosol. *Filt. Separat.*, pp. 473–479, 499.

Decker, H. M., Citek, F. J., Harstad, J. B., Gross, N. H., and Piper, F. J. 1954. Time temperature studies of spore penetration through an electric air sterilizer. *Appl. Microbiol.*, **2:** 33–36.

Decker, H. M., Harstad, J. B., and Lense, F. T. 1957. Removal of bacteria from air streams by glass fiber filters. *J. Air Pollut. Contr. Assoc.*, **7:** 15–16.

Dimitriades, B. 1967. Methodology in air pollution studies using irradiation chambers. *J. Air Pollut. Contr. Assoc.*, **17:** 460–466.

Dimmick, R. L. 1960. Delayed recovery of airborne *Serratia marcescens* after short-time exposure to ultra-violet irradiation. *Nature*, **187:** 251–252.

Dimmick, R. L., Hatch, M. T., and Ng, J. 1958. A particle sizing method for aerosols and fine powders. *A.M.A. Arch. Ind. Health*, **18:** 23–29.

Dorman, R. G., Sergison, P. F., and Yeates, L. E. J. 1965. An apparatus for the measurement of particulate penetration through high-efficiency air-conditioning filters: *J. Ind. Heating Ventilating Engineers*, **32:** 390–396.

Doxie, F. T. and Ullom, K. J. 1967. Human factors in designing controlled ambient systems. Western Electric. *The Engineer*, Vol. XI, No. 1.

Elsworth, R., Morris, E. J., and East, D. N. 1961. The heat sterilization of spore infected air. *The Chem. Engr.*, A47–A52.

Elsworth, R., Telling, R. C., and Ford, J. W. S. 1955. Sterilization of air by heat. *J. Hyg.*, **53:** 445–457.

Favero, M. S., and Berquist, K. R. 1968. Use of laminar air-flow equipment in microbiology. *Appl. Microbiol.*, **16:** 182–183.

Ferry, R. M., Brown, W. F., and Damon, E. B. 1958. Studies of the loss of viability of stored bacterial aerosols. II. Death rates of several non-pathogenic organisms in relation to biological and structural characteristics. *J. Hyg.*, **56:** 125–150.

Fiese, M. J. 1958. *"Coccidioidomycosis."* p. 86. Charles C Thomas, Publisher, Springfield, Ill.

Fraser, D. A., Bales, R. E., Lippman, M., and Stokinger, H. E. 1959. Exposure chambers for research in animal inhalation: Design, construction, operation and performance. Public Health Monograph No. 57, pp. 1–54.

Glauser, E. M., and Glauser, S. C. 1966. Environmental chamber for the study of respiratory stress in small animals. *Arch. Environ. Health*, **13:** 61–65.

Gogolak, F. M. 1953. A quantitative study of the infectivity of murine pneumonitis virus in mice infected in a cloud chamber of improved design. *J. Infect. Dis.*, **92:** 240–247.

Goldberg, L. J. 1968. Application of the microaerofluorometer to the study of dispersion of a fluorescent aerosol into a selected atmosphere. *J. Appl. Meteorol.,* 7: 68–72.

Goldberg, L. J., Watkins, H. M. S., Boerke, E. E., and Chatigny, M. A. 1958. The use of a rotating drum for the study of aerosols over extended periods of time. *Am. J. Hyg.,* 68: 85–93.

Griffin, C. W., Kantzes, H. L., Ludford, P. M., and Pelczar, M. J., Jr. 1956. Studies of aerosols with a simple cloud-chamber technic. I. The evaluation of a technic for the rapid and convenient determination of the survival of air-borne microorganisms. *Appl. Microbiol.,* 4: 17–20.

Griffith, W. R. 1964. A mobile laboratory unit for exposure of animals and human volunteers to bacterial and viral aerosols. *Am. Rev. Resp. Diseases,* 89: 240–249.

Hanson, R. P. Sulkin, S. E., Buescher, E. L., Hammon, W. McD., McKinney, R. W., and Work, T. H. 1967. Arbovirus infections of laboratory workers. *Science,* 158: 1283–1286.

Harper, G. J., Hood, A. M., and Morton, J. D. 1958. Airborne micro-organisms: A technique for studying their survival. *J. Hyg.,* 56: 364–370.

Harstad, J. B., Decker, H. M., Buchanan, L. M., and Filler, M. E. 1967. Air filtration of submicron virus aerosols. *Am. J. Public Health,* 57: 2186–2193.

Hassler, J. W. 1963. *Activated Carbon.* Chemical Publishing Co., New York.

Hatch, M. T., and Dimmick, R. L. 1965. A study of dynamic aerosols of bacteria subjected to rapid changes in relative humidity. In: *A Symposium on Aerobiology* (R. L. Dimmick, Ed.), p. 265–281 Univ. of Calif., Nav. Biol. Lab., Naval Supply Center, Oakland, Calif.

Henderson, D. W. 1952. An apparatus for the study of airborne infection. *J. Hyg.,* 50: 53–68.

Hinners, R. G., Burkart, J. K., and Contner, G. L. 1966. Animal exposure chambers in air pollution studies. *Arch. Environ. Health,* 13: 609–615.

Horsfall, F. L., Jr., and Bauer, J. H. 1940. Individual isolation of infected animals in a single room. *J. Bacteriol.,* 40: 569–580.

Jemski, J. V., and Phillips, G. B. 1965. Aerosol challenge of animals In: *Methods of Animal Experimentation* (W. I. Gay, Ed.), Vol. I, Chap. 8. Academic Press, New York.

Johansson, K. R., and Ferris, D. H. 1946. Photography of airborne particles during bacteriological plating operations. *J. Infect. Dis.,* 78: 238–252.

Kethley, T. W. 1963. Air: Its importance and control. *Proc. Nat'l. Conf. Institutionally Acquired Infect.,* U.S. Dept. of Health, Education, and Welfare, PHS Bulletin No. 1188, pp. 35–46, U.S. Government Printing Office, Washington, D.C.

Kirchheimer, W. F., Jemski, J. V., and Phillips, G. B. 1961. Cross-infection among experimental animals by organisms infectious for man. *Proc. Animal Care Panel,* 11: 83–92.

Kraft, L. M. 1958. Observations in the control and natural history of epidemic diarrhea of infant mice (EDIM). *Yale J. Biol. Med.,* 31: 121–137.

Kruse, R. H. 1962. Potential aerogenic laboratory hazards of *Coccidioides immitis.* *Am. J. Clin. Pathol.,* 37: 150–158.

Mantel, C. L. 1945. *Adsorption.* McGraw-Hill Book Co., New York.

McDade, J. J., Whitcomb, J. G., Rypka, E. W., Whitfield, W. J., and Franklin C. M. 1968. Microbiological studies conducted in a vertical laminar airflow surgery. *J. Am. Med. Assoc.,* 203: 125–130.

McPhee, C. W. 1966. (Ed.) *ASHRAE Handbook of Fundamentals.* Am. Soc. of Heating, Refrigeration and Air-conditioning Engineers, Inc., New York.

Mercer, T. T. 1964. Aerosol production and characterization: Some considerations for improving correlation of field and laboratory derived data. *Health Phys.*, **10**: 873–887.

Middlebrook, G. 1952. An apparatus for airborne infection of mice. *Proc. Soc. Exptl. Biol. Med.*, **80**: 105–110.

Norseth, H. G., and Mitchell, R. I. 1963. The dynamic dilution of aerosols by mixing with clean air. *Ann. N.Y. Acad. Sci.*, **105**: 88–133.

Perry, R. H., Chilton, C. H., and Kirkpatrick, S. D. (Eds.). 1963. *Chemical Engineers' Handbook.* (4th ed.) McGraw-Hill Book Co., New York.

Phillips, G. B. 1965a. Safety in the chemical laboratory. XIII. Microbiological hazards in the laboratory. Part One-Control. *J. Chem. Education,* **42(1)**: A43–A44, A46–A48.

Phillips, G. B. 1965b. Safety in the chemical laboratory. XIII. Microbiological hazards in the laboratory. Part Two-Prevention. *J. Chem. Education,* **42(2)**: A117–A120, A122, A124, A126, A128, A130.

Phillips, G. B., Jemski, J. V., and Brant, H. G. 1956a. Cross infection among animals challenged with *Bacillus anthracis. J. Infect. Dis.*, **99**: 222–226.

Phillips, G. B., Broadwater, G. C., Reitman, M., and Alg, R. L. 1956b. Cross infections among brucella infected guinea pigs. *J. Infect. Dis.*, **99**: 56–59.

Phillips, G. B., Reitman, M., Mullican, C. L., and Gardner, G. D., Jr. 1957. Applications of germicidal ultraviolet in infectious disease laboratories. III. The use of ultraviolet barriers on animal cage racks. *Proc. Animal Care Panel,* **7**: 235–244.

Piggott, W. R., and Emmons, C. W. 1960. Device for inhalation exposure of animals to spores. *Proc. Soc. Exptl. Biol. Med.*, **103**: 805–806.

Reitman, M., and Wedum, A. G. 1956. Microbiological safety. *Public Health Repts.*, **71**: 659–665.

Reyniers, J. A. 1959. Design and operation of apparatus for rearing germfree animals. *Ann. N.Y. Acad. Sci.*, **78**: 47–79.

Roesler, J. F. 1966. Application of polyurethane foam filters for respirable dust separation. *J. Air Pollut. Control Assoc.*, **16**: 30–34.

Roessler, W. G., and Kautter, D. A. 1962. Modifications to the Henderson apparatus for studying air-borne infections. Evaluations using aerosols of *Listeria monocytogenes. J. Infect. Dis.*, **110**: 17–22.

Saslaw, S., Eigelsbach, H. T., Wilson, H. E., Prior, J. A., and Carhart, S. 1961. Tularemia vaccine study. I. Intracutaneous challenge. II. Respiratory challenge. *Arch. Int. Med.*, **107**: 689–714.

Schwartz, W. B., and Silverman, L. 1965. A large environmental chamber for the study of hypercapnia and hypoxia. *J. Appl. Physiol.*, **20**: 767–774.

Schechmeister, I. L. 1957. In: *Antiseptics, Disinfectants, Fungicides, and Chemical and Physical Sterilization* (G. F. Reddish, Ed.), pp. 928–952. Lea and Febiger, Philadelphia, Pa.

Silver, S. D. 1946. Constant flow gassing chambers: Principles influencing design and operation. *J. Lab. Clin. Med.*, **31**: 1153–1161.

Sivinski, H. D., (Ed.). 1968. *NASA Contamination Control Handbook.* SANDIA Laboratory Report No. SC-M-68-370, SANDIA Corp., Albuquerque, N.M.

Southern, P. M., Jr., Pierce, A. K., and Sanford, J. P. 1968. Exposure chamber for 66 mice suitable for use with the Henderson aerosol apparatus. *Appl. Microbiol.*, **16**: 540–542.

Spiegl, C. J., Leach, L. J., Lauterbach, K. E., Wilson, R., and Laskin, S. 1953. Small chamber for studying test atmospheres. *A.M.A. Arch. Indust. Hyg. Occup. Med.*, **8**: 286–288.

Sulkin, S. E., and Pike, R. M. 1951a. Survey of laboratory-acquired infections. *Am. J. Public Health*, **41**: 769–781.

Sulkin, S. E., and Pike, R. M. 1951b. Laboratory-acquired infections. *J. Am. Med. Assoc.*, **147**: 1740–1745.

Sullivan, J. F., Songer, J. R., and Mathis, R. G. 1967. Fluorometric method for determining the efficiency of spun glass air filtration media. *Appl. Microbiol.*, **15**: 191–196.

Thiéblemont, P., Marblé, G., Perrault, G., and Pasquier, Ch. 1965. Technique d'-administration d'aérosols radioactifs liquides par voie respiratorie au singe. *Phys. Med. Biol.*, **11**: 307–312.

Tomlinson, A. J. H. 1957. Infected air-borne particles liberated on opening screw-capped bottles. *Brit. Med. J.* **II** (5035): 15–17.

Trexler, P. C. 1959. The use of plastics in the design of isolator systems. *Ann. N.Y. Acad. Sci.*, **78**: 29–36.

Urban, E. C. J. 1954. Two chambers of use in exposing laboratory animals to the inhalation of aerosols. *A.M.A. Arch. Ind. Hyg. Occup. Med.*, **9**: 62–68.

Walter, C. W. 1966. Cross-infection in hospitals. *ASHRAE J.*, **8**: 41–45.

Wedum, A. G. 1957. In: *Fourth National Conference on Campus Safety. Safety Monographs for Colleges and Universities.* No. 7, pp. 15–20. Joint project of Purdue University and The National Safety Council, Chicago, Ill.

Wedum, A. G. 1964. Laboratory safety in research with infectious aerosols. *Public Health Repts.*, **79**: 619–633.

Wedum, A. G., Hanel, E., Jr., and Phillips, G. B. 1956. Ultraviolet sterilization in microbiological laboratories. *Public Health Repts.*, **71**: 331–336.

Weiss, E., and Segeler, J. C. 1952. A cloud chamber for the uniform air-borne inoculation of mice. *J. Infect. Dis.*, **90**: 13–20.

Whitby, K. T., Lundgren, D. A., McFarland, A. R., and Jordan, R. C. 1961. Evaluation of air cleaners for occupied spaces. *J. Air Pollut. Control Assoc.*, **11**: 503–515.

Whitby, K. T., McFarland, A. R., and Lundgren, D. A. 1960. Generator for producing high concentrations of small ions. Tech. Rept. No. 12, Mechanical Engr. Dept., University of Minnesota to U.S. Public Health Service.

Whitby, K. T., Liu, B. Y. H., and Peterson, C. M. 1965. Charging and decay of monodispersed aerosols in the presence of unipolar ion sources. *J. Colloid Sci.* **20**: 585–601.

Whitfield, W. J., Mashburn, J. C., and Neitzel, W. E. 1962. A new principle for airborne contamination control in clean rooms and work stations. ASTM Special Technical Publication #342. Am. Soc. for Testing and Materials, Philadelphia, Pa.

Wolfe, E. K., Jr. 1961. Quantitative characterization of aerosols. *Bacteriol. Rev.*, **25**: 194–202.

Wolochow, H., Chatigny, M., and Speck, R. S. 1957. Studies on the experimental epidemiology of respiratory infections. VII. Apparatus for the exposure of monkeys to infectious aerosols. *J. Infect. Dis.*, **100**: 48–57.

Young, G. A., Jr., Zelle, M. R., and Lincoln, R. E. 1946. Respiratory pathogenicity of *Bacillus anthracis* spores: I. Methods of study and observations of pathogenesis. *J. Infect. Dis.*, **79**: 233–246.

PART 3

ANALYSIS OF CONCEPTS
AND RESULTS

11

BACTERIAL SURVIVAL: CONSEQUENCES OF THE AIRBORNE STATE

Melvin T. Hatch and H. Wolochow

NAVAL BIOLOGICAL LABORATORY
SCHOOL OF PUBLIC HEALTH
UNIVERSITY OF CALIFORNIA, BERKELEY, CALIFORNIA

INTRODUCTION

The natural dispersal of microorganisms into the atmosphere was demonstrated many years ago. In the laboratory, studies have shown that there are three consequences of rendering microorganisms airborne: 1, no effect on the microorganism; 2, lethality; 3, nonlethal effects. It is only recently that the third effect has been differentiated from the first.

The literature concerning survival of airborne microorganisms is voluminous. The reader will find suitable literature reviews in the paper by Anderson and Cox (1967) or in the book by Wells (1955). Perusal of this literature reveals many discordant results. It is our belief that these differences are referable to differences in experimental technique. The problem is serious enough to have prompted several laboratories to initiate a reference standard testing procedure (Wolfe, 1961).

Microbial aerosols differ from other aerosols only in that some of the airborne particles contain living microorganisms; the presence of these life forms in no way invalidates the laws governing aerosol behavior discussed in previous chapters. Although an understanding of physical principles underlying general properties of aerosols is pre-requisite to a discussion of survival of microorganisms in aerosols, these principles, alone, are of no avail in furnishing a comprehensive theory concerning mechanisms leading to the death of airborne microorganisms.

Let us define the term *survival*. For our purposes, the one general attribute most useful in defining life is the ability of the individual microbe to multiply or reproduce. In effect, this is an all-or-nothing criterion. For nonpathogenic bacteria or molds, we define survival as the ability of a single cell (or a group of cells contained in a single aerosol particle) to initiate growth of a colony eventually visible to the eye. For pathogenic species, the attribute of initiation of a disease process in a susceptible host is also survival. As we shall see, a given cell may not contain mechanisms to initiate growth processes on both a lifeless medium and in a living host. Viruses and rickettsiae require living cells for multiplication. A living bacterium is required by bacterio-phage, whereas a tissue culture cell or intact host cell, is required by many viruses and rickettsiae. For viruses and rickettsiae, manifesta-tions of reproduction will depend on the test "host"—plaque formation for phage and some viruses, cytopathogenic effects in tissue culture, ob-servable disease signs and symptoms of plants or animals, or antibody production.

A commonly used term (to which a purist in English would certainly object) is *viable* or *viability*. As used here, these terms have both qualita-tive and quantitative connotations. A viable culture contains living microorganisms. A viable count is a quantitative estimate of the number of living microorganisms in a measured sample (air or liquid). We speak, colloquially, of differences in viability. A moment's reflection leads us to the realization that this term can apply only to *populations* of micro-organisms. The individual microbial particle is either living or dead, by the criteria used, whereas a population of such particles might, for example, appear to be half dead (50% loss in viability) following some treatment. In other words, some microbes may have reacted to the treat-ment in such a way that half can no longer be considered living.

The results from studies on survival of airborne microorganisms—studies of survivor curves, changes in morphology, physiology, growth rates—can be applied generally to the field of microbiology as well as specifically to the field of aerobiology. In these studies, the investigator may vary parameters such as particle size, total-solids content and their qualitative nature, the rate and extent of drying, etc., and can ascertain their effect on survival. Wells (1955) has suggested that the term *lethes* be utilized as a measure of the extent of microbial death.

It is convenient to group factors known to influence survival of microorganisms into three major categories:

1. The specific microorganism (affected by culture medium, growth temperature, suspending medium, age) ;
2. Aerosol manipulation (affected by aerosol generation techniques, suspending fluid, and sampling techniques) ;
3. Environment (affected by temperature, humidity, radiation, oxygen, ions, and trace substances).

Of course, these may not be the only factors involved in the survival of airborne microbes, but they have received the most attention to date. Each of these factors implicated in survival (or apparent survival) of airborne microbes will be evaluated in this chapter.

FACTORS AFFECTING THE SURVIVAL OF AIRBORNE MICROORGANISMS

The Microorganism

Species

Few generalizations can be made in comparing capacity for survival of the various classes of microorganisms; each species, or even strain, of microorganism must be treated as an entity unto itself. Survival is dependent in large measure on experimental conditions—generation, suspending fluids, collection techniques, and other environmental conditions. Anderson and Cox (1967) have listed various genera and species that have been studied; the original references to any species should be consulted for particulars.

Whereas bacterial spores survive better than vegetative cells, recent work has shown that spores are not totally immune to trauma in the aerosol state (Anderson, 1966). Conversely, a nonspore-forming species, tentatively identified as a *Flavobacterium*, has been shown by Leif and Hebert (personal communication) to maintain viability as an aerosol to almost the same extent as spores of *Bacillus subtilis* var. *niger*. *Micrococcus radiodurans* is in the same category (Dimmick, personal communication).

Cultural Conditions

Several comparisons have been made of the effect on survival of growth medium and age of culture. *Escherichia coli*, for example, was shown by Brown (1953) to survive better in aerosol when cells were grown on agar than when grown in a liquid medium. Both *E. coli* and *Serratia marcescens* (Goodlow and Leonard, 1961), suffered maximal death rates when cells to be atomized were in the lag-to-logarithmic-transition phase of growth as compared to stationary phase cultures. Thus it would seem that the physiological age of the culture rather than its temporal age is the governing factor. From the few data available it is tempting to implicate metabolic systems concerned in nuclear material synthesis and organization as being the sensitive loci.

Culture Treatment

There is no doubt that changes in airborne survival capabilities do occur as a result of cultural treatment, and that such variations, interacting with sampling problems, interfere with our ability to interpret results in a meaningful way. As an example, temperature shifts imposed on cultures of *Serratia marcescens* prior to atomization caused extensive and unusual changes in the ability of cells to survive in air (Dimmick and Heckly, 1965). Changes in buffer molarity (at pH 7.0) also caused previously washed cells to react differently to relative humidity.

Hatch and Dimmick (1966) have shown that when cells from the same inoculum are grown on two different media they react differently to shifts in humidity. Anderson and Cox (1967) list numerous other examples of similar phenomena.

Storage Conditions

In publications, pre-experimental treatment of cells is usually reported as "cells were washed and stored for some time at usual refrigeration temperatures," or they were "grown in such and such a manner" before they were tested. However, one might as well not mention such manipulation unless experimental results are to be compared with cells treated in some "standard" manner—there is no other way the reader can judge the importance of the cell pre-treatment. However, there are a few recent studies on the effects of prolonged storage on subsequent airborne survival (Sawyer *et al.*, 1966; Schlamm, 1960), and if one considers the freeze-dry state a storage condition, then data of Dimmick (1960) show that profound changes in survival capacity occur after periods of storage. Storage in buffer at 4°C decreases resistance of cells to subsequent thermal stress at 52°C, indicating that the physiological state of the stored cells differed from that of the original culture. Again, the problem of interacting parameters—storage being another possible factor—often prevents meaningful analysis of survival data.

Bacteriophage

Whereas the number of papers on survival of airborne bacteria is voluminous, there is a paucity of papers concerning survival of phage.

Hemmes (1959) reported on survival of T5 phage; several papers concerning sampling, or the use of phage as a tracer, have appeared (Harstad, 1965; Cox and Baldwin, 1966); Erlich *et al.* (1964) studied T3 phage and Songer (1967) studied T3 and Newcastle disease virus. A satisfying picture of the relationship of survival to humidity has not been established. One unusual phenomenon, however, seems evident; the initial loss, or atomizer loss, apparently increases as humidity is decreased. In fact, data from Ehrlich and Songer (above) indicate that a linear relationship exists between log loss and decrease of humidity. As with bacteria, survival after the initial loss seems to have little direct relationship to humidity.

Using the Dual Aerosol Transport Apparatus (DATA), we have studied the effects of shifts of humidity on airborne T3 and *Pasteurella pestis* phage and have found that, when humidity is shifted upward (e.g., 33% RH to 66% RH), the number of apparently surviving phage particles increased to the level they would have been at the higher humidity; as high as a four-log increase has been noted. We also found that drawing a sample of bacteriophage aerosol through a small chamber where moisture could be added (i.e., shifting the humidity to about 99% immediately prior to impingement) caused a significant increase in apparent phage concentration compared with using no chamber. In fact, the increase is sufficient to suggest that phage held in the airborne state at less than 100% RH do not die, but become sensitive to collection via direct impingement.

If the same situation should be shown to obtain with mammalian viruses, then the implication to viral respiratory diseases is clear. Usual field sampling collects particles at ambient RH; when microbes are breathed, however, they encounter an increased humidity condition before deposition. Hence, the apparent concentration of airborne particles, and the apparent respiratory dosage, could be grossly underestimated. We must emphasize that, currently, data with virus are presumptive; increased effort toward clarifying this finding is mandatory. Use of phage as a model system makes it possible to conduct research in this area without highly complex equipment. Too often, however, model systems (eventually) have been shown to be generally not applicable, and studies of the effect of humidity on specific, pathogenic viruses will have to be conducted.

The tentative conclusion is that drying *per se* does not "kill" phage, but makes it sensitive to instantaneous addition of moisture, although not to slow rehydration. It is also possible that addition of moisture before sampling causes particles to enlarge and therefore enhances sampling efficiency; the true cause remains to be determined.

Aerosol Techniques

Generation

Production of an aerosol either from a liquid or a dry powder, requires energy. This energy appears in the form of new surfaces, in heat, and in the rupture of chemical bonds. By mechanisms not clearly understood, some of this energy may "damage" a cell, with the consequence that when airborne, the cell is unable to cope with additional stresses imposed either by the atmospheric environment, collection procedures, or subsequent assay. Whether or not this trauma is sufficient to destroy cellular integrity is dependent on characteristics of each preparation; i.e., properties of suspending medium and species in question. The extent of cellular damage that occurs during aerosol generation in the natural environment is unknown and can only be inferred from experimental production of microbial aerosols (see Chapter 2).

Several reported examples of the importance of the generator, its mode of operation and subsequent survival of the aerosolized bacteria may be cited.

Webb (1959) demonstrated that S. marcescens, on being aerosolized from a reflux atomizer driven by dry air, died at a faster rate than when humid air was used. Furthermore, this difference in viability was noticed only when the relative humidity of the resultant aerosol was 75% or greater.

Other recent studies of aerosol generation have indicated that very little damage occurs to cells disseminated into air by a vibrating reed apparatus (Wolfe, 1961) and only minimal amount of damage when an ultrasonic aerosol generator was used.

Ferry et al. (1951), showed that E. coli may be gradually destroyed during atomization whereas Micrococcus candidus is comparatively resistant. An apparent increase in the number of viable cells during atomization of Streptococcus zooepidemicus was observed by Shechmeister and Goldberg (1950). Wright et al. (1968) have reported that reflux atomization of Mycoplasma pneumoniae for 5 minutes caused a tenfold increase in apparent viability. These data are most easily explained on the basis of deagglomeration or breaking apart of chains of cells. Alternately, one may find no appreciable alteration in numbers of viable cells in the suspension during early periods of atomization, as illustrated in Chapter 2.

Viability of microbial cells has been discussed by Rosebury (1947) who noted that "recovery of S. marcescens with direct-spray atomizers increases as input air pressure on the atomizer is increased; by this means recovery values with S. marcescens as high as 58.8% have been

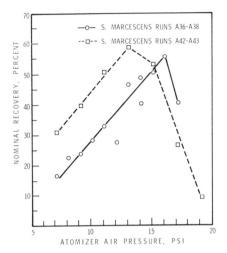

FIGURE 11-1. The atomizer air pressure effect with *Serratia marcescens*. (Rosebury, T. 1947. *Experimental Air-Borne Infection*, Fig. 39, p. 100.) Reprinted with permission of The Williams and Wilkins Co., Baltimore, Md.

obtained compared with 73.4% for *Bacillus globigii* at lower pressures" (Figure 11-1).

The shape of typical survivor curves of bacterial suspensions undergoing constant aerosolization (Figure 11-1) suggests that there are two reactions involved, each with characteristic rates. As pressure on the atomizer is increased, spraying efficiency increases; at the same time, cells are being subjected to shear forces, and as pressure is increased further, death of the cells resulting from shear stress increases at a greater rate than does spray efficiency.

Although interactions between all forces at play during aerosol generation are unknown, it is reasonable to assume the combined effects of all forces can, and do in many instances, produce a profound effect. It is quite likely that this initial trauma is a deciding factor as to whether a given cell can or cannot withstand subsequent stresses incurred in the aerosol. One of the problems involved in such measurements is the difficulty of obtaining a representative sample of aerosol prior to particle evaporation; most data include combined effects of atomization and evaporation. We are not aware of this problem having been resolved.

Suspending Media

Unless a well-washed suspension in water is used for aerosol production (or from which dried powders are prepared) the resulting aerosol particles contain not only microbial cells, but nonvolatile constituents of the suspending medium. These constituents, interacting with the environment, influence survival of the contained microbes.

The suspending medium may contain organic and/or inorganic substances. After leaving the spray device, the moisture content of droplets rapidly equilibrates with the ambient atmosphere. Even if air is saturated with water (100% RH) the droplet will lose water and shrink in size. All nonvolatile medium constituents increase in concentration during this drying to a final concentration limited by both the relative humidity and the nature of the constituents. Both of these factors are amenable to experimental manipulation, and both have received considerable attention in terms of survival of microorganisms in aerosols, as well as in bulk drying (by freeze-drying in particular). In general, results obtained in freeze-dry studies seem to carry over into the aerosol field.

Anderson and Cox (1967) list, and reference, a number of microorganisms that have been used in attempts to modify survival by manipulation of suspending media.

Inorganic Additives

This class of additives includes various salts, of mono-. di-, and trivalent metals in the form of chlorides, phosphates, sulfates, bromides, etc. If any generalization can be made, it is that inorganic additives, as a class, tend to depress survival of airborne microorganisms—particularly at low initial concentrations. Zimmerman (1965) has shown that cells of airborne *S. marcescens* aerosolized from 1% NaCl died faster than when 5% or 10% NaCl was used. The effect of other monovalent metal salts appear to be similar to that of NaCl. There is some evidence that airborne microbes may be protected by certain divalent metal salts, but the extent is so small that firm conclusions cannot be drawn.

Organic Additives

Organic compounds as moderators of survival of airborne microorganisms have received more attention than inorganic substances. Simple di-, tri-, and polysaccharides, sugar alcohols, proteins, polypeptides, organic acid salts, antibiotics, chelating agents, culture supernatant fluid, and other more esoteric mixtures have been examined. Most of the organic compounds shown to be the best protective agents (of which inositol appears to be particularly effective) are sugars or polyhydric alcohols. Some produce a time-dependant response (Heckly *et al.*, 1967). However, simpler compounds, such as mixtures of glycerol and thiourea, are also effective.

Silver (1965) described the use of an aerosol-simulating technique wherein microdrops of suspension are held under controlled atmospheric conditions on glass fibres and examined microscopically. When various organic compounds were added to *E. coli* suspensions, Silver's results are similar to those obtained in aerosols after organic substances were added either to the suspension or to collecting fluids. In general, compounds that did not penetrate cells seemed to protect cells—the greatest apparent protection being afforded when both suspension and collecting fluids contained these additives (such as sucrose, raffinose or sucrose and dextran).

Silver's technique also permits observations to be made on the physical state of the droplets. Some additives (1M mannitol) crystallized in some droplets when equilibrated with air at 70% RH. In these droplets recovery was poor compared with those where crystallization did not occur. Furthermore, slow rehydration of droplets by exposure to saturated air, compared with fast rehydration by immersing droplets in collecting fluid, tended to increase survival. Both steps, dehydration and rehydration, seem to influence survival of microorganisms initially suspended in a given medium.

Combination of Organic and Inorganic Additives

The toxic effect of salts, such as NaCl, at low initial concentration tends to be counteracted by organic media constituents (Anderson and Cox, 1967), at least for *E. coli*. Growth medium may also be considered under this heading. Unwashed cells of *S. marcescens* survived markedly less than did cells washed in, and dispersed from, water (Hess, 1965).

The Mechanism of Protective Agents

At present no single hypothesis adequately serves to unify all observations concerning the protection of airborne microorganisms by additives in the spray suspension. Perhaps this state of affairs is an expression of the multicomponent nature of the stresses imposed on microorganisms while in the airborne state. A cell probably dies not as a consequence of a single, unique event, but as a result of one of several possible events; that event with the highest reaction rate would be considered the lethal one, and all cells may not be alike in this one respect. Three explanations that have been offered, none of which is mutually exclusive, are:

1. *Plasmolysis.* This hypothesis may be stated briefly as follows: as a droplet evaporates, dissolved solids increase in concentration to exert an osmotic effect on the cell. Monosaccharides have been

shown to be indifferent in modifying survival (Zimmerman, 1965), whereas the di- and trisaccharides enhance survival. The former sugars apparently penetrate into the cell whereas the latter do not. The inference is that nonpenetrating solutes plasmolyze the cell by withdrawing moisture; the cell contents recede from the cell wall, and metabolic activity is interrupted. Interruption of glucose oxidation has been shown for plasmolytic concentrations of sucrose or NaCl. Plasmolysis, however, is not always lethal.

Cox (1965) has elaborated a series of expressions to explain events occurring during the evaporative (and rehydration during collection) process. His treatment of the subject involves not only rates of water movement as a function of relative humidity, but also temperature, latent heats, surface tension, solute solubility and concentration, equilibrium moisture values, droplet diameter, and relative air movement past the droplet.

2. *The Water Replacement Hypothesis.* Studies on nucleic acid and protein structure show that water molecules, held by hydrogen bonding, are involved in the stereochemical configuration of these macromolecules. Webb (1965) explains the protective effect of hydroxyl-type compounds (such as inositol) by their ability to replace water molecules as structural components in maintaining the spatial integrity of proteins and nucleic acids, thus allowing them to maintain their metabolic and genetic functions. Inositol was found to minimize, or even abolish, the sensitivity of *S. marcescens* to particular humidities. Furthermore, lethal effects of ultraviolet or X-irradiation were diminished in the presence of inositol, and the typical RH-dependent lethality of these energy sources was minimized.

Webb explained the protection afforded by water at high RH, or inositol at all humidities, by assuming that both compounds are absorbers and transformers of energy, thereby "protecting" the macromolecule. Webb's replacement hypothesis may be open to question as an outcome of studies wherein oxygen content of the atmosphere is controlled.

3. *Enzyme Stabilization Hypothesis.* Protective agents, such as metabolic inhibitors, chelating agents, or antibiotics, are thought to act by "slowing down" the metabolism of airborne microorganisms, thereby minimizing the accumulation of toxic by-products that could harm the cell after it is returned to a liquid medium. Experiments of Heckly and Dimmick (1965) suggest that free radicals could be the "toxic" substances, and that protective agents might act as substrate for these radicals.

In summary, the medium in which microbes are suspended prior to, and during, the production of aerosols influences the subsequent behavior of cells in the airborne state. The content of this medium has been extensively studied in an attempt to elucidate mechanisms or factors that allow airborne bacteria to live, or cause them to die. None of the current hypotheses is intellectually satisfying, so the subject remains open. Since little is known about the "suspending medium" from which *naturally occurring* microbes arise, we must simply confess our ignorance as well as our failure to supply those who need to know (e.g., the public health officer) with useful data. For those seeking a challenge the door remains open.

Sampling

Comparative trials of various methods for obtaining samples of aerosols have demonstrated that the mode of collection has an effect on the *apparent* survival of airborne microorganisms. The adjective "apparent" is essential in this context, for it is obvious that the act of collection, occurring as it does *after* the establishment of the aerosol, can have no bearing on what actually takes place during the time cells are airborne. (For a discussion of various sampling devices that have been used, see Chapter 4.)

The experimentalist is cautioned, when investigating survival characteristics of any microorganism, to use and compare several sampling techniques. Under given conditions, that technique yielding highest numbers of recovered organisms should be the method of choice. At the same time it must be understood that even this optimal technique may, in fact, be less than 100% efficient both from the physical and biological viewpoints.

The difficuty that arises is that "optimal" media or techniques are seldom universally optimal. Sometimes the medium that collects most efficiently during the initial existence af the aerosol is least effective for collecting viable microbes that have been airborne for extended time periods. (This effect, for freeze-dried bacteria, is illustrated in Chapter 14.)

There is no known way of collecting aerosols and subsequently assaying them for the number of viable cells or particles that does not involve the transition of particles through a water phase—a water phase with an activity close to unity. This transition occurs rapidly enough to permit osmotic shock as a result of the transient rise in osmotic pressure in the interior of the cell. Cox (1965, 1966) has shown for several bacterial species that there is an interaction between temperature, humidity, additives, age of culture, and osmotic shock resulting from transition of particles through a water phase. All species do not react equally to such shock, so the previous statement that no one mechanism can be responsible for the general phenomenon of death of airborne bacteria is again applicable.

In this chapter we shall assume that sampling techniques can be employed on a comparative basis, but that such data are valid only for specified situations; the reader may extrapolate to the general case, if he chooses, at his own peril.

The Environment

Temperature

It is evident that cellular inactivation in the airborne state is influenced by atmospheric temperature, and that there are other interdependent factors that affect biological properties related to survival (Kethley et al., 1957; Brown, 1954). However, few data are available wherein temperature effects have been separated from other effects. Obviously, rates of chemical and biochemical processes are regulated by temperature. It often appears that a relationship exists between temperature coefficients, death rates, and humidity. A typical Arrhenius plot, showing activation energies for airborne S. marcescens, is illustrated in Figure 11-2. These kinds of data illustrate a possible association of temperature with thermodynamic functions that characterize an activated state. Apparently, greater amounts of energy are required to inactivate bacteria as they progressively age in the airborne state—an interesting concept.

It is customary to determine influence of temperature on airborne microorganisms by conducting tests at a given RH and at varied temperatures. Because droplets dehydrate rapidly at low humidity, drying temperature has less chance to influence biological than physical properties; biological parameters are more likely to exert influence at the midrange levels or higher. Stated another way, rapid and momentary changes in the temperature of a droplet (e.g., < 0.1 sec) might not harm the cell, or cause biological mechanisms to change, but slower changes might trigger adaptive mechanisms that would have to be readjusted by the cell after evaporation (and the cooling therefrom) had ceased.

Although survival appears to decrease with increase in temperature at a constant RH, lethal processes attributable to effects of a temperature rise in the range of 14 to 33°C have not been clearly demonstrated for aerosols of pneumococci held below 30% RH. Above this humidity, survival decreases significantly with increasing temperature. Studies of airborne S. marcescens, carried out by Kethley et al. (1957), over a

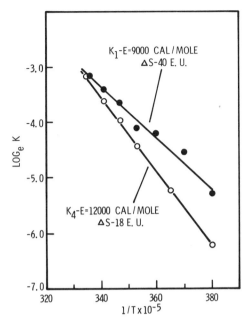

FIGURE 11-2. Effect of temperature on K_1 and K_4 for *Serratia marcescens*. Reproduced by permission of the National Research Council of Canada, from the *Canadian Journal of Microbiology*, Vol. 5, p. 658 (1959).

temperature range of -40 to $32°C$ and 20 to 80% RH, have generally indicated an increase in death rates with increasing temperature.

At either end of this temperature range, humidity appeared to exert little effect. At the higher temperatures studied, the death-rate constant was 0.05 to 0.1, decreasing by a factor of 10 at temperatures below $-20°C$. Between 30 and $-20°C$, changes in reaction rates were humidity-dependent, in a complex manner. In general, low humidity at a given temperature was detrimental to survival; i.e., the death-rate constant was higher.

Similar findings have been reported for other bacteria, as well as some viruses. At humidities in the range of 30 to 66% RH, decay rates of *E. coli*, *Staphylococcus aureus*, and *Streptococcus salivarius* were essentially unchanged by a rise in temperature from 24 to 30°C.

Humidity

Constant Conditions. Aerosolized microorganisms are influenced most dramatically by water content of the air; the most noticeable effect occurs immediately after particles are formed. Years ago, investigators realized that survival and infectivity of microbial cells in the aerosol were RH-dependent. Since then, almost without exception, aerobiologists have characterized behavior of airborne cells by determining survival of microbes at varied humidity levels—and at constant temperature where possible.

A brief account of results of earlier work will be considered now for two reasons: (a) the reader should be cognizant of the general views and different opinions in the older literature concerning the influence of RH (particularly at the critical humidity levels) on death rates of some representative microorganisms, and (b) such information is the best available way to predict persistence and infectivity of microbial agents in air environments, as well as to give possible insights into inter-actions of water, solutes, and structural alterations of cells in the per-plexing problem of cellular behavior in the aerosol.

When E. coli is aerosolized into air humidified in the range of 45 to 90% RH, the survival rate increases with an increase in humidity. Humidity levels less than 40% RH favor greater numbers of survivors of airborne S. marcescens, S. albus, S. aureus and S. salivarius (dispersed in distilled water) than at higher levels. The survival rate of S. pullorum increases as humidity is raised from 15 to 80% RH. The influenza virus was thought to survive better in dry air than in humid environments. On the other hand, hemolytic streptococci cannot survive well in dry air but are relatively stable in moist atmospheres.

In more sophisticated studies, however, the most lethal humidity range for aerosols of Pneumococcus type 1 and hemolytic Streptococcus group C (dispersed in saliva, 0.5% saline or broth medium) is the mid-range; above or below this level, cells are capable of surviving for longer periods. Many other microorganisms have been studied to determine the influence of RH and other factors on survival and the general rule seems to be that mid-range humidities are most lethal. Reports by investigators, such as Dunklin and Puck (1948) and Ferry and Maple (1954), who initiated the types of studies wherein the whole range of humidities was included, established without doubt the presence of critical RH ranges.

At the present time, we still acknowledge that RH is a contributing factor in survival of airborne cells, but we are not as sure of the extent of the effect as were our immediate predecessors. At a critical degree of cellular dehydration, which sometimes occurs in rather narrow hu-midity zones near the mid-range and elsewhere, airborne microorganisms may be quite susceptible to lesions apparently caused by the concentra-tion of certain ionized chemical substances within the particle micro-environment. That rapid decay does not necessarily take place at critical humidities in the absence of such ions may be shown by aerosolizing bacteria suspended in ion-free medium at varied RH.

In such studies Goetz (1951) clearly demonstrated that survival of airborne S. marcescens was not affected by RH, an observation seemingly in direct contrast to the customary alteration of decay rates of this organism as a function of RH in ion-containing medium. In light of this, it may be conjectured that the presence of water available in airborne particles possibly controls the movement as well as site of deposition of intracellular ions. It may be that an ion-protein interaction can affect cellular structure in a way that causes a cell to either succumb or survive in a given air environment.

Variable Conditions of Humidity. If moisture equilibrium causes airborne bacteria to die at some specific rate, then one could ask whether a shift in humidity, imposed on cells after equilibrium has been established, will change the rate. One might suppose, for example, that once a critical humidity level has been established, shifting to a more favorable humidity should certainly not decrease the death rate because a death mechanism has already been triggered. If removal of moisture reversibly altered an essential protein or nucleic acid (or their relationship), but added moisture could restore it, then sampling should be an act of restoration, and death would not be observed.

Tests of some of these suppositions have been conducted at the Naval Biological Laboratory where dynamic aerosols of bacteria, equilibrated in a given air atmosphere for about 6 minutes, were manipulated so that RH could be abruptly shifted up or down, at constant temperature, in the DATA described in Chapter 9. We found that a sudden shift in moisture content of an aerosol may produce a pronounced change in rate of killing of airborne microorganisms, the extent of which depends on the range and direction of the shift. As an example, a loss in viability of airborne S. marcescens may be about fortyfold greater than physical loss when the humidity is abruptly shifted from 24 to 51% RH (Figure 11-3). However, increased death rates seem to be associated only with cultures grown or atomized from suspensions of an enriched medium; cells grown and atomized from suspensions of a defined, minimal medium are seemingly not so affected.

Studies were initiated to determine whether enriched medium may have been toxic and defined medium protective, or whether cells grown or suspended in enriched medium were more sensitive to "sorbed water" (Monk and McCaffrey, 1957) than cells from defined medium. Cells were removed from the medium in which they were grown, washed once, and resuspended in the other medium. The removal of enriched medium from cells lowered susceptibility to sorbed water death, but failed to

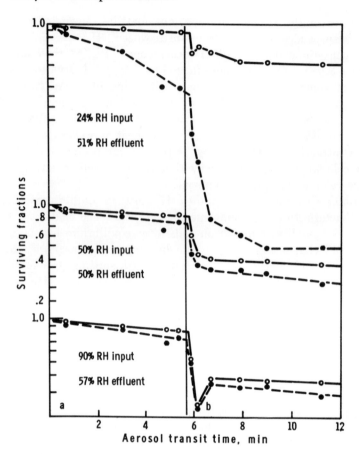

FIGURE 11-3. The effect of rapid changes in relative humidity on airborne *Serratia marcescens* grown in an enriched medium. Physical (O) and total (●) decay of primary (*a*) and diluted (*b*) aerosols.

eliminate completely the apparent toxic effect of enriched medium; no change in relative sensitivity occurred with the reverse procedure. On the other hand, if cells were cultivated in the defined medium and then held at 4°C for varied intervals, followed by an equilibration period of 30 min in enriched medium at 4°C, cells became increasingly resistant to sorbed water as the initial storage time increased (Figure 11-4). Apparently enriched medium was not always "toxic."

It seems reasonable to suppose that cellular sensitivity to stresses of aerosolization and sorbed-water killing was directly related to metabolic activities, because cellular resistance to the above stress environment was shown to increase as metabolic processes were suppressed by low temperature. In order to substantiate this notion, we cooled the atomizer fluid to 4°C throughout the spray operation by using a non-refluxing procedure. Cells aerosolized by this process were found to be basically unaffected by a rapid shift-up in RH, although they had been suspended in the enriched medium (Figure 11-5). Additional evidence that alterations in cellular metabolism are implicated in survival of airborne *S. marcescens* is presented in Figure 11-6. In these studies, cell suspensions were held at 21.6°C throughout the spray operation, then atomized with a nonrefluxing atomizer. Few cells in these spray

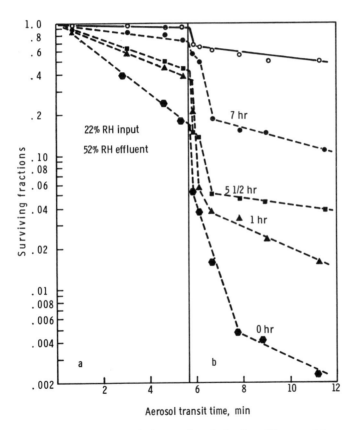

FIGURE 11-4. The effect of rapid changes in relative humidity on airborne populations of *Serratia marcescens* grown in a chemically defined medium, aged for varying periods at 4°C and aerosolized in an enriched medium after a 30-min equilibration period. Physical (○) and total (●) decay of primary (*a*) and diluted (*b*) aerosols.

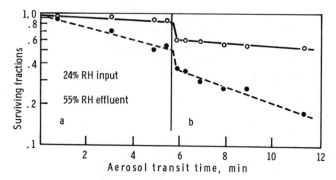

FIGURE 11-5. The effect of rapid changes in relative humidity on airborne populations of *Serratia marcescens* grown in defined medium and aerosolized in enriched
medium. The spray suspension was held at 4°C prior to and during atomization
with a Wells nonrefluxing atomizer. Physical (○) and total (●) decay of primary
(*a*) and diluted (*b*) aerosols.

suspensions survived after a shift-up in humidity. Effects of sorbed-water
death were almost eliminated when chloramphenicol, a substance that
inhibits protein synthesis, was added to these suspensions.

Two entirely different phenomena were observed as a result of an
abrupt shift-down in RH. One was an unchanged rate of biological decay
together with an instantaneous 90% loss of those airborne *P. pestis*
cells subjected to a change from 87 to 61% RH (Figure 11-7). The
other was a decreased rate of decay, but with no instantaneous loss
of cells, from the same suspension subjected to a shift from 39 to 26%
RH (Figure 11-8). It is of interest to note that surviving cells in the
first instance were quite sensitive to inactivation during dilution and
plating. For example, more than 1,000 viable cells were estimated to
be in a 0.1-ml portion of the undiluted impinger fluid, whereas the number of viable cells estimated in a similar portion of a 1:10 dilution
was significantly less than 100; frequently a two-log difference was
found. This phenomenon was also observed during assay of certain bacterial populations collected from aerosols at higher humidities, indicating
that an additional impairment of cellular function had occurred as a
result of the dilution procedure. Addition of whole blood to the assay
medium and storage for two hours at 4°C prior to incubation nullified
to some extent the detrimental effects of dilution shock.

It is suggested that airborne bacteria are metabolically active and
that apparent effects of movement of water includes the influence of
the subsequent physiological responses of injured cells in given growth
environments. Presumably, the majority of cells in the process of adapt-

ing to the initially dry environment are unable to adjust all internal mechanisms sufficiently to accommodate the sudden increase in moisture followed by the added insult of sampling, and consequently are unable to immediately produce a viable colony. The capability of cellular systems to respond quickly and successfully to sequential injuries is influenced by the particular species and the pre- and post-aerosol treatments of the culture.

Similar responses of freeze-dried cells have been reported (Heckly et al., 1967). Cox and Baldwin (1966) present evidence that cells in the airborne state can maintain sufficient metabolic functions to support phage replication.

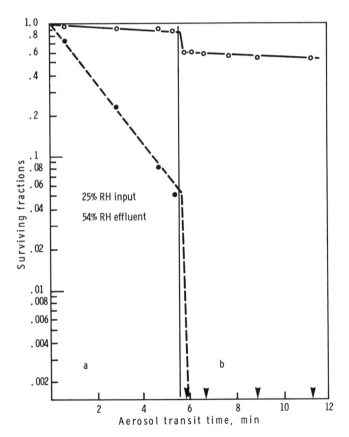

FIGURE 11-6. The effect of rapid changes in relative humidity on airborne populations of Serratia marcescens grown in defined medium and aerosolized in enriched medium. The spray suspension was held at 21.6°C for 30 min prior to and during atomization with a Wells nonrefluxing atomizer. Physical (○) and total (●) decay of primary (a) and diluted (b) aerosols.

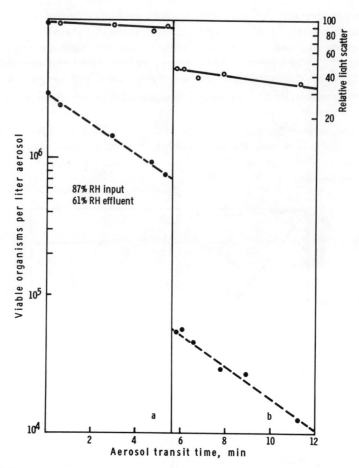

FIGURE 11-7. The effect of rapid changes in relative humidity on survival of airborne populations of *Pasteurella pestis* A1122 in the DATA. Physical (○) and total (●) decay of primary (*a*) and diluted (*b*) aerosols.

Oxygen and Other Gases

The toxicity of air has been attributed to a "toxic" action of oxygen. Data in support of this notion have been described in what may be termed a "classical" paper by Hess (1965). He demonstrated that the concentration of oxygen markedly affected the ratio between number of viable *S. marcescens* cells and *B. subtilis* spores in a mixed aerosol. The viability of both organisms was nearly equal only at the minimal oxygen concentration; as the concentration of oxygen was increased, viability of *S. marcescens* decreased. When the time of addition of 5%

concentrations of oxygen to a nitrogen atmosphere was varied (and when the survival of airborne *S. marcescens* was never greater than 30% in 5 hours in the absence of oxygen), approximately 80% of the cells were inactivated in 30 min after oxygen addition. A comparative study of these cells in air, with and without added oxygen, has convincingly demonstrated the significance of the lethal action of oxygen on airborne cells.

Heckly and Dimmick (1968) have shown that oxygen concentrations as low as 0.001% are sufficient to decrease survival of freeze-dried *S.*

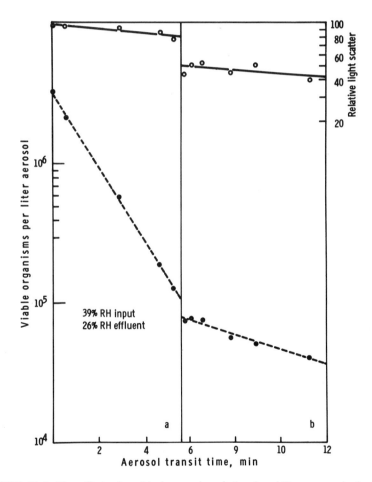

FIGURE 11-8. The effect of rapid changes in relative humidity on survival of airborne populations of *Pasteurella pestis* A1122 in the DATA. Physical (○) and total (●) decay of primary (a) and diluted (b) aerosols.

marcescens. For a more detailed treatment of this subject, the reader is referred to experiments discussed by Zentner (1966). There is no doubt that oxygen harms "dry" microbes and that the extent of damage varies with moisture content. It is of utmost importance that any theory of death mechanisms included the combined effects of both moisture and oxygen.

Helium, argon, and nitrogen atmospheres, with varying water content were used by Cox (1968) to hold aerosols of *E. coli B.* After 30 min in the aerosol state, survival at low RH (defined from psychrometric constants experimentally determined) was high in all three gases, whereas at high RH, survival was markedly reduced and varied from gas to gas, indicative of "biological" activity of the gases. A double minimum in survival was noted at 82 to 87% RH. Cox ruled out differences in rates of initial water evaporation from droplets as factors influencing survival of *E. coli B.* He suggests that RH operates by controlling the water content of the bacterium which, in turn, is a more immediate factor in the death mechanism; the role played by water movement in the intracellular environment, and hence in DNA or RNA synthesis, is not clear.

Irradiation

Past investigators convincingly demonstrated the bactericidal properties of solar irradiation when they exposed various bacterial species to sunlight and found that many cultures became sterilized in a relatively short time. The lethal effects of sunlight were associated chiefly with the ultraviolet (UV) region of the solar spectrum, whereas wavelengths in the red and infrared regions were generally considered to have little effect other than heating. More recent studies have indicated that visible light up to 5,800 Å will actually inactivate bacteria in the aerosol (Webb, 1961). Airborne *S. marcescens* may be inactivated at varied rates by incident wavelengths of 3,400 to 4,500 Å. As a result of these observations, attempts have been made to sterilize air by means of radiant energy (Jensen, 1965).

It is generally considered that most cells behave similarly to the same specific wavelength, although the action spectrum, or relative bactericidal effectiveness, varies somewhat with microbial species. For example, the dosage required to inactivate 90% of various bacterial species may vary by as much as fiftyfold. Some investigators have reported that only a single quantum of radiant energy absorbed in a vital structure is sufficient to inactivate a sensitive cell, whereas other workers believe that one in about 4 million quanta absorbed by a given cell is effective.

There is some evidence that bacteria are more sensitive to UV irradiation in air than in liquid suspension or on nutrient agar surfaces. In some situations there is little difference in sensitivity of airborne bacteria to irradiation, regardless of RH, whereas in others there is a tendency for bacteria to be more resistant at high than at low RH. The viability of airborne *Francisella tularensis*, for example, has been shown to be affected noticeably by radiant energy in the range of 30 to 95 mw/cm² (Beebe and Pirsch, 1958). The extent of damage may be modified by the addition of various compounds to spray suspensions (Webb and Dumasia, 1964).

Trace Substances in Air

Evidence is now mounting to indicate that, in natural air at least, air pollutants may be detrimental to the airborne microbe. Several laboratories are studying the effects of contaminants in the atmosphere on survival of airborne bacteria. An instance has been observed where vaporized oil from an oil-sealed pump produced a detrimental effect on bacteria; at least, the effect ceased when a carbon-vane pump was substituted. "Smog" components, specifically the 2-pentene-ozone complex, were found by Druett and Packman (1968) to be markedly lethal to *E. coli* cells suspended on spider web threads (May and Druett, 1968). One part of the complex per thousand million of air appeared to be lethal. Neither the olefin or ozone, separately, at ten times this concentration was lethal.

It is questionable whether quantitative assay of chemical substances in amounts capable of influencing survival of airborne bacteria is possible. Actually changes in survival might be a more sensitive indicator of pollution than currently available chemical or physical methods. Indeed, Druett and Packman (1968) have so reported for their biological test as a detector of the pentene-ozone complex. Other instances in which living systems have been used to detect and measure air pollution are shown in the table in Chapter 6. Effects of air ions as trace substances are discussed in Chapter 5.

NONLETHAL EVENTS

Some of the events we have included here are really misplaced. Perhaps a better heading would read, "pre-lethal," since some of the events appear to precede the loss of viability, in that they occur at a faster rate than does the loss in viability. This reasoning, then, represents a projection of a "population phenomenon" to that which occurs in an individual within the population.

Phenotypic variation—mutation

There is reasonable evidence (Webb, 1965, 1968), that desiccation causes phenotypic changes or even mutation of airborne cells exposed to 55% RH or less, with maximum numbers of mutants appearing in the 40 to 50% humidity range. Some of the specific nutritional requirements and physiological properties of *E. coli* and *S. marcescens* mutants are summarized in Table 11-1. These data provide evidence that mutations can be either forward or backward, depending on the RH. Certain metabolites of low molecular weight are said to be able to stabilize the genetic material in these cells. In view of the present lack of knowledge about the stability of such mutants, it is difficult to decide whether new phenotypes were the product of alterations of hereditary material or of extrinsic environmental influences. The situation is obviously com-

Table 11-1   The effect of bound water content on the production of bacterial mutants.

Mutant type		30	40	50	60	70	80
				Relative humidity			
White *Serratia marcescens*		316[a]	342	404	204	1	2
Amino-acid-requiring		62	173	120	15	0	0
Lactose negative		72	156	148	50	0	0
Chloramphenicol sensitive from resistance		94	318	331	67	5	4
Streptomycin independent from streptomycin dependent		103	216	208	62	1	0
Tryptophan independent from	A	433	755	273	42	2	1
dependent	B	10	8	4	1	0	2
Arginine independent from	A	284	522	322	28	3	0
dependent	B	4	7	3	1	2	1
Methionine independent from	A	208	396	201	5	3	1
dependent	B	7	8	1	0	2	0

[a] Calculated on the basis of approximately 10^6 viable cells.
A = Aerosolized from water.
B = Aerosolized from water plus the metabolite.
(From Webb, S. J. 1965. *Bound Water in Biological Integrity*, p. 81. Reprinted with permission of Charles C Thomas, Publisher, Springfield Ill.)

plicated by many other factors. As Webb points out, "results . . . are not understood and will require much more experimentation for a reasonable explanation to be offered." Nonetheless, the general inference from these kinds of data is that desiccation may be mutagenic as well as produce other changes in bacterial cells.

Metabolic Events

When staphylococci that survive desiccation and rehydration are cultivated under suitable conditions, evidence for metabolic and/or structural damage is evident by a noticeable increase in the lag period, a reduction in rate of enzyme synthesis, and a greater instability to environmental conditions (Maltman et al., 1960). Such physiological effects of drying are greater when cells in the log phase are compared with cells in the stationary phase. Although the former are the most sensitive, these log-phase cells recover more quickly than other cells. Damage becomes more pronounced with increased drying time, which in turn results in the progressive lengthening of the lag period.

Recently, a technique using bacteriophage to differentiate between damage to productive and reproductive systems of airborne bacteria was described by Cox and Baldwin (1966). Cultures of E. coli infected with T7 phage were subsequently rendered airborne. Some cells incapable of colony production were, nonetheless, capable of producing at least one replicative cycle of phage growth; hence some cellular systems were not only intact, but possessed sufficient enzymatic activity to synthesize phage-required proteins and nucleic acids. In other words, the inability of injured cells to form colonies when placed in a given growth environment was insufficient criterion that the cells were "dead" in all respects. A complex relationship existed between humidity and aerosol age as environmental factors, and lysis and colony formation as criteria of injury or death.

Anderson (1966) showed that populations recovered from aerosols of E. coli B suffered no predictable damage to pre-formed beta-galactosidase. However, the recovered bacteria were shown to be unable to synthesize the enzyme de novo for a period of time thereafter. This technique represents another way in which processes occurring in airborne cells can be analyzed.

One of the events that takes place in a bacterial cell in an aerosol has been shown by Anderson and Dark (1967) and Anderson, Dark, and Peto (1968) to be the loss of intracellular potassium ions. The magnitude and rate of loss was humidity dependent and was found for three strains of E. coli, Aerobacter aerogens, S. marcescens, and Staphylococcus epidermidis. Loss of phosphate by E. coli B was not associated with loss in viability. Whereas the correlation between loss of potassium ions and of viability was high, it is not clear that this is indeed the lethal step; it may only reflect a more generalized loss in control of permeability of other ions and substrates.

One of the ambiguities that confuses our thinking is that all events occurring in bacteria may or may not be directed by cellular control, and we do not really know whether airborne cells are metabolically active or are in a temporary quasi-quiescent metabolic state. Any change in atmospheric conditions (i.e., fluctuations in temperature or relative humidity) could conceivably trigger alterations in the cellular micro-environment. Throughout this interval, cells probably attempt to maintain their integrity and vital functions, otherwise they would cease to exist as a living entity. The exact physical stresses and conditions which are sufficient to trigger a biological response have not been adequately elucidated. We are reasonably sure, however, that airborne cells behave as dynamically active systems, and that bacteria are very responsive to environmental changes occurring within the aerosol.

SUMMARY

We can summarize observations that lead us to believe that dynamic, physiological phenomena, whether directly or indirectly controlled by genetic mechanisms, are involved in survival processes of bacteria:

1. Nonlogarithmic survival curves; nonhomogeneous populations; enhanced survival rates when humidity is shifted.
2. Survival curves differ when different sampling methods or media are employed.
   a. Differences not constant as a function of age of culture, or humidity.
   b. "Best" medium not always best as a function of cell age or aerosol age.
   c. "Best" medium for uninjured cell not always best for stressed cell.
3. Frequent lack of agreement between numbers of colonies arising from one dilution compared to the number from one above or below that dilution (dissonance).
4. Frequent observation of petite colonies arising from airborne cells, a characteristic that is apparently not transmissible.
5. Differences in numbers of colonies on duplicate samples as a function of pre-incubation treatment (cooling for example).
6. Observations of occasional intervals where the apparent number of survivors increases (recuperation) as a function of aerosol age.

7. Instances where more than one maximal sensitivity to humidity occurs.

8. Influence of oxygen on the apparent reaction of airborne cells to humidity.

9. Effects of age of culture, or cultural pre-treatment, on survival.

10. Differences between rates of loss of ability to support phage lysis and loss of ability to form colonies.

Any unified hypothesis for the mechanism of death of airborne bacteria must consider all of these phenomena; any unified hypothesis for *a* mechanism of death must show all were considered, and did not contradict the data upon which the hypothesis is built. The problem is not just an academic one, for if we know these many factors operate in the laboratory then they must certainly operate under field conditions. In studies of survival of bacteria, one can, at present, correlate laboratory and field data only within broad limits. Perhaps the creation of new approaches to experimental design and execution are needed for both the laboratory and field.

REFERENCES

Anderson, J. D. 1966. Biochemical studies of lethal processes in aerosols of *Escherichia coli. J. Gen. Microbiol.*, **45**: 303–313.

Anderson, J. D., and Cox, C. S. 1967. Microbial survival. *Airborne Microbes*, pp. 203–226. 17th Symp. Soc. Gen. Microbiol. Cambridge University Press, London.

Anderson, J. D., and Dark, F. A. 1967. Studies on the effects of aerosolization on the rates of efflux of ions from populations of *Escherichia coli* Strain B. *J. Gen. Microbiol.*, **46**: 95–105.

Anderson, J. D., Dark, F. A., and Peto, S. 1968. The effect of aerosolization upon survival and potassium retention by various bacteria. *J. Gen. Microbiol.*, **52**: 99–105.

Beebe, J. M., and Pirsch, G. W. 1958. Response of airborne species of *Pasteurella* to artificial radiation simulating sunlight under different conditions of relative humidity. *Appl. Microbiol.*, **6**: 127–138.

Brown, A. D. 1953. The survival of airborne microorganisms. I. Experiments with *Escherichia coli* near 0°. *Aust. J. Biol. Sci.*, **6**: 470–480.

Brown, A. D. 1954. The survival of airborne microorganisms. III. Effects of temperature. *Aust. J. Biol. Sci.*, **7**: 444–451.

Cox, C. S. 1965. Protecting agents and their mode of action. In *A Symposium on Aerobiology, 1963* (R. L. Dimmick, Ed.), pp. 345–368. Nav. Biol. Lab., Nav. Supply Center, Oakland, Calif.

Cox, C. S. 1966. The survival of *Escherichia coli* sprayed into air and into nitrogen from distilled water and from solutions of protecting agents, as a function of relative humidity. *J. Gen. Microbiol.*, **43**: 383–399.

Cox, C. S. 1968. The aerosol survival of *Escherichia coli* B in nitrogen, argon and helium atmospheres and the influence of relative humidity. *J. Gen. Microbiol.* **50**: 139–147.

294   Analysis of Concepts and Results

Cox, C. S., and Baldwin, F. 1966. The use of phage to study causes of loss of viability of *Escherichia coli* in aerosols. *J. Gen. Microbiol.*, 44: 15–22.
Dimmick, R. L. 1960. Characteristics of dried *Serratia marcescens* in the airborne state. *J. Bacteriol.*, 80: 289–296.
Dimmick, R. L. 1965. Rhythmic response of *Serratia marcescens* to elevated temperature. *J. Bacteriol.*, 89: 791–798.
Dimmick, R. L., and Heckly, R. J. 1965. An approach to study of microflora in atmosphere. *Proc Atmos. Bio. Conference* (Tsuchiya, H. M., and Brown, A. H., Eds.), pp. 187–197. Univ. of Minnesota, Minneapolis, Minn. (Lib. Cong. Cat. #65-22526).
Druett, H. A., and Packman, L. P. 1968. Sensitive microbiological detector for air pollution. *Nature*, 218: 699.
Dunklin, E. W., and Puck, T. T. 1948. The lethal effects of relative humidity on air-borne bacteria. *J. Exptl. Med.*, 87: 87–101.
Ehrlich, R., Miller, S., and Idoine, L. S. 1964. Effect of environmental factors on the survival of airborne T-3 coliphage. *Appl. Microbiol.*, 12: 479–482.
Ferry, R. M., Farr, L. E., Rose, J., and Blau, M. R. 1951. A study of freshly generated bacterial aerosols of *Micrococcus candidus* and *Escherichia coli*. *J. Infect. Diseases*, 88: 256–271.
Ferry, R. M., and Maple, T. G. 1954. Studies on the loss of viability of stored bacterial aerosols. I. *Micrococcus candidus*. *J. Infect. Diseases*, 95: 142–159.
Goetz, A. 1954. Early detection of bacterial growth. Parameters for biocolloidal matter in the atmosphere. *Proc. Atmos. Biol. Conference* (Tsuchiya, H. M. and Brown, A H., Eds.). University of Minnesota, Minneapolis, Minn.
Goodlow, R. J., and Leonard, F. A. 1961. Viability and infectivity of microorganisms in experimental airborne infection. *Bacteriol. Rev.*, 25: 182–187.
Harstad, J. B. 1965. Sampling submicron T1 bacteriophage aerosols. *Appl. Microbiol.*, 13: 899–908.
Hatch, M. T., and Dimmick, R. L. 1966. Physiological responses of airborne bacteria to shifts in relative humidity. *Bacteriol. Rev.*, 30: 597–602.
Heckly, R. J., and Dimmick, R. L. 1965. Survival of lyophilized bacteria during storage. In *A Symposium on Aerobiology, 1963* (R. L. Dimmick, Ed.), p. 305–318. Nav. Biol. Lab., Nav. Supply Center, Oakland, Calif.
Heckly, R. J., and Dimmick, R. L. 1968. Correlations between free radical production and viability of lyophilized bacteria. *Appl. Microbiol.*, 16: 1081–1085.
Heckly, R. J., Dimmick, R. L., and Guard, N. 1967. Studies on survival of bacteria: Rhythmic response of microorganisms to freeze-drying additives. *Appl. Microbiol.*, 15: 1235–1239.
Hemmes, J. H. 1959. *De overleving Van Microorganismen in Lucht* (The survival of airborne microorganisms). Labor, Utrecht, The Netherlands.
Hess, G. E. 1965. Effects of oxygen on aerosolized *Serratia marcescens*. *Appl. Microbiol.*, 13: 781–787.
Jensen, M. M. 1965. Inactivation of virus aerosols by ultraviolet light in a helical baffle chamber. In *A Symposium on Aerobiology, 1963* (R. L. Dimmick, Ed.), pp. 219–226. Nav. Biol. Lab., Nav. Supply Center, Oakland, Calif.
Kethley, T. W., Fincher, E. L., and Cown, E. B. 1957. The effect of sampling method upon the apparent response of airborne bacteria to temperature and relative humidity. *J. Infect. Diseases*, 100: 97–102.
Maltman, J. R., Orr, J. H., and Hinton, N. A. 1960. The effect of desiccation on *Staphylococcus pyogenes* with special reference to implications concerning virulence. *Amer. J. Hyg.*, 72: 335.

May, K. R., and Druett, H. A. 1968. A microthread technique for studying the viability of microbes in a simulated airborne state. *J. Gen. Microbiol.,* **51**: 353–366.

Monk, G. W., and McCaffrey, P. A. 1957. Effect of sorbed water on the death rate of washed *Serratia marcescens. J. Bacteriol.,* **73**: 85–88.

Rosebury, T. 1947. *Experimental Air-Borne Infection.* The Williams and Wilkins Company, Baltimore, Md.

Sawyer, W. D., Jemski, J. V., Hogge, A. L., Jr., Eigelsbach, H. T., Wolfe, E. K., Dangerfield, H. G., Gochenour, W. S., Jr., and Crozier, D. 1966. Effect of aerosol age on the infectivity of airborne *Pasteurella tularensis* for *Macaca mulatta* and man. *J. Bacteriol.,* **91**: 2180–2184.

Schechmeister, I. L., and Goldberg, L. J. 1950. Studies on the experimental epidemiology of respiratory infections. II. Observations on the behaviour of aerosols of *Streptococcus zooepidemicus. J. Infect. Diseases,* **87**: 117–127.

Schlamm, N. A. 1960. Detection of viability in aged or injured *Pasteurella tularensis. J. Bacteriol.,* **80**: 818–822.

Silver, I. H. 1965. Viability of microbes using a suspended droplet technique. In *A Symposium on Aerobiology, 1963* (R. L. Dimmick, Ed.), pp. 319–333. Nav. Biol. Lab., Nav. Supply Center, Oakland, Calif.

Songer, J. R. 1967. Influence of relative humidity on the survival of some airborne viruses. *Appl. Microbiol.,* **15**: 35–42.

Webb, S. J. 1959. Factors affecting the viability of air-borne bacteria. I. Bacteria aerosolized from distilled water. *Can. J. Microbiol.,* **5**: 649–669.

Webb, S. J. 1961. Factors affecting the viability of air-borne bacteria. IV. The inactivation and reactivation of air-borne *Serratia marcescens* by ultraviolet and visible light. *Can. J. Microbiol.,* **7**: 607–619.

Webb, S. J. 1965. *Bound Water in Biological Integrity.* Charles C Thomas, Publisher, Springfield, Ill.

Webb, S. J. 1968. Effect of dehydration on bacterial recombination. *Nature,* **217**: 1231–1234.

Webb, S. J., and Dumasia, M. D. 1964. Bound water, inositol, and the effect of X-rays on *Escherichia coli. Can. J. Microbiol.,* **10**: 877–885.

Wells, W. F. 1955. *Airborne Contagion and Air Hygiene.* Harvard University Press, Cambridge, Mass.

Wolfe, E. K. 1961. Quantitative characterization of aerosols. *Bacteriol. Rev.,* **25**: 194–202.

Wright, D. N., Bailey, G. D., and Hatch, M. T. 1958. Role of relative humidity in the survival of airborne *Mycoplasma pneumoniae. J. Bacteriol.,* **96**: 970–974.

Zentner, R. J. 1966. Physical and chemical stresses of aerosolization. *Bacteriol. Rev.,* **30**: 551–557.

Zimmerman, L. 1965. Additives to increase aerosol stability. In *A Symposium on Aerobiology, 1963* (R. L. Dimmick, Ed.), pp. 285–289. Nav. Biol. Lab., Nav. Supply Center, Oakland, Calif.

12

SURVIVAL OF AIRBORNE VIRUS, PHAGE AND OTHER MINUTE MICROBES

Thomas G. Akers

NAVAL MEDICAL RESEARCH UNIT NO. 1
UNIVERSITY OF CALIFORNIA, BERKELEY

The advent of antibiotics has reduced the importance of bacteria to a secondary role compared with that of viruses as agents causing respiratory disease. Unfortunately, studies of virus in the airborne state have lagged behind studies of airborne bacteria, but knowledge gained from studies of bacterial aerosols has permitted faster progress than if extensive technology developed previously for bacteria had not been available.

The major discussion will involve certain representative viruses as well as some rickettsia, psittacosis, phage, and mycoplasma employed in airborne stability and/or transmission studies as listed in Table 12-1. The literature on airborne microbes, except for bacteria and viruses, is sparse, but available information will be included in the final paragraphs.

VIRUSES

Early Studies on Influenza

Historically, the World War I pandemic of influenza was the impetus for initiating virus aerosol studies, and if one traces the progression

296

Table 12-1 Viruses, rickettsiae and phage employed for aerosol stability of airborne transmission studies.

Virus	Aerosol stability	Airborne transmission
West Nile (WN)	. . .	Nir *et al.*, 1965
Yellow fever	Miller *et al.*, 1963	Miller *et al.*, 1963
Rift Valley fever	Miller *et al.*, 1963	Miller *et al.*, 1963
Tick-borne encephalitis	. . .	Danes *et al.*, 1962
Venezuelan equine encephalitis (VEE)	Harper, 1961	Kuehne *et al.*, 1962
Colorado tick fever (CTF)	Watkins *et al.*, 1965	. . .
Sindbis	Jensen, 1964	. . .
Sendai (parainfluenza)	. . .	Jakab *et al.*, 1967
Influenza	Shechmeister, 1950	Loosli, 1949
Rhinoviruses	. . .	Cate *et al.*, 1964
Adenoviruses	Miller & Artenstein, 1967	Couch *et al.*, 1966
Measles	De Jong *et al.*, 1965	. . .
Coxsackie A$_{21}$	Couch *et al.*, 1965	Couch *et al.*, 1966
Coxsackie B	Jensen, 1964	. . .
Polio	Harper, 1961	. . .
Encephalomyocarditis (EMC) group	Akers *et al.*, 1966	Akers *et al.*, 1966
Foot & mouth disease (FMV)	. . .	Hyslop, 1965
Hog cholera (HC)	. . .	Beard and Easterday, 1965
Vesicular exanthema (VESV)	Lief and Beam, 1964	. . .
Newcastle disease (NDV)	Songer, 1967	Hitchner *et al.*, 1952
Infectious bovine rhinotracheitis (IBR)	Songer, 1967	. . .
Vesicular stomatitis (VSV)	Watkins *et al.*, 1963	. . .
Vaccinia	Harper, 1961	Hahon *et al.*, 1961
Neurovaccinia	Watkins *et al.*, 1963	. . .
Pigeon pox	Webb *et al.*, 1963	. . .
Miscellaneous pox viruses	. . .	Hahon *et al.*, 1961
Monkey B.	. . .	Chappel, 1960
Herpes	. . .	Edward *et al.*, 1943
Ectromelia	. . .	Edward *et al.*, 1943
Mouse hepatitis	. . .	White and Madin, 1964
Rous sarcoma (RSV)	Webb *et al.*, 1963	. . .
Phage	**Aerosol stability**	**Airborne transmission**
T1	Harstad, 1965	. . .
T2	Webb *et al.*, 1965	. . .
T3	Ehrlich *et al.*, 1964	Buckland and Tyrrell, 1964
T4	Webb *et al.*, 1965	. . .
T5	Hemmes *et al.*, 1962	. . .
T7	Cox and Baldwin, 1964	. . .
Psittacosis	**Aerosol stability**	**Airborne transmission**
Psittacosis (Lori strain)	. . .	Bolotovskii, 1959
Psittacosis (Borg strain)	. . .	McGavran *et al.*, 1962
Ricksettsia	**Aerosol stability**	**Airborne transmission**
Q-fever (*R. burnetii*)	. . .	Tigertt *et al.*, 1961
Rocky Mountain spotted fever (*R. rickettsii*)	. . .	Saslow *et al.*, 1966
Mycoplasma	**Aerosol stability**	**Airborne transmission**
M. hominis	Kundsin, 1966	. . .
M. pharyngis	Kundsin, 1966	. . .
M. pneumoniae	Kundsin, 1966	. . .
M. meleagridis	Beard and Anderson, 1967	. . .
M. laidlawii	Wright *et al.*, 1968	. . .
M. gallisepticum	Wright *et al.*, 1968	. . .

from the initial isolation of the virus to later sophisticated airborne studies, the evolution of the role of viruses in aerobiology becomes evident. The first significant step was the isolation of influenza virus. Smith *et al.* (1933) employed filtrates of garglings from influenza patients and inoculated this material intranasally into ferrets. The spectrum of disease observed in the infected ferret closely resembled the various clinical forms of influenza observed in man.

Adaptation of influenza to mice (Andrewes *et al.*, 1934) provided a means by which lung lesions (or fatality) could provide end points for either virus-neutralization tests or for studies of immunity and pathogenesis. This information, coupled with isolation techniques using the amniotic sac of 13- to 14-day-old chick embryos (Burnet, 1940) and the agglutination of virus by erythrocytes (a simplified serological test; Hirst, 1941), completed the necessary prerequisites for airborne studies. These "prerequisites" include:

(*a*) A method of producing large quantitites of virus that can be preserved in the frozen state. This is necessary if replicate studies are to be conducted.

(*b*) An efficient isolation and assay methodology.

(*c*) A serological method to test immune response in cases where aerosols were shown to have aerogenic immunization properties.

(*d*) A method to transmit the virus to several varieties of laboratory animals by the airborne route, and thereby to cause infection resembling the clinical disease observed in man.

In 1937, Smorodintseff *et al.* exposed 72 volunteers to an aerosol sprayed from a 10% mouse-lung suspension of the Leningrad W.S. strain of influenza. Each volunteer inhaled approximately 10^5 to 10^6 mouse minimal lethal doses (MLD) during exposure times varying from 15 to 60 minutes. Twenty percent of the volunteers, who were also those with low initial anti-influenza antibody titers, showed symptoms of influenza. Chalkina (1938) exposed 272 men to varying concentrations and doses of mouse-lung virus suspensions. By controlling the quantity of virus inhaled, an antibody response could be elicited without undesirable symptoms. Twenty percent of volunteers exposed to the highest virus concentration for the longest periods did develop influenza symptoms.

On the other hand, none of the 150 volunteers Burnet and Lush (1938) inoculated intranasally with an egg-passage strain of the Melbourne virus, which is avirulent for ferrets and mice, developed influenza symptoms or produced antibody.

In 1944, Francis and associates exposed volunteers to aerosols of the Lee strain of influenza B. This particular strain was aerosolized by a nebulizer equipped with glass nasal adapters inserted into the nostrils of the human subjects. After $2\frac{1}{2}$ minutes' exposure via one nostril, the adapter was shifted and exposure continued for another $2\frac{1}{2}$ minutes. As a result of intranasal spray, mild symptoms resulted. Most of the subjects developed a fourfold rise in antibody titer and 11 of 30 controls residing in the same quarters also evidenced serological changes. When this experiment was repeated 4 months later, the same symptoms were observed again (chills, aches, nausea, loss of appetite). Volunteers subjected to reinoculation showed considerably less antibody response to the second exposure than to the first. No virus was isolated from throat washings collected at 24 hours. However, mice became ill when placed in the spray room and the day room occupied by the subjects, indicating that airborne virus was in the rooms. No demonstrable relationship between height of antibody titers and clinical response to infection was observed.

Personnel of the Naval Laboratory Research Unit No. 1 at the University of California (1944) exposed volunteers to aerosols of influenza A (F-99 strain). This virus had been passaged 4 times in embryonated chicken eggs. Allantoic fluid, titered at $10^{6.6}$ 50% egg lethal doses (egg LD_{50})/ml, was sprayed from a modified Wells atomizer, and groups of 6 men were exposed to aerosols for 1, 3, 6, and 12 minutes. No significant results were observed. When PR8 was substituted for the F-99 strain, and with a group of 6 men exposed for 6 minutes, 5 for 10 minutes, and 6 for 12 minutes, influenza symptoms were observed in 10/17 of the subjects and 16/17 responded with a rise in antibody titer.

In 1949 Loosli reported on the pathogenesis and pathology of experimental airborne influenza in mice. Forty to fifty mice were exposed in a 60-liter static chamber for 40 minutes to an aerosol sprayed from 2 to 2.5 ml of PR8 virus dilution. After 72 hours, the mice were sacrificed their lungs were homogenized and prepared in broth and 10% horse serum to serve as a stock virus suspension. Groups of 40 to 70 mice were then exposed to aerosols of 0.26 to 0.28 ml of a 10^{-3} dilution and with 0.30 ml of a 10^{-4} dilution of the stock virus for 15 minutes. Animals exposed to lethal doses showed illness by the second day after exposure, whereas sublethal concentrations produced varying degrees of clinical signs. The significant aspect of this study was the pathogenesis. It was noted that the infectious process was confined to the lower trachea, bronchial tree, and lungs. Histopathologically, the bronchial epithelial cells, although not destroyed, shed that portion of the cytoplasm associated with virus growth. The remaining tissue underwent proliferation, obstructing the terminal brochiole, as well as growing peripherally into collapsed alveolar ducts and alveoli. The regenerated stratified squamous epithelium then degenerated, leaving permanent cystic spaces with non-respiratory functions.

Davenport (1961), reporting on the pathogenesis of influenza, notes that the virus first settled on the mucous film covering the respiratory epithelium of the nasopharynx, bronchi, or large bronchioles. The ciliary action tends to move the virus particle outward prior to swallowing. Meanwhile, the neuraminadase activity of influenza virus rapidly lowers the viscosity of the mucus which subsequently exposes cellular receptors. Pathology, when cellular invasion is successful, is seen to include necropsis and degeneration of respiratory epithelium.

Characterization of Airborne Influenza

Concomitant with the employment of airborne viruses to study infection processes in man or animals, reports appeared in the literature concerning biological and physical characterizations of influenza virus. Wells and Brown (1936) reported on the stability of airborne influenza virus and its susceptibility to ultraviolet irradiation, and Shechmeister (1950) showed that survival was related to relative humidity (RH). Alantoic fluid containing W.S. strain of type A influenza buffered to pH 7.1 was aerosolized with a modified Wells atomizer (DeOme et al., 1944) and capillary impingers were used to collect aerosol samples. It was reported that (a) the mean particle diameter was less than 0.5 μ; (b) the virus could withstand 15 minutes of physical agitation commensurate with atomization in the Wells atomizer (the production of virus aerosol did not affect the titer); (c) the percent reactivity was independent of the initial concentration of the atomized suspension and later was shown to be related to the humidity of the secondary mixing air stream. Virus aerosol concentrations were determined by titrating sampled impinger fluid in mice or eggs. From this it was concluded that recovery of infectious airborne virus was minimal at 60% RH and greater at 32 and 68% RH. Influenza infectivity was found to be greater when the virus was introduced by an airborne rather than by an intranasal route. Additional studies of the airborne virus by Hemmes et al. (1960) and Harper (1961) confirmed Shechmeister's report.

Hood (1963) reported on the infectivity of influenza virus aerosols in relation to aerosol age. This work was most informative because by 1963: (a) The in vitro assay system, using the egg membrane piece technique, was applicable; (b) the toroid drum designed by Goldberg et al. (1958) to hold aerosols, often over prolonged periods, was available, and (c) physical decay could be followed by using radioactive isotopes. PR8 aerosols were studied in a 500-liter toroid drum and effects of 20, 50, and 80% RH at 23°C were studied. It was observed that the mouse respiratory LD_{50} for aerosols aged 3 seconds was not significantly different from those at 2 hours (50% RH), 4 hours (80% RH), and 20 hours (20% RH). Moreover, any change in viability of virus in aerosols (as measured by membrane piece titrations) was a direct indication of the subsequent respiratory infectivity and virulence for mice; i.e., *in vivo* and *in vitro* qualities did not change independently during the aging of virus aerosols.

While it may seem that early research with airborne influenza viruses was rather extensive, the impetus for this work was based on the fact that influenza in the early thirties was an important public health problem. Moreover, the use of human volunteers was possible because of the low mortality potential.

Natural Environmental Effects

The association of seasonal virus epidemics with changes in humidity in the natural environment evolved from the earlier influenza aerosol studies. Results from studies on influenza, polio, and measles outbreaks revealed a correlation between environmental conditions and increased morbidity.

Hemmes and associates (1960) were of the opinion that naturally occurring changes in humidity, associated with seasonal and climatic changes, accounted for influenza's winter and polio's summer prevalence. Their opinion was based on studies conducted to determine airborne inactivation rates for influenza and poliomyelitis. For poliovirus, it was observed that inactivation was slow at high humidities and very rapid below 45% RH. Inactivation rates for influenza were high at 50 to 90% RH and low in the range of 15 to 50% RH. In temperate climates, survival of airborne influenza virus would then be best in the winter when indoor relative humidity would be low. In contrast, poliovirus survival could possibly be enhanced by higher indoor humidites which occur in the summer months (Figures 12-1 and 12-2).

In Nigeria, epidemics of measles usually occur during the end of the summer's dry season and decline with the onset of July rains (Morley, 1962), whereas in India (Delhi) the peak incidence of measles occurred between March and June, but declined abruptly with the onset of the monsoons in July (Taneja et al., 1962). In Chile the environment varies to such an extent that correlations are nebulous. Ristori et al. (1962) reported that population in Northern Chile, living in an arid and warm climate with little temperature variation, has a low measles mortality. The population of Central Chile, where cold, humid winters and warm,

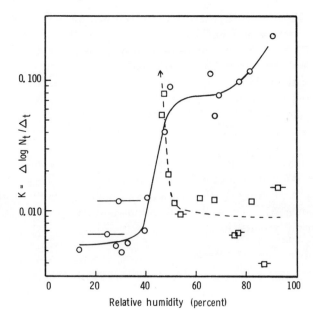

FIGURE 12-1. Death rates for influenza virus (○) and poliomyelitis virus (□) in air plotted against relative humidity. (Hemmes, Winkler and Kool. 1960. *Nature,* 188(4748) : 431.) Reprinted with permission of Macmillan (Journals) Ltd., London.

dry summers occur, also has a low mortality rate. However, the death rate from measles in Southern Chile, with its heavy rainfall, prolonged winter, and low temperatures, is high in the winter and spring with lowest rates reported for summer and autumn. De Jong (1965) pointed out that the fluctuation in seasonal variation may depend on the variation in the susceptibility of the host and incidence of virus, as well as virus survival in air. With the Edmonston strain of measles, survival of airborne virus was high at low relative humidities and low at mid-range values (50 to 70%). The author concluded that seasonal fluctuation in the morbidity of measles was largely caused by seasonal variation in indoor relative humidity rather than outdoor climate (Figures 12-3 and 12-4).

Andrewes (1964) believes that epidemiology of respiratory viruses varies with abrupt meteorological changes or other climatic-derived stresses which affect host susceptibility, and that climatic seasonal changes do not influence viral disease incidence. Spicer (1959) was unable to relate meteorological data with polio incidence. Lidwell *et al.* (1965) noted that the incidence of the common cold was not related to sunshine, pollution, day minimum minus night maximum temperatures, anti-

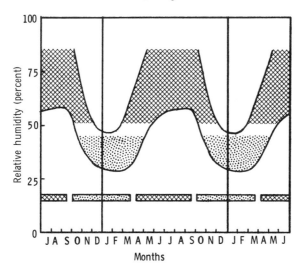

FIGURE 12-2. Seasonal variation of relative humidity indoors in the Netherlands. The upper and lower curves represent mean maximal and minimal relative humidity as calculated from temperature and absolute humidity. The range for optimal virus survival is stippled for influenza virus and hatched for polio virus. The period of increasing morbidity of influenza (data for England and Wales) and for poliomyelitis (Dutch data) are given at the bottom of the figure as stippled and hatched bars. (Hemmes, Winkler and Kool. 1960. *Nature,* **188**(4748): 431.) Reprinted with permission of Macmillan (Journals) Ltd., London.

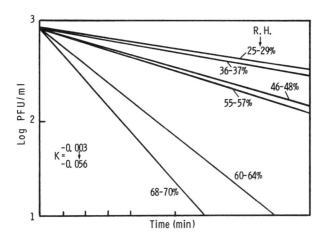

FIGURE 12-3. Survival of measles virus at varying relative humidity. Virus survival was examined at 20°C. (J. G. de Jong. 1965, *Arch. fur die Gesamte Virusforschung,* **16:** 99.) Reprinted with permission of Springer-Verlag, Vienna.

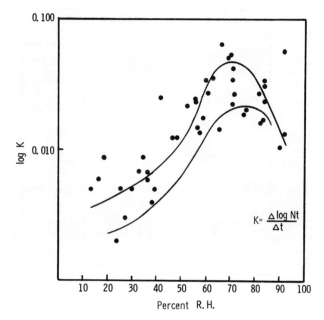

FIGURE 12-4. Survival of measles at varying relative humidity. $K = \dfrac{\Delta \log Nt}{\Delta t}$
●———● 20°C and ○———○ 15°C. (J. G. de Jong, 1965. *Arch. fur die Gesamte Virusforschung,* **16:** 99.) Reprinted with permission of Spring-Verlag, Vienna.

cyclonic or cyclonic conditions, relative humidity, or rainfall. Only two weather variables existing on the day, and/or each of the preceding 9 days, were shown to be significantly correlated. These were the mean temperature and water vapor pressure at 9 A.M. The strongest correlation was with a lowered mean temperature that occurred 2 to 4 days before the onset of symptoms. The results of this study suggested that either low outdoor temperatures, or the absolute humidity at low temperature, promotes transmission or development of the disease.

Other Airborne Viruses

Aerosol studies with viruses other than influenza were based on analysis of occurrences of laboratory accidents and circumstances which pointed toward airborne transmission within the confines of the laboratory itself. It is noteworthy that in most cases the outcomes were serious enough to preclude human transmission studies, and even work with laboratory animals was conducted under closely controlled environments. Human viruses for which airborne transmission is implicated to date are numerous. Lennette and Koprowski (1943) reported on human infections of Venezuelan equine encephalomyelitis (VEE) virus. The source of the virus was conjectured to be contaminated dust raising from mouse cages holding VEE-infected mice, and infection was supposed to be via the respiratory route.

Nir (1959) reported on a laboratory technician who became infected with West Nile (WN) virus after operating a Waring blender to grind up infected mouse brains. Circumstantial evidence again pointed to the airborne route of infection. The author attempted to show the potential infectivity of WN aerosols. Monkeys, hamsters, and mice were exposed for 15 minutes to aerosols of a 20% mouse-brain virus suspension in 10% rabbit serum sprayed from a DeVilbis type 40 atomizer. Aerosols were collected in glass impingers and blood was collected for serum-virus neutralization tests. It was estimated that each monkey inhaled 1.5×10^4 mouse intracerebral LD_{50} ($MICLD_{50}$) doses. Although no deaths occurred and antibody levels were negligible after 2 weeks, at 6 weeks a neutralization index (NI) of 3.0 logs was obtained in the exposed monkeys. The 12 hamsters, which had each inhaled approximately 1.4×10^3 $MICLD_{50}$, were dead after 6 days. Mice exposed under similar conditions (inhaling between 4×10^2 and 1.4×10^3 $MICLD_{50}$) were all dead by the seventh day. The infectious airborne potential of WN virus was thereby confirmed.

In a review article, Sulkin (1961) pointed out that since 1932 there have been more than 16 laboratory-acquired infections with monkey B virus, and that circumstantial evidence pointed to aerogenic spread of the virus.

There are additional references to a number of different viruses associated with aerogenic laboratory infections. Briefly, Beard and Easterday (1965) infected swine with aerosols of hog cholera virus. Hyslop (1965) was able to transmit foot-and-mouth disease virus (FMV) to cattle by means of FMV aerosols. The latter confirmed earlier reports which inferred that the virus had been disseminated in air currents.

Experimental Conditions during Aerosolization

As emphasis on *in vitro* techniques developed, the following questions emerged: What was the effect of the suspending fluids? What was the effect of stabilizing additives? And what was the effect of various impinger fluids on apparent airborne stability? As indicated by Nobel (1967), many of the aerosol methods for handling virus were modifications of those devised for bacteria. However, the diversity in the number of culture systems, which include embryonating eggs, chicken embryo membrane pieces, suckling mice, and different tissue culture cell lines and systems, has made direct comparison of results very difficult. To date, very few laboratories have equipment that is similar in design. In addition, the virus methodology (e.g., assay techniques) may not be uniform in laboratories working independently.

Take, for example, the matter of containers or chambers in which virus aerosols have been stored during airborne stability studies. Although this has been discussed in another chapter, it is important to reiterate that variations in equipment and mechanical procedures preclude comparison of results from different laboratories. For instance, Harper (1961) employed a 75-liter drum system, Songer (1967) a 140-liter drum, Miller and Artenstein (1967) a 500-liter drum, Webb et al. (1963) a 1,100-liter drum, and Watkins et al. (1965) and Akers et al. (1966) a 1,200-liter drum—all require a different filling and "secondary air" apparatus, and impose different physical characteristics on the aerosol. These variables are seldom reported.

A variety of atomizers have been employed—the Collison and Wells types being the most popular. If one considers mechanical variations in drum-filling techniques in conjunction with different drum volumes and atomizer delivery rates, it becomes apparent that direct comparison of past virus aerosol studies to current undertakings is very tenuous indeed. Therefore, before one can evaluate the present literature, one must know the means by which the virus suspension was obtained, aerosolized, stored, collected, and assayed.

Suspending Fluids

Harper (1961) was the first to note the effect of different suspending fluids on virus survival after atomization. Table 12-2 is a composite of data presented by Harper (1965) and shows the effects of different suspending fluids on poliovirus. It is plain that by using different fluids a series of different results could be obtained. With vaccinia, in contrast to poliovirus, survival of 1-second aged aerosols was independent of the suspending fluid. At 23 hours, however, NaCl (0.5%) exceeded both water and crude infected tissue culture supernatant as the most effective suspending fluid (Table 12-3). It is apparent that an all-purpose suspension fluid has not been developed, and that, with each virus employed, one has to screen several suspending fluids until a maximal recovery is achieved. When one attempts to consider possible effects of different suspending fluids on virus airborne stability, or what is meant by "maximal," it is impossible to draw definite conclusions.

Influenza virus, which is readily propagated in embryonating chicken eggs, is usually suspended in allantoic fluid (Shechmeister, 1950; Hemmes et al., 1960; Harper, 1961; Hood, 1963). Allantoic fluid was also used for Newcastle disease virus (NDV) and Sindbis virus (Jensen,

Table 12-2   Poliovirus (Brunhilde strain). 20–21°C.

	% Viable 1 sec after spraying			% Viable 23 hr after spraying		
	% Relative humidity			% Relative humidity		
Suspending fluid	20	50	80	20	50	80
Crude	19[a]	66[a]	120[a]	1.1[a]	Trace[a]	85[a]
Water	1[a]	85[a]	97[a]	0[a]	0[a]	50[a]
Phosphate Buffer 0.02 M	0.6	102	110	0	0	23
Cysteine 0.1%	0.8	60	118	0	0	32
Gelatin 0.5%	2.4	116	69	0	Trace	89
Calf serum 1%	0.2			0		
Tissue culture S.N.F.	14			3.7		
NaCl 0.5%	11[a]	(5.2)	110[a]	1.2[a]	(0.9)	0.2[a]
KCl 0.65%	7.1	18[a]	91	4.3	2.2[a]	9.1
K$_2$SO$_4$ 0.78%	4.7	11	123	3.1	2.9	9.6
Na$_2$SO$_4$ 0.63%	8.9	33	106	0	0	3.9
CaCl$_2$ 0.47%	0		95	0		15

[a] Arithmetic means of between two and five tests. Other values are the results of single tests.
(From Harper, G. J. 1965. In A Symposium on Aerobiology, 1963 (R. L. Dimmick, Ed.), pp 337–340. Nav. Biol. Lab., Nav. Supply Center, Oakland, Calif.)

1964; Songer, 1967). With vesicular stomatitis virus (VSV), chorioallantoic and amniotic membrane suspensions have been used (Songer, 1967), whereas other workers preferred suckling-mouse brain suspensions (Watkins et al., 1965). Vaccinia (Harper, 1965) and pigeon pox (Webb et al., 1963) suspensions have been prepared from chorioallantoic membranes. With Rous Sarcoma virus (RSV), the tumor tissue was homogenized (Webb et al., 1963), and with VEE, the suspensions were chick embryo preparations (Harper, 1965).

Not only is a diversity of egg materials employed as suspending fluids, but additives such as McIlvaine's citric acid/disodium phosphate buffer with 1% dialyzed horse serum (Harper, 1965), 0.2% gelatin (Hood, 1963), peptone (Hemmes et al., 1960), and bactotryptose broth (Beard and Easterday, 1967) have been employed.

In contrast to the findings of Hood (1963) that infectivity of influenza PR8 did not change during aging of virus aerosols, Webb et al., (1963) reported that virulence of RSV for embryonating chicken eggs was increased when aerosols were held at 80% RH in the presence of 6% inositol (Figures 12-5 and 12-6). Without inositol, maximal RSV survival

Table 12-3   Vaccinia virus. 20–21°C.

Suspending fluid	% Viable 1 sec after spraying			% Viable 23 hr after spraying		
	% Relative humidity			% Relative humidity		
	20	50	80	20	50	80
Crude	97[a]	93[a]	112[a]	15[a]	12[a]	0.02[a]
Water	100	102	52	38	4.5	0.6
McIlvaine buffer (0.002 M)			111			5
Water + 1% serum			110			20
McIlvaine buffer + 1% serum			89[a]			18[a]
NaCl 0.5%	96	102	114[a]	25	31	32[a]
KCl 0.65%	135		82	30		10

[a] Arithmetic means of between two and five tests. Other values are the results of single tests.

(From Harper, G. J. 1965. In *A Symposium on Aerobiology*, 1963 (R. L. Dimmick, Ed.), pp 337–340. Nav. Biol. Lab., Nav. Supply Center, Oakland, Calif.)

occurred at relative humidities above 70%. Citrate buffers were found to enhance recovery of RSV at 40% RH and below. The medium from which virus aerosols were generated also affected ultimate airborne survival. Addition of organic salts (as with airborne bacteria) resulted in increased survival at lower humidities (30% RH) and reduced survival at higher humidities (60% RH); the addition of proteins apparently reverses this effect. However, the possible effect of inositol in the presence of citrate buffers or of protein solutions was not reported.

Hearn et al. (1966) reported that with the Asibi strain of yellow fever virus (serially passaged in HeLa cells) variations of *in vivo* characteristics such as attenuation of virulence for monkeys, and of *in vitro* characteristics such as aerosol stability, occurred with changes in relative humidity. Aerosols were less stable and less infectious at 50% RH than at 80% RH. This leads one to suspect that the preatomization history of the virus (growth medium, for example) may affect aerosol infectivity.

Many viruses have been prepared in suckling-mouse brain tissue suspensions, i.e., VSV; Columbia-SK (Col-SK); Colorado Tick fever (CTF) and neurovaccinia (Watkins et al., 1965); and West Nile (WN) (Nir et al., 1965).

Viruses which can be easily propagated in tissue culture are usually suspended in supernatant fluid of the infected tissue culture. However,

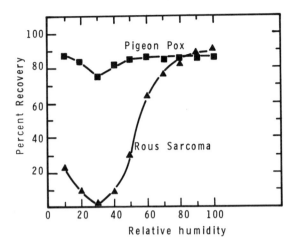

FIGURE 12-5. The effect of relative humidity on the survival of Pigeon pox and Rous sarcoma viruses during 5 hours of storage in aerosols. (Webb, Bather, and Hodges. 1963. *Canadian J. Microbiol.*, **9**: 90.) Reproduced by permission of the National Research Council of Canada.

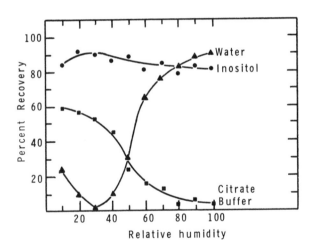

FIGURE 12-6. The effect of relative humidity and chemicals on the 5-hour survival of Rous sarcoma virus in an aerosol. (Webb, Bather, and Hodges. 1963. *Canadian J. Microbiol.*, **9**: 90.) Reproduced by permission of the National Research Council of Canada.

there is a diversity of suspending fluids, depending on the growth media used for the cell line employed for virus preparation. Harper (1965) separated his virus by ultracentrifugation prior to suspension in a defined diluent. With polio propagated in cell culture, the suspending fluid was balanced salt solution (BSS) with lactalbumin hydrolysate, 5% horse serum, and peptone (Hemmes *et al.*, 1962). For Coxsackie B, the suspending fluid was Eagle's medium with 20% agamma calf serum (Jensen, 1964); with encephalomyocarditis (EMC) group viruses, Eagle's medium only (Akers *et al.*, 1966); for polio, Earle's BSS only (Harper, 1961). For adenoviruses, Jensen employed 20% tryptose-phosphate broth, 0.1% yeast extract, and 1% agamma calf serum, whereas Miller and Artenstein (1967) used Eagle's medium with 2% chick or fetal bovine serum.

Temperature and Relative Humidity

The most complete study on the effect of temperature and relative humidity on airborne virus survival was conducted by Watkins *et al.* (1965) who studied airborne VSV at 50, 70, 80, and 90°F and at relative humidity values ranging from 20 to 80% (Figure 12-7). With VSV,

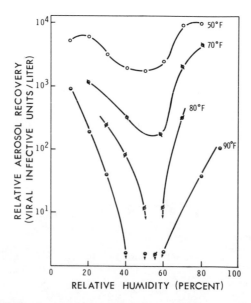

FIGURE 12-7. Effect of temperature and relative humidity on vesicular stomatitis virus aerosols aged 5 to 6 hr. (Watkins, Goldberg, Deig, and Leif. In *A Symposium on Aerobiology, 1963* (R. L. Dimmick, Ed.), p. 386. Nav. Biol. Lab., Nav. Supply Center, Oakland, Calif.)

maximal stability occurred in the range of 20% and 80% RH and minimal stability in the neighborhood of 50% RH. With successive increases in temperature the rate of aerosol decay was increased, whereas the response to humidity remained essentially the same. In this report the authors compared aerosols aged 5 to 6 hours; no comparison was made between young and old aerosols.

Harper (1961) studied airborne stability of vaccinia, influenza, and VEE at temperature ranges of 7 to 12°C, 21 to 24°C, and 32 to 34°C. He concluded that VEE, vaccinia, and influenza survive better at the lowest temperature range. Songer (1967) reported on the airborne stability of NDV virus aerosols held at 4, 23, and 37°C. At 23°C survival was better at 10% RH than at 35% or 90% RH. At 10% RH, airborne NDV survived equally well at 4, 23, and 37°C, whereas at higher humidities, lower temperature favored survival.

Akers et al. (1966) observed that humidity-dependent inactivation patterns of the EMC group of viruses was not affected by temperature. One can see by comparison of Figures 12-8 and 12-9 that at both 26 and 16°C there was an immediate decrease in survival; this decay constant is referred to as K1. At the lower temperature studied (16°C), there was little viable loss after the first 5-minute reduction (2 logs at 60% RH). However, at 26°C, and particularly at mid-range humidities, there was considerable virus inactivation (decay constant, K2) throughout the 6-hr study period. The effect of relative humidity on the initial and final decay pattern is seen in Figure 12-10. Virus particles that survive the initial stress of dehydration are inactivated at a much slower rate than those inactivated initially. An exception occurs at the mid-range (40 to 60% RH).

In conclusion, humidity-dependent inactivation of airborne viruses is established immediately after atomization and, once established, does not drastically change with the aging of the aerosol. The effect of temperature is apparently secondary. The initial inactivation expressed by K1 occurs immediately post-nebulization and within limits is not temperature dependent. However, the population that survives aerosolization with concomitant dehydration reflects a temperature-dependent decay, K2.

The question then arises as to the relation between concentration of virus suspension atomized, ultimate aerosol concentration, and the effect of relative humidity on the initial (immediately post-atomization) aerosol concentration. Harper (1965) studied aerosols 1 second after spraying and Hemmes et al. (1960) initiated studies at 30 seconds. The fact that virus concentrations in the pre-atomization suspension can influence initial aerosol concentration was evident.

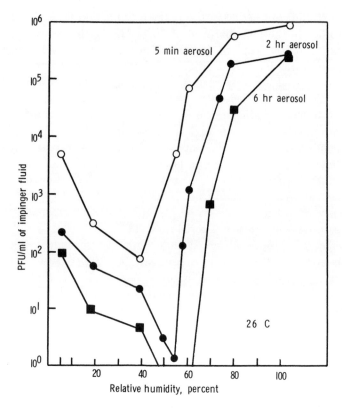

FIGURE 12-8. Airborne Maus Elberfeld virus at 26°C; effect of time and various relative humidities on virus survival. Concentration of virus suspension before atomization = 1×10^9 plaque-forming units/ml. (Akers, Bond, and Goldberg. 1966. *Appl. Microbiol.*, **14**: 363.) Reprinted with permission of the American Society for Microbiology.

Coxsackie A21 aerosols were generated in a 7-foot by 6-inch tubular chamber and studied at 25°C and 50 to 60% RH (Couch *et al.*, 1965). Virus suspensions of 10^6 to 10^{11} 50% tissue culture infective doses ($TCID_{50}$) per liter of fluid were atomized and virus aerosol concentration was found to be $10^{0.3}$ to $10^{4.3}$ $TCID_{50}$ per liter of air (Figure 12-11). The average difference in concentration of virus in the suspension and in the aerosol was $10^{6.3}$ $TCID_{50}$ per liter. This difference was attributed to the dilution of the aerosol particles in air, loss of virus viability, slight inefficiency of sampling, and sedimentation and impaction of particles in the chamber.

If one determines virus survival over a broad spectrum of humidities, as was done with members of the EMC group of viruses, one finds

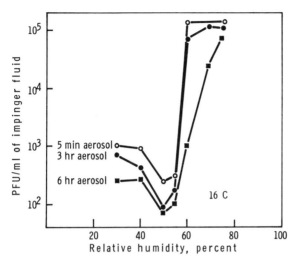

FIGURE 12-9. Airborne Maus Elberfeld virus at 16°C; effect of time and various relative humidities on virus survival. Concentration of virus suspension before atomization = 1×10^9 plaque-forming units/ml. (Akers, Bond, and Goldberg, 1966. *Appl. Microbiol.*, **14**: 363.) Reprinted with permission of the American Society for Microbiology.

that initial aerosol concentration is altered, depending on the concentration of virus in the suspension atomized and on the humidity. A Mengo-37A virus suspension of 1×10^9 plaque-forming units (PFU) per ml yielded aerosol concentrations, at humidities greater than 60% RH, of at least 2.8×10^5 PFU/ml of impinger fluid when sampled at 5 minutes post-atomization. This figure can be converted to PFU/liter of air sampled as follows;

$$\frac{(2.8 \times 10^5 \,\text{PFU/ml})(20 \,\text{ml per impinger})}{(12.5 \,\text{liters/min AGI sampling rate})(2 \,\text{min})} = 2.2 \times 10^5 \,\text{PFU/liter of air}$$

Note that the difference between pre-atomized virus suspension and the 60% RH aerosol concentration is approximately 6.3 logs, in good agreement with the results of Couch *et al.* (1965). However, it should also be noted that this is only true for relative humidity values of 60% or higher. At 50% RH and at 11% RH, the differences are approximately 8.7 logs and at 40% RH a difference of 9.3 logs is found. It is thus apparent that final aerosol concentration is determined by two factors; (1) the concentration of virus in the suspension to be atomized, and (2) the relative humidity of the secondary mixing air.

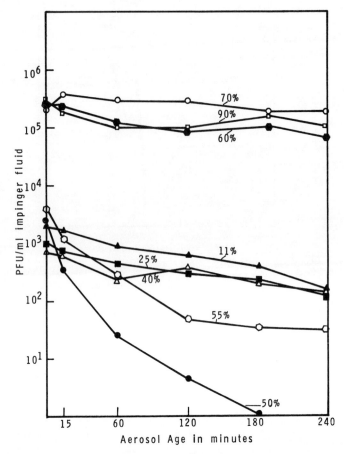

FIGURE 12-10. Stability of airborne Mengo 37A virus at 26°C. Concentration of virus suspension before atomization = 1×10^9 plaque-forming units/ml.

Particle Size

One of the important problems facing the aerobiologist is that of the effect of size of virus-laden particles on the physical decay of the particles and on inhalation efficiency in the experimental animal. Methods perfected in the evolution of bacterial aerosol techniques have been applied to the study of viruses; these are discussed elsewhere. However, since virus particles *per se* are almost a thousand times smaller than most bacteria, they could theoretically form much smaller airborne droplets than other microbes.

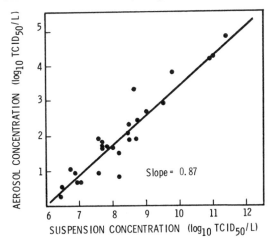

FIGURE 12-11. Relation of virus concentration in atomization suspension to virus concentration in aerosol. Each point represents an aerosol experiment and the line represents the slope that best fits these points. (Couch, Gerone, Cate, Griffith, Alling, and Knight. 1965. *Proc. Soc. Exptl. Biol. Med.*, 118: 820.) Reprinted with permission of the Society for Experimental Biology and Medicine.

Employing static chambers, Edward *et al.* (1943) passed aerosols of influenza virus through dry membranes of known porosity and observed that particles penetrating membranes with pores of 4.2 μ, were infectious for mice. The authors estimated that under dry conditions the mean particle size was 1.3 μ and under moist conditions was 2.3 μ. Thus the relationship of particle size to the relative humidity of the environment was indicated. Moreover, dry aerosols were found to be more persistent than those held at high relative humidities; the lack of persistence at high humidities was thought to be largely due to settling.

Shechmeister reported that the mean diameter of influenza virus droplets encountered in his aerosol study was less than 0.5 μ. Particle size was determined by the particle size analyzer described by Goldberg (1950).

Some investigators have simply employed the Andersen sampler for particle size determination. Jensen (1964) passed aerosols of Coxsackie B_1, Influenza A, Sindbis, and vaccinia through an Andersen sampler in which each petri plate was filled with 2% agar overlayed with 0.3 ml of a 20% skim milk suspension. After aerosol sampling, the skim milk was resuspended in Hank's BSS and assayed for virus. The highest concentrations of virus were found in stages 3 and 4 of the Andersen sampler (Table 12-4). However, this is misleading. For instance, a particle trapped on stage 3 will have approximately 15 times the volume of a particle trapped on stage 5 ($3.9^3/1.5^3$). Based on this difference in volume particles trapped on stage 3 will contain more infectious units (PFU) per particle than those trapped on stage 5. In order to correct for this the total number of PFU's per stage is divided by the cube of the characteristic diameter for that particular stage. Using this correction, I have included in Table 12-4 the approximate percentage

Table 12-4 Stage distribution of viral aerosols collected in the Andersen sampler[a]

	Plaque-forming units per stage of sampler					
Stage	1	2	3	4	5	6
Particle size, μ	>9.2	5.5–9.2	3.3–5.5	2–3.3	1–2	<1
Virus						
Coxsackie B_1	10,500	17,750	21,750	26,750	20,250	11,250
	(<1)[b]	(<1)	(1)	(7)	(23)	(69)
Vaccinia	8,775	30,480	42,930	141,750	49,500	1,785
	(<1)	(<1)	(3)	(32)	(54)	(11)
Sindbis	1,830	4,000	15,750	14,100	8,850	840
	(<1)	(<1)	(6)	(17)	(50)	(26)
Influenza A	150	1,200	4,200	3,300	450	...
	(<1)	(1)	(20)	(48)	(31)	...

[a] From Jensen, M. M. 1965. In *A Symposium on Aerobiology, 1963* (R. L. Dimmick, Ed.), p. 222. Nav. Biol. Lab., Nav. Supply Center, Oakland, Calif.
[b] Numbers in parentheses are percentages of total particles per stage (see text).

of particles collected per stage. It is obvious that the largest numbers of particles were collected by stages 5 and 6 and not by stages 3 and 4.

Guerin and Mitchell (1964) collected airborne virus in a 3% solution of gelatin in tissue culture medium. They dispensed 5 ml into each Andersen sampler petri plate; the airborne viruses were impacted onto the gelatin cushion, and the gelatin was then liquified by heating the plates to 37°C. When concentrations of virus are very small, virus samples can be concentrated by centrifugation at high speeds. Otherwise, serial dilutions can be made from samples collected at each stage. With poliovirus, 75 to 90% of the droplets passing to the fourth through sixth stages (up to 3.3 μ) contained infectious virus, whereas only 10% of particles remaining on stages 1 through 3 (3.3 μ and larger) were infectious. With influenza (PR8 strain) virus, whereas all stages gave positive results for infectious virus, only stages 4 through 6 showed a positive direct hemagglutination (HA) test. These results are inconclusive, however, because larger particles may have contained viral aggregates.

Beard and Easterday (1965) exposed chickens to a Newcastle disease virus aerosol which was then passed through a membrane filter of 0.8-μ pore size. The size of the particles trapped by the membrane (and therefore infecting the chickens) was estimated by light microscopy and found to be less than 1–2 μ.

Couch et al. (1965), studying the properties of Coxsackie A_{21} aerosols held in a 64-oz Mason jar, allowed the aerosol to settle on a glass slide coated with Permount. The particle diameter was then measured by light microscopy. Eighty-five percent of the particles were less than 1 μ in diameter; the mean particle diameter was 0.656 μ. When an Andersen sampler was employed, 68% of the virus was recovered from stage 5 (1 to 2 μ) and 90% was recovered on stages 4 through 6 (less than 3.3 μ).

Physical Fallout

Tracer materials added to spray suspensions are an important adjunct in determination of physical fallout. The decay of tracer material in sequential samples is a measurement of physical behavior and the ratio of tracer to biological activity measures the biological behavior.

Harper (1961) used a suspension of formalin-killed *Pasteurella tularensis* cells, previously labeled by growth in a medium containing P^{32}, as a tracer to be added to a suspension of VEE which was subsequently aerosolized. In tracer studies with vaccinia, influenza, and polio, P^{32} was added to the virus suspension so that the final isotope concentration was about 10 μc/ml. The original virus/tracer ratio was regarded as 100% and sample ratios thereafter were expressed as percentages of this original number.

Songer (1967) employed a 0.1% solution of Rhodamine B as a tracer for his virus aerosols. An aliquot of each sample was assayed for fluorescence, and change in concentration of tracer was used to determine physical decay. Total decay was determined by virus assay, and by subtracting physical decay from total decay, biological decay was determined for each time interval (Figure 12-12). Miller and Artenstein (1967) used Uranine dye (soluble sodium fluorescein) and related loss of infectivity of aerosols of adenoviruses 4 and 7 and parainfluenza to physical fallout.

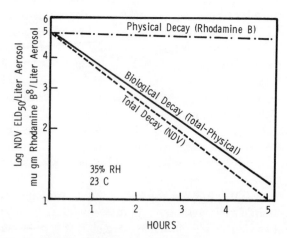

FIGURE 12-12. Comparison of physical, biological and total aerosol decay in the rotating drum chamber. (Songer, 1967. *J. Appl. Microbiol.*, **15**: 38.) Reprinted with permission of the American Society for Microbiology.

To date the most sensitive tracer for determining physical fallout is Calcafluor White BGT. The compound is added in minute amounts to a viral or bacterial suspension and the actual number of fluorescent particles are counted with the microaerofluorometer designed by Goldberg (1968).

In conclusion, comparison of results of the airborne stability of viruses under various conditions is difficult if not impossible because, even in laboratories employing the same equipment, different investigators working with the same strain of virus report different results. It is obvious that many factors affecting survival in the airborne state are as yet not understood.

Pathogenesis

We shall now turn our attention to viral pathogenesis in animals and to aerogenic immunization studies. Pathogenesis of an infectious agent is defined as the movement or spread of the infecting agent through the host from the portal of entry to its portal of exit, including its lodgment, multiplication, and tissue invasion. Concomitant with the above is the correlation with clinical and histopathological observations.

To study pathogenesis of a virus infection in a laboratory host, the investigator may employ the intraperitoneal (ip), intracranial (ic), subcutaneous, and other inoculation routes; he is then able to initiate infection with a known quantity of inoculum. However, with aerosol exposures there exists an inherently greater margin of error that significantly affects dosage calculations. For example, to initiate infections of the respiratory tract, aerosols must consist of small particles (1 to 5 μ). Larger airborne particles invariably are trapped in the upper respiratory system and are swallowed along with a certain proportion of the smaller ones. Furthermore, inherent variabilities of the exposure equipment must be measured and included in calculations. For instance, impingers have a certain slippage rate; i.e., they operate with an efficiency in the neighborhood of 90% for particles 1 μ and above—below this size, efficiency decreases; when air is expanded, as from tubing into a chamber, moisture equilibrium changes; the height of an exposure chamber influences the physical fallout rate. The retention rate for inhaled 1 μ (assuming one had a monodisperse aerosol) particles in mice varies from 30% (Goldberg and Lief, 1950), 36% (Harper and Morton, 1953), to 65% (Rosebury, 1947). At best, one can consider the respiratory dosage to be approximately one-half the inhaled dosage. Therefore, variability in impinger collecting efficiency coupled with fluctuations in small particle retention by the lungs reduces the accuracy of dose calculations. In addition, swallowed virus particles may initiate infection in the intestinal tract, thus confusing the airborne-respiratory nature of the infection. This is particularly true with airborne picornaviruses which simultaneously initiate intestinal and respiratory involvement in mice.

With the above points in mind, one generally calculates the inhaled dose according to Guyton's (1947) formula:

$$D = (A)(w)(c)(t)$$

which for mice becomes

$$D = (0.00125)(w)(c)(t)$$

where D = dose of virus inhaled in PFU's or LD_{50},
A = respiratory volume in liters/gm body weight/ minute = 0.00125 for mice,
w = weight of test animal in grams,
c = aerosol concentration in PFU's or LD_{50}/liter of air, and
t = time of aerosol exposure in minutes.

Guyton's formula is easy to use and interpret when impinger fluid infectivity (as is done with bacteria and certain viruses) is expressed as colonies or plaque forming units (PFU). After animals are exposed, one can express the inhaled dose in terms of a given number of PFU's, etc. However, when serial dilutions of impinger fluid are tested for virus content by calculating a 50% infectious endpoint of that series of dilutions, the results may be somewhat confusing. Where laboratory animals or embryonating chicken eggs are used, the endpoint may be based on mortality (a 50% lethal dose LD_{50}) or infectivity (a 50% infectious dose ID_{50}). When tissue cultures are employed for assay, 50% endpoints are expressed as tissue culture infective doses ($TCID_{50}$), or as cytopathic doses. One is then confronted by experimental results in which mice were reported to have inhaled so many egg LD_{50}'s, or exposed monkeys have inhaled a quantitated $MICLD_{50}$ of virus. Therefore, one must be alert to the fact that (a) less than half of the inhaled particles may be retained in the lungs after inhalation, and (b) it must be assumed that the host used for assay is equally, if not more, susceptible to the virus than the animal exposed to the aerosol.

Danes et al. (1962) reported on the experimental airborne infection of mice with tick-borne encephalitis (TBE) virus. The estimated inhaled lethal dose was between 10 and 40 $MICLD_{50}$. Mice which had inhaled approximately 1,000 $MICLD_{50}$ first evidenced virus multiplication in lung tissues and showed increases in virus titers from 1.5 $MICLD_{50}/0.03$ ml at 24 hours to nearly 4 $MICLD_{50}/0.03$ ml at 8 days. After appearance of virus in this initial target organ, viremia became evident at 72 hours post exposure and involved the central nervous system (CNS) at 96 hours.

Benda et al. (1962) exposed monkeys to TBE virus aerosols and observed no clinical signs indicative of CNS involvement, although the inhaled dose was estimated to be 5,000 to 50,000 $MICLD_{50}$. Pathological changes consisted of irregular and limited microscopic changes in the brain and spinal cord tissues only. Elevated temperature, and viremia, aside from antibody formation, were the only noteworthy events. It is of interest that at this dosage level death would have ensued if virus had been introduced via the ip or ic routes.

Hahon and McGavran (1961) studied airborne infectivity of the variola-vaccinia group of pox viruses (variola, vaccinia, rabbit-pox, alastrim, monkey pox, and cow pox) for the cynomologus monkey. The pattern of disease consisted of a febrile reaction, constitutional disturbances, variable mortalities, and an immune response. The basic histopathologic changes consisted of ulcerative bronchitis, bronchitis, and peribronchitis. Virus was recovered from the lung tissues of some monkeys exposed to aerosols. The calculated aerosol doses ranged from 1.7×10^5 to 5.8×10^7 egg infectious units, depending on the virus strain aerosolized. This study showed the potential airborne infectivity of the pox-virus group for monkeys.

White and Madin (1964) were able to infect the C3hf/CRGL strain of mice with murine hepatitis virus by means of aerosols. Feeding or inoculation of virus into the stomach by means of a gavage did not produce infection. After aerosol exposure, virus was not recovered from the lungs until after 24 hours—at least 1 day earlier than other organs. Histopathological changes, indicative of a bronchogenic pneumonitis, were evident. Following initial multiplication in the lungs, virus was carried to the liver, kidney, and spleen by the blood. It was concluded, based on positive aerosol and negative oral transmission results, that the respiratory tract was a natural route of infection.

Nir *et al.* (1965) followed the course of West Nile pathogenesis in mice after aerosol exposure. Test mice were estimated to have inhaled between 2.4-16 $\times$ 10^4 MICLD$_{50}$. Virus was recovered from lung specimens immediately after aerosol exposure, but decreased during the next 4 hours. Lung virus titers began to increase above the inhaled dose 24 hours after exposure (Table 12-5).

Table 12-5   Distribution of West Nile virus in selected tissues of mice at various intervals following exposure to the virus in aerosol form.

Tissue	Time (hr)							
	0	4	24	48	72	96	120	144
Blood	0[a]	0	0	0.2	0	0.6	0.5	0.2
Liver	0	0	0	0	0	0.2	0.3	1.1
Spleen	0	0	0	1.0	0.9	0.5	0.6	1.7
Kidneys	0	0	0	0	0	0	0.5	1.8
Adrenals	0	0	0	0	0	0	1.7	3.0
Cervical lymph nodes	0	0	0	0	0	0	0	0.3
Lung	1.0	0.6	2.6	4.0	4.5	4.1	4.2	4.1
Nasal mucosa	0	0	0	0	0.1	0.4	1.4	1.4
Brain	0	0	0	0.1	1.7	3.7	8.0	7.6

Titers of virus expressed as the log of the reciprocals of the LD$_{50}$ endpoint dilution. Means of titrations from 4–6 mice.
[a] 0 = No virus detected in highest concentration tested.
(From Nir, Beemer, and Goldwasser, *Brit. J. Exptl. Pathol.*, **46**: 446, August 1965. Reprinted with permission of H. K. Lewis & Co. Ltd., London.)

Most fluorescent antibody slides prepared from liver, spleen, and kidney tissues taken from 0 hours through 144 hours indicated the absence of viral antigen. Antigen was located in the cytoplasm of macrophages in lung sections collected at post-24 hours and thereafter. Whereas lung, liver, spleen, and kidney tissues showed no histopathological changes following aerosol exposure, the CNS pathology was typical for neuro-tropic viruses. It was noteworthy that virus multiplication in the lungs was not accompanied by pathological changes.

Akers *et al.* (1968), reporting on the pathogenicity in mice of aerosols of EMC group viruses or their infectious nucleic acids, observed that when mice were exposed to aerosols of different strains the observed differences in pathogenicity were correlated with plaque size, state of the virus [intact or infectious ribonucleic acid (RNA)] and the presence or absence of circulating antibodies. With aerosols of mengo-37A, a small plaque-forming (Spf) immunogenic strain, virus was recovered from the lung, intestines, spleen, liver, and blood (Figure 12-13); pathological changes occurred in the lungs and heart. Mice exposed to lethal aerosols of Col-SK or mengovirus yielded virus from every organ (Figure 12-14). With mengo RNA, however, there was a delayed appearance of virus in the intestinal tract. This delay was a result of the inactivation of the initially swallowed RNA; the complete virus which appeared in

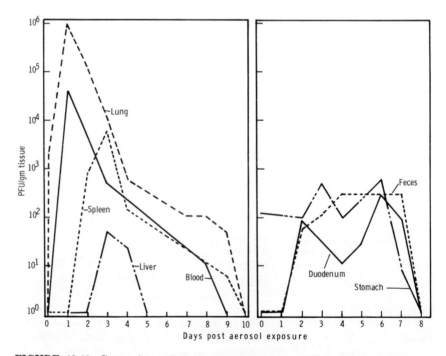

FIGURE 12-13. Comparison of virus content of tissues after exposure of mice to aerosols of Mengo 37A. (Akers, Madin, and Schaffer. 1968. *J. Immunol.*, **100:** 122.) Reprinted with permission of the Williams and Wilkins Co., Baltimore, Md.

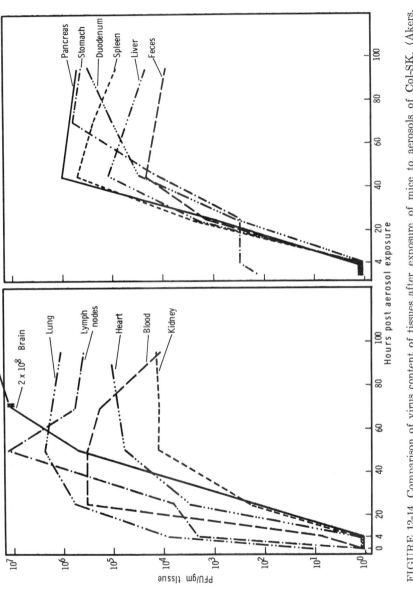

FIGURE 12-14. Comparison of virus content of tissues after exposure of mice to aerosols of Col-SK. (Akers, Madin, and Schaffer. 1968. *J. Immunol.*, **100**: 121.) Reprinted with permission of the Williams and Wilkins Co., Baltimore, Md.

the stomach 48 hours later was synthesized in either the respiratory tract or the oral-pharyngeal area. Mice exposed to lethal mengo RNA, Col-SK, or mengovirus aerosols exhibited similar pathological changes, which occurred only in brain and liver tissues. No deaths were observed when mengo-37A-immunized mice were challenged with lethal Col-SK or mengovirus aerosols. However, challenge virus was recovered from the lungs, intestinal tract, spleen, liver, and blood. Pathological changes were observed in lung and liver tissues. When mengo-37A-immunized mice were exposed to mengo RNA aerosols, virus was isolated from lung tissues only.

Henderson et al. (1967) exposed hamsters to aerosols of Semliki Forest virus (SFV). Virus multiplication occurred in the respiratory and olfactory mucosa, then sequentially invaded the olfactory region of the CNS. Virus was not present in the feces or intestinal tissue, but was detected in the buccal cavity. Histological changes were observed in the CNS, liver (acute hepatitis), and olfactory mucosa.

In summary, airborne virus invades the respiratory tract as a primary target area; swallowing of virus to cause intestinal involvement depends on the nature of the virus itself. Picornaviruses can infect the gastrointestinal (GI) tract, but myxoviruses and arboviruses have limited intestinal invasiveness.

Aerogenic Immunization

The technique of immunization by exposure to a virus aerosol has not reached the stage where large scale use is practicable. Earlier studies with influenza showed that the airborne route of exposure to certain strains of influenza produced immunity. Except for a recent report in the literature pertaining to measles, emphasis on aerogenic immunization has been directed towards diseases of commercial animals and flocks. Studies employing NDV aerosols for flock vaccination (Hitchner and Reisling, 1952), distemper aerosols with ferrets and mink (Gorham et al., 1954), and the use of aerosols to immunize piglets against swine fever (Kulesko et al., 1963) have been reported. However, Beard and Easterday (1965) were unable to immunize swine with aerosols of undiluted hog cholera virus vaccine.

Earlier aerosol techniques were rather primitive. Hitchner and Reisling (1952) simply atomized a predetermined amount of virus suspension (20 ml/1000 birds) and later bled a number of the exposed fowl to determine antibody response. No attempts were made to quantitate aerosol concentration or droplet size. Nevertheless, this empirical approach was successful on the basis of serological results. Later work by Beard and Easterday (1967) confirmed this earlier NDV study; chickens vaccinated with aerosols of NDV virus developed higher levels of hemagglutination inhibition (HI) and neutralization antibodies than chickens vaccinated with larger amounts of the same vaccine injected via the intramuscular (im), conjunctival, or intranasal route. Moreover, aerosol vaccinees were resistant to both aerosol and intramuscular challenge. Chickens vaccinated intramuscularly, while resistant to intramuscular challenge, were not resistant to aerosol challenge. Viral aerosol particles probably became widespread throughout the respiratory tract and were able to elicit a more effective antibody response. Also, chickens that had initially received either aerosol or intramuscular inoculations when later challenged by either route yielded virus from tracheal swabs. This would tend to indicate the circulating NDV antibodies could not prevent infection of the respiratory epithelium, but were effective in confining the infection to the general area of the respiratory tract.

Kuehne et al. (1962) showed that mice, guinea pigs, and Rhesus monkeys could be immunized by exposure to aerosols of living attenuated VEE virus. Immunization attained was comparable to that observed following parental inoculation. The fact that aerogenic immunization against the highly infectious strain of VEE is possible with attenuated strains lends encouragement to the search for possible respiratory immunization methods for other viruses.

The immunization efficiency of attenuated virus strains depends on their genetic histories. In general, the use of inactivated virus suspensions will not suffice for aerogenic immunization. Schulman and Kilbourne (1962) showed that aerosols of inactivated influenza virus (influenza A) produced in mice a transient (3-day) period of protection as noted by reduction in pulmonary virus titers and reduced occurrence of gross lesions. The effect was attributed to viral interference.

Thus, the best aerogenic immunization results are attainable with a living attenuated vaccine. Association of plaque size (a genetic marker) with virulence or immunogenicity is sometimes possible. For instance, Hearn (1961) observed that the attenuated VEE strain, which produced numerous small plaques, was also associated with the absence of mouse ip virulence. The parent VEE strain, a large plaque-former (Lpf) was lethal for mice when inoculated ip. Furthermore, the small plaque-former (Spf), instead of being lethal when airborne, was capable of aerogenic immunization. With the EMC group viruses we have observed the same phenomenon; i.e., small plaque-forming strains (administered via the respiratory route) were capable of immunizing mice against lethal aerosols of the large plaque-forming strains. When mice inhaled 1.1 to 16×10^3 PFU's of Col-SK or mengovirus, mortality was 100%; when mice inhaled 6×10^3 PFU's of either Maus Elberfeld (ME) or Mengo-

37A (both Spf strains) no deaths occurred, neutralization antibodies appeared, and mice were resistant to a later challenge with a lethal Col-SK aerosol.

When mice inhaled 42 to 50 PFU's of either Lpf strains, approximately 40% of the exposed mice survived. None were resistant to a later lethal aerosol challenge; i.e., no immunity developed in the survivors. With Lpf infection, results were absolute; i.e., either the animal died or it survived with no immunity, implying that no infection occurred in the survivors. On the other hand, 42 to 45% of mice that inhaled 15 to 28 PFU's of Spf strains were resistant to later lethal aerosol challenge. This indicates that Spf strains can initiate nonlethal infections and immune responses. Mice immunized with Spf and challenged with lethal Lpf doses supported virus replication in the respiratory tract, the GI tract, and in certain other organs. Mice exposed to Spf aerosols and Spf-immunized mice challenged with lethal Lpf aerosols—except for time of occurrence—show almost identical pathogenesis. Probably, barriers effectively limiting the spread of Spf strains within the mouse, and the site of antibody-virus reaction, are the same. Further credence for this probability is found in results of experiments in which Spf-immunized mice were exposed to lethal aerosols of mengo RNA. Virus was recovered only from lung tissues—all other tissues tested were negative. The inference is that neutralizing antibodies were of sufficient quality and quantity to limit virus replication to the respiratory tract. The swallowed portion of the RNA aerosol did not survive passage in the intestinal tract.

Thus one can see that with certain arboviruses and picornaviruses, aerogenic immunizations can produce interesting laboratory model systems. Only one recent report on application of aerogenic immunization to human populations is relevant. Okano (1965) successfully immunized 313 of 318 (98.4%) persons against measles virus by having them inhale a vaccine strain (the attenuated Toyashima strain) of measles virus. Marked increases in neutralizing antibodies were noted in the vaccinees; responses obtained with the inhalatory method of immunization were comparable to results obtained with vaccine injection.

One ought to be aware of factors limiting the employment of aerosols for large scale immunization:

(a) The simple dispensing of an aerosol into a large room containing the subjects has many pitfalls. Temperature and humidity may affect the stability of the airborne virus, and hence influence dosage. Particles may adhere to the clothing, hair, skin, etc., of the vaccinee and be subjected simply to mechanical transfer outside of the confines of the exposure area. Vaccinees should be selected for similar breathing patterns to provide for dosage adjustments; e.g., children should not be exposed to the same aerosol concentration as adults.

(b) The virus may produce a transient viremia in the subjects and potentiate an arthropod transmission to a species in which the attenuated strain might be lethal.

(c) The aerosol may travel beyond the confines of the exposure chamber and spread to the local wildlife to establish a natural reservoir of infection. In this case, the attenuated virus strain may mutate or revert to a more virulent form.

Aerogenic immunization might be feasible in a military recruit camp where facilities, such as ultraviolet light barriers, disinfectant showers, negative air pressures, and, probably most important, effective equipment maintenance and operation together with personnel discipline, could be maintained.

BACTERIOPHAGE

The potential role in aerobiology of bacteriophage, as approximate models for viral systems, is just being realized. Besides being so easy to work with in the laboratory, their variety of shape and size is such that phage can be employed to simulate pathogenic mammalian viruses. Moreover, the very nature of the differences in nucleic acid structure of different phage—single- or double-stranded deoxyribonucleic acid (DNA) or ribonucleic acid (RNA), as the case may be—makes phage an ideal model system for determining the role of nucleic acid structure relative to airborne stability.

If one considers methods by which phage are prepared for aerosolization, collection, and assay, it is evident that there are no novel techniques specific for phage alone. For instance, most coliphage suspending fluids consisted of the growth media for *Escherichia coli* in which the phage was replicated. This situation is the same as for viruses that are produced in tissue culture, where the atomizer fluid was that medium employed to grow cell cultures prior to virus inoculation. As with viruses, phage may be concentrated by ultracentrifugation and resuspended as desired. The intricacies of sampling phage aerosols are discussed in another chapter.

Aerosol Stability

To date, there is a dearth of literature on airborne phage; at best, it can be stated that phage aerosol research is in a developing state.

Hemmes *et al.* (1962) were the first to effectively study the airborne stability of a T5 phage. Using a static chamber system, these investigators observed that the death rate of airborne, T5 was critically influenced by humidity. Although airborne survival was modified by the suspension fluid used, simiiar humidity-dependent inactivation patterns were observed irrespective of suspending fluid (Figure 12-15). As can be seen, 1% peptone provided a stabilizing effect. When different temperatures were employed (10, 20, 30°C), the effect of humidity was still manifest, regardless of temperature. In each case a rather sharp transition from high recovery at low humidity to low recovery at high humidity occurred. The increase in death rate with temperature (Q_{10}) was 2–3. The authors concluded that the temperature at which T5 phage aerosols were held was of minor influence on death when compared with humidity effects.

Ehrlich *et al.* (1964) studied the effect of environmental factors on airborne stability of T3 coliphage. Freshly prepared or stored T3 suspen-

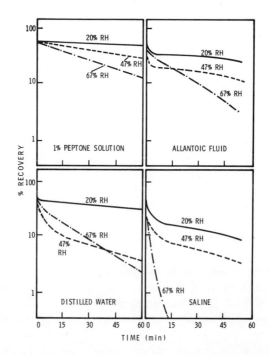

FIGURE 12-15. The influence of relative humidity on the survival of bacteriophage T5 aerosolized in media of different composition at 24°C. (Hemmes, Winkler, and Kool. 1962. Antonie van Leeuwenhoek. *J. Microbiol., Serol.,* **28:** 224.) Reprinted with permission of Swets and Zeitlinger, Amsterdam, The Netherlands.

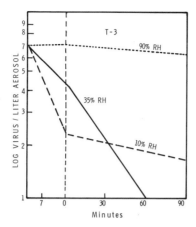

FIGURE 12-16. Decay of T3 phage aerosols at 23°C and at 35, 90, and 100% relative humidity. Each line represents three tests. (Songer. 1967. *Appl. Microbiol.,* **15**: 39.) Reprinted with permission of the American Society for Microbiology.

sions possessed similar aerosol properties and were therefore suitable for replicate studies. Using a 1600-liter static chamber, the authors found the highest decay rate at 55% RH (12.8%/min), followed by 30% RH (9.7%/min); greatest stability was at 85% RH (decay rate of 5.0%/min.) This sensitivity of T3 phage to midrange humidities is in sharp contrast to the airborne behavior of T5 phage. When assorted chemicals were added to the phage suspensions prior to aerosolization, it was noted that 0.1 M dextrose solution reduced aerosol decay at 50% but not at 85% RH. Spermine, spermidine phosphate, thiourea, galacturonic acid, and glucosaminic acid had no effect on decay rates. Survival data derived from the T3 studies indicated that this airborne coliphage behaves similarly to VEE, polio, and Rous sarcoma viruses.

Further studies with airborne T3 coliphage by Songer (1967), who used a 140-liter capacity torpid drum, confirmed the sensitivity at midrange humidities of this coliphage. T3 aerosols survived best during storage at 90% RH, least at 35% RH, and were fairly stable at 10% RH. As seen in Figure 12-16, the "generation and stability" loss was low at 90% RH, but high at 10% RH. After this initial period, the decay was more gradual compared to T3 aerosols at 35% RH. This characteristic of stability of phage at low relative humidities, seen after some loss induced by dilution of the newly generated aerosol with a dry, secondary air mixture, closely resembles that observed for several viruses.

Air Ion Effects

Because of their general ease of handling, their nonpathogenicity for man, and the fact that they are unique in relation to bacteria phage have been employed for divergent aerosol model systems. For instance, Happ et al. (1966) studied the effect of air ions on aerosols of T1 phage Recoveries of all ionized aerosols were less than the recoveries from nonionized aerosols, indicating that air ions affected stability of the aerosols of T1 phage. With the AGI-4 liquid impinger, the mean recovery at 22°C and approximately 29% RH was 48.3% for mixed ions, 43.6% for negative ions, and 11.4% when positive ions were added to the aerosol. It was also reported, as indicated by electron micrographs, that there were no significant changes in particle size or shape between airborne phage exposed to either ionized or controlled environments.

Mechanisms of Death

Phage have also been employed to investigate death mechanisms of airborne bacteria. Cox and Baldwin (1964) examined aerosol samples of E. coli, or E. coli infected with T7 phage aged up to 2 hours. They reported that in samples collected after 30 minutes, more E. coli cells were capable of supporting phage replication than were capable of colony formation. After 2 hours, the proportion of cells capable of supporting phage replication declined. However, it is significant that T7 DNA remained intact within the airborne cell and even while its own reproductivity diminished, the ability of E. coli to support phage replication remained active. Later, Cox and Baldwin (1966) reported that aerosol stability of T7 phage exposed to E. coli was dependent on the stage of development in the host bacterium. They reasoned that this aerosol stability was related to the initial process of adsorption, and injection of phage DNA, as well as to processes which occur later in the replication cycle.

Webb et al. (1965) reported on effects of relative humidity and inositol on the ability of airborne stressed E. coli to produce phage. Whereas T1, T3, and T7 phage replication in stressed E. coli was affected by relative humidity, T2 and T4 were not. For instance, E. coli aerosolized and held at 60% RH possessed a reduced T1 phage replication capacity that was even more evident at 40% RH. Two points from this study warrant mentioning: First, inositol prevented loss of replication capacity; and second, loss of the host cell's ability to support phage replication was never as great as the loss of cell viability.

Further discussion of this type of phage experiment would lead into the area of possible mechanisms of death that are discussed elswhere.

However, it is worth pointing out not only that phage systems are unique, but also how they can be employed in aerobiology.

Buckland and Tyrrell (1964) employed T3 phage as a biological tracer in studying airborne dispersal of nasal secretions. After instilling known amounts of phage in the nose, conjunctiva, or mouth, they observed that the biological half-life in the nose was between 3 and 7 minutes, and in the mouth, between 2 and 6 minutes. Tracers placed in the nose passed rapidly down the throat and were found only in small amounts in the saliva. The most efficient means of dispersal was by sneezing, and even then, only about 0.1% of the droplets were small enough to remain airborne. This study is another example of the use of phage in airborne particle studies.

PSITTACOSIS, RICKETTSIAE, AND MYCOPLASMA

To date, there are no reports of studies on the airborne stability of either psittacosis or rickettsiae. In part this is due to the highly infectious nature of these airborne organisms and to the fact that few laboratories are suitably equipped to safely conduct airborne studies.

Psittacosis

Airborne-initiated epidemics of psittacosis have been associated with laboratory accidents (Rosebury et al., 1947), poultry dressing establishments (Irons et al., 1955), and contact with psittacine bird populations (Meyers and Eddie, 1958). A most significant contribution was the report of Bolotovskii (1959), who studied experimental infections initiated in mice by psittacosis aerosols. Starting with a suspension of a Lori strain, titering between 10^7 and 10^8 $MICLD_{50}$ approximately 1 ml of a 10% mouse lung suspension was atomized in a volume of 126 to 378 liters of air; 100% of the test mice exposed to such aerosols succumbed. Interestingly, it was noted that those test mice exposed to aerosols at 80% RH exhibited survival times of 9 to 11 days, whereas mice exposed to aerosols held at 95% RH survived only 5 to 8 days. It was concluded that 90% RH, or higher, was an optimal humidity for holding airborne psittacosis for the greatest efficiency of infection. The minimal infective dose for a 5-minute aerosol exposure (100% mortality) was 100 $MICLD_{50}$.

McGavran et al. (1962) exposed 24 Rhesus monkeys to aerosols of a Borg strain of psittacosis. The aerosol was disseminated into a 543-liter static chamber held at 21 to 22°C with the relative humidity of the aerosol ranging from 26 to 30%. The median diameter of the airborne particles was 1 to 1.5 μ. Anesthetized monkeys were exposed for 5 minutes. The calculated inhaled dose of agent was between 4,000 and 5,000 $MICLD_{50}$. None of the test monkeys became ill after exposure. The initial site of infection was in the respiratory bronchiole, spreading so as to result in lobular pneumonia. Anatomic evidence of infection was scant in organs other than the liver, although recovery of virus was achieved from the blood, liver, and spleen, which emphasized the systemic nature of the infection. It was also reported that the organism persisted for more than two weeks after the appearance of serum antibodies.

Rickettsiae

Because of the stability and general resistance to environmental stresses of *Rickettsia burnetii*, coupled with the fact that a single organism can initiate infection in man (Tigertt *et al.*, 1961), the natural method of initiation of Q-fever is thought to be by inhalation. The incrimination of the inhalation route becomes more tangible when one notes that this is the only rickettsial disease naturally transmitted in the absence of an arthropod vector. Although there are reports pertaining to epidemics of Q-fever among packing-house workers, and outbreaks among laboratory personnel, the study of Wellock (1960) showed the true airborne potential of the disease. Aerosols of Q-fever, generated in an animal-fat rendering plant that had received potentially infected ewes, infected numerous individuals, the majority of cases residing within a fan-shaped area downwind from the suspected plant. Twenty cases or more were located within 10 city blocks of the point of origin. In fact, the concentration of case residence was higher in the upwind portion of the case distribution pattern. Windborne dissemination was successful in infecting cases as far as 10 miles from the source of origin.

When guinea pigs were exposed to Q-fever aerosols with particles in the 1-μ diameter range, it was concluded that a single particle inhaled was capable of initiating infection (Tigertt *et al.*, 1961). A single particle could also initiate infection via the i.p. inoculation route. Results of this study also indicate that in man, one organism can initiate infection via the inhalation route, and although the lung may be the primary target organ after aerosol exposure, pathologic manifestations usually occur elsewhere.

Saslaw *et al.* (1966) exposed monkeys to aerosols of Rocky Mountain spotted fever (*Rickettsia rickettsii*). Fifty-six of 60 monkeys became ill after exposure to aerosols containing 1 to 6,000 yolk sac LD_{50}'s (YLD_{50}) per liter. Forty of these 56 monkeys died between 7 and 24 days following exposure. No primary rickettsial pneumonia was observed, and it thus appeared that the lungs served mainly as a portal of entry.

The authors concluded that the disease produced in monkeys clearly resembles that observed in naturally occurring infections in man, and that the potential for airborne laboratory infections should not be underestimated.

Other rickettsial diseases, aside from Q-fever and Rocky Mountain spotted fever, have been associated with laboratory infections resulting from generated aerosols. In fact, as with psittacosis, rickettsia must be handled with a high degree of awareness. Although known to survive the rigors of the natural environment, no aerosol stability studies have been reported to confirm this.

Mycoplasma

The last of the submicron organisms to be discussed are the mycoplasma. These organisms, which have a minimal reproductive unit of approximately 125 mμ in diameter, are one of the smallest living forms that can be cultured in a cell-free system. To date, the only study of airborne mycoplasma over a wide range of relative humidities is the recent report of Wright et al. (1968). Previously, Kundsin (1966) and Beard and Andersen (1967) reported on airborne mycoplasma, but not at a variety of humidities. Kundsin (1966), employing an Andersen sampler, observed the medium diameter of droplet nuclei for aerosols of M. hominis, M. pharyngis, and M. pneumoniae to be from 1.5 to 3.1 μ. Using a static chamber, and aging the aerosols at 23% RH, there was a reduction in colony-forming units (CFU) of less than 1 log after 45 minutes; the resultant decay constant, K (0.008), indicated a survival potential of approximately 6 hours. Beard and Andersen (1967) studied M. gallisepticum and M. meleagridis aged in a rotating 55-gallon steel drum at 25°C and at 40 to 50% relative humidity. One percent and 0.1% of the original recoveries were mean values obtained for these two species after 6 hours in the airborne state.

Wright et al. (1968) aerosolized suspensions of either M. laidlawii or M. gallisepticum into rotating drums. Their results are shown in Figure 12-17. M. laidlawii was very stable at 25% RH or less, and at 75% RH or greater. However, at midrange relative humidities, airborne decay was enhanced. As can be seen, M. gallisepticum in the airborne state did not survive well except at 10% RH. As with rickettsiae and psittacosis, mycoplasma methodology is still too undeveloped to warrant discussion of the effect of suspending fluids, etc., on airborne stability.

It is important to note that mycoplasma, with no rigid cell wall, exhibit midrange RH sensitivity, as do many bacteria, phage, and viruses. Moreover, viruses (excluding the rickettsia and psittacosis

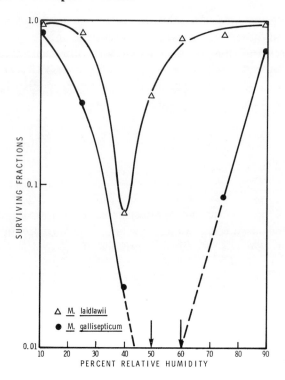

FIGURE 12-17. Effect of relative humidity on survival of *Mycoplasma* species *laidlawii* and *gallisepticum* in aerosols at 27°C for 60 minutes. Data shown are corrected for physical loss and represent biological decay. This figure is taken from decay-rate values obtained by best-fit techniques, representing 4 or 6 experiments at each relative humidity. No single value deviated more than 10% from those shown. (Wright, Bailey, and Hatch. 1968. *J. Bacteriol.*, **95**: 252.) Reprinted with permission of the American Society for Microbiology.

group) are even more simple than the mycoplasma, in that they possess no known enzymatic systems. It would therefore appear possible that one can eliminate the role of both cell walls and enzyme systems as having a major effect on airborne stability. This emphasizes the attractiveness of pointing to nucleic acids and their related proteins as having a leading role in mechanisms of airborne stability of microorganisms.

REFERENCES

Akers, T. G., Bond, S., and Goldberg, L. J. 1966. Effect of temperature and relative humidity on survival of airborne Columbia-SK group viruses. *Appl. Microbiol.*, **14**: 361–364.

Akers, T. G., Madin, S. H., and Schaffer, F. L. 1968. The pathogenicity in mice of aerosols of encephalomyocarditis group viruses or their infectious nucleic acids. *J. Immunol.,* **100**: 2–18.

Akers, T. G., Bond, S. B., Papke, C., and Leif, W. R. 1966. Virulence and immunogenicity in mice of airborne encephalomyocarditis viruses and their infectious nucleic acids. *J. Immunol.,* **97**: 379–385.

Andrewes, C. H. 1964. The complex epidemiology of respiratory virus infections. *Science,* **146**: 1274–1277.

Andrews, C. H., Laidlaw, P. P., and Smith, W. 1934. The susceptibility of mice to the viruses of human and swine influenza. *Lancet,* **2**: 859–862.

Beard, C. W., and Anderson, D. P. 1967. Aerosol studies with avian mycoplasma. 1. Survival in air. *Avian. Diseases,* **11**: 54–57.

Beard, C. W., and Easterday, B. C. 1965a. Aerosol transmission of hog cholera. *Cornell Vet.,* **55**: 630–636.

Beard, C. W., and Easterday, B. C. 1965b. An aerosol apparatus for the exposure of large and small animals: description and operating characteristics. *Am. J. Vet. Research,* **26**: 174–182.

Beard, C. W., and Easterday, B. C. 1967. The influence of the route of administration of Newcastle disease virus on host response. 1. Aerological and virus isolation stuides. *J. Infect. Diseases,* **117**: 55–61.

Benda, R., Fuchsova, M., and Danes, L. 1962. Experimental airborne infection of monkeys with tick-borne encephalitis. *Acta Virol.,* **6**: 46–52.

Bolotovskii, V. M. 1959. Necessary conditions for the successful production of experimental infection with ornithosis virus aerosol. *Problems Virol.,* **4**: 102–106.

Buckland, F. E., and Tyrrell, D. A. J. 1964. Experiments on spread of colds. 1. Laboratory studies on the dispersal of nasal secretions. *J. Hyg.,* **62**: 365–377.

Burnet, F. M. 1940. Influenza virus infections of the chick embryo by the amniotic route. 1. General character of the infection. *Australian J. Exptl. Biol. Med. Sci.,* **18**: 353–360.

Burnet, F. M., and Lush, Dora. 1938. Influenza virus on the developing egg: The antibodies of experimental and human sera. *Brit. J. Exptl. Pathol.,* **19**(1): 17–29.

Cate, T. R., Couch, R. C., and Johnson, K. M. 1964. Studies with rhinovirus in volunteers: Production of illness, effect of naturally acquired antibody, and demonstration of a protective effect not associated with serum antibody. *J. Clin. Invest.,* **43**: 56–67.

Chalkina, O. M. 1938. Immunological changes in the blood of men vaccinated against the virus of epidemic influenza. *Arch. Sci. Biol.,* **52**: 126–131.

Chappal, W. A. 1960. Animal infectivity of aerosols of monkey B virus. *Ann. N.Y. Acad. Sci.,* **85**: 931–934.

Couch, R. B., Cate, T. C., Fleet, W. F., Gerone, P. J., and Knight, V. 1966. Aerosol induced adenovirus illness resembling the naturally occurring illness in military recruits. *Am. Rev. Respirat. Diseases,* **93**: 529–535.

Couch, R. B., Gerone, P. J., Cate, T. R., Griffith, W. R., Alling, D. W., and Knight, V. 1965. Preparation and properties of a small particle aerosol of coxsackie A_{21}. *Proc. Soc. Exptl. Biol. Med.,* **118**: 818–822.

Cox, C. S., and Baldwin, F. 1964. A method for investigating the cause of death of airborne bacteria. *Nature,* **202**: 1135.

Cox, C. S., and Baldwin, F. 1966. The use of phage to study causes of loss of viability of *Escherichia coli* in aerosols. *J. Gen. Microbiol.,* **44**: 15–22.

Danes, L., Libich, J., and Benda, R. 1962. Experimental airborne infection of mice with tick-borne encephalitis virus. *Acta. Virol.*, **6**: 37–45.

Davenport, F. M. 1961. Pathogenesis of influenza. *Bacteriol. Rev.*, **25**: 294–299.

de Jong, J. G. 1965. The survival of measles virus in air in relation to the epidemology of measles. *Arch. Ges. Virusforsch.*, **16**: 97–102.

De Ome, K. B. and The Personnel, U.S. Navy Medical Research Unit No. 1, Berkeley, Calif. 1944. The effect of temperature, humidity, and glycol vapor on the viability of airborne bacteria. *Am. J. Hyg.*, **40**(3) : 239–250.

Edward, D. G., Elford, W. J., and Laidlaw, P. P. 1943. Studies on airborne virus infections. I. Experimental technique and preliminary observations on influenza and infectious ectromelia. *J. Hyg.*, **43**: 1–10.

Ehrlich, R., Miller, W. S., and Idoine, L. S. 1964. Effects of environmental factors on the survival of airborne T3 coliphage. *Appl. Microbiol.*, **12**: 479–482.

Francis, T. J., Pearson, H. E., Salk, J. E., and Brown, P. N. 1944. Immunization in human subjects artificially infected with influenza virus type B. *Am. J. Public Health*, **34**: 317–334.

Gerone, P. J., Couch, R. B., Keefer, G. V., Douglas, R. G., Derrenbacher, E. B., and Knight, V. 1966. Assessment of experimental and natural viral aerosols. *Bacteriol. Rev.*, **30**: 576–583.

Goldberg, L. J. 1950. Studies on the experimental epidemiology of respiratory infections. IV. A particle size analyzer applied to the measurement of viable airborne bacteria. *J. Infect. Diseases*, **87**: 133–141.

Goldberg, L. J. 1968. Application of the Microaerofluorometer to the study of dispersion of a fluorescent aerosol into a selected atmosphere. *J. Appl. Meteorol.*, **7**: 68–72.

Goldberg, L. J., and Leif, W. R. 1950. The use of a radioactive isotope in determining the retention and initial distribution of airborne bacteria in the mouse. *Science*, **112**: 299–300.

Goldberg, L. J., Watkins, H. M. S., Boerke, E. E., and Chatigny, M. A. 1958. The use of a rotating drum for the study of aerosols over extended periods of time. *Am. J. Hyg.*, **68**: 85–93.

Gorham, J. R., Leader, R. W., and Gutierrez, J. C. 1954. Distemper immunization of ferrets by nebulization with egg adapted virus. *Science*, **119**: 125–126.

Guerin, L. F., and Mitchell, C. A. 1964. A method for determining the concentration of airborne virus and size droplet nuclei containing the agent. *Can. J. Comp. Med. Vet. Sci.*, **28**: 283–287.

Guyton, A. A. 1947. Measurement of respiratory volumes of laboratory animals. *Am. J. Physiol.*, **150**: 70–77.

Hahon, N., and McGavran, M. H. 1961. Airborne infectivity of the variola-vaccinia group of poxviruses for the cynomolgus monkey *Macaca irus*. *J. Infect. Diseases*, **109**: 294–298.

Happ, J. W., Harstad, J. B., and Buchanan, L. M. 1966. Effects of air ions on submicron T_1 bacteriophage aerosols. *Appl. Microbiol.*, **14**: 888–891.

Harper, G. J. 1961. Airborne microorganisms survival test with four viruses. *J. Hyg.*, **59**: 479–486.

Harper, G. J. 1965. Some observations on the influence of suspending fluids on the survival of airborne viruses. In *A Symposium on Aerobiology*, 1963 (R. L. Dimmick, Ed.), pp. 335–343. Nav. Biol. Lab., Nav. Supply Center, Oakland, Calif.

Harper, G. J., and Morton, J. D. 1953. The respiratory retention of bacterial aerosols; experiments with radioactive spores. *J. Hyg.*, **51**: 372–385.

Harstad, J. B. 1965. Sampling submicron T_1 bacteriophage aerosols. *Appl. Microbiol.*, **13**: 899–908.

Hearn, H. J., Jr. 1961. Differences among virus populations recovered from mice vaccinated with an attenuated strain of Venezuelan equine encephalomyelitis virus. *J. Immunol.*, **87**: 573–577.

Hearn, H. J., Jr., Chappell, W. A., Demchak, P., and Dominik, J. W. 1966. Attenuation of aerosolized yellow fever virus after passage in cell culture. *Bacteriol. Rev.*, **30**: 615–623.

Hemmes, J. H., Winkler, K. C., and Kool, S. M. 1960. Virus survival as a seasonal factor in influenza and poliomyelitis. *Nature*, **188**(4748) : 430–431.

Hemmes, J. H., Winkler, K. C., and Kool, S. M. 1962. Virus survival as a seasonal factor in influenza and poliomyelitis. *Antonie van Leeuwenhoek J. Microbiol. Serol.*, **28**: 221–233.

Henderson, D. W., Peacock, S., and Randles, W. J. 1967. On the pathogenesis of Simliki forest virus (SFV) infection in the hamster. *Brit. J. Exptl. Pathol.*, **48**: 228–234.

Hirst, G. K. 1941. The agglutination of red cells by allantoic fluid of chick embryos infected with influenza virus. *Science*, **94**: 22–23.

Hitchner, S. B., and Reisling, G. 1952. Flock vaccination for Newcastle disease by atomization of B1 strain of virus. *Proc. Am. Vet. Med. Assoc.*, **89**: 258–264.

Hood, A. M. 1963. Infectivity of influenza virus aerosols. *J. Hyg.*, **61**: 331–335.

Hyslop, N. St. G. 1965. Airborne infection with the virus of foot and mouth disease. *J. Comp. Pathol.*, **75**: 119–126.

Irons, J. B., Denlay, M. L., and Sullivan, T. D. 1955. Psittacosis in turkeys and fowls as a source of human infection. In *Psittacosis, Diagnosis, Epidemiology and Control* (E. R. Beaudette, Ed.), pp. 44–65. Rutgers Univ. Press, New Brunswick, N.J.

Jakab, G. J., and Dick, E. C. 1967. Experimental parainfluenza (Sendai) virus infection of mice by aerosol. *Bacteriol. Proc.*, Paper # V. 76 p. 147.

Jensen, M. M., 1964. Inactivation of airborne viruses by ultraviolet irradiation. *Appl. Microbiol.*, **12**: 418–420.

Jensen, M. M. 1965. Inactivation of virus aerosols by ultraviolet light in a helical baffle chamber. In *A symposium on Aerobiology, 1963* (R. L. Dimmick, Ed.), pp. 219–226. Nav. Biol. Lab., Nav. Supply Center, Oakland, Calif.

Kuehne, R. W., Sawyer, W. D., and Gochenour, W. S., Jr. 1962. Infection with aerosolized attenuated Venezuelan equine encephalomyelitis. *Am. J. Hyg.*, **75**: 347–350.

Kulesko, I. I., Shikov, A. T., and Yarnykh, Y. S. 1963. Aerosol immunization of piglets against swine fever. *Veterinariya*, 5:30. (English summary in *Vet. Bull.* 33:23 (#3927), 1963).

Kundsin, R. B. 1966. Characterization of mycoplasma aerosols as to viability, particle size, and lethality of ultraviolet irradiation. *J. Bacteriol.*, **91**: 942–944.

Lennette, E. H., and Koprowski, H. 1943. Human infection with Venezuelan equine encephalomyelitis virus: Report of eight cases of infection acquired in the laboratory. *J. Am. Med. Assoc.*, **123**: 1088–1095.

Lidwell, O. M., Morgan, R. W., and Williams, R. E. O. 1965. The epidemiology of the common cold. IV. The effect of weather. *J. Hyg.*, **63**: 427–439.

338   Analysis of Concepts and Results

Loosli, C. G. 1949. The pathogenesis of experimental airborne influenza virus A infection in mice. *J. Infect. Diseases*, 84(2) : 153–168.

McGavran, M. H., Beard, C. W., Berendt, R. F., and Nakamura, R. M. 1962. The pathogenesis of psittacosis: Serial study of Rhesus monkeys exposed to a small particle aerosol of the Borg strain. *Am. J. Pathol.*, 40: 653–670.

Meyers, K. F., and Eddie, B. 1958. Ecology of avian psittacosis, particularly in parakeets. In *Progress in Psittacosis Research and Control*, pp. 52–79. Rutgers University Press, New Brunswick, N.J.

Miller, W. S., and Artenstein, M. S. 1967. Aerosol stability of three acute respiratory disease viruses. *Proc. Soc. Exptl. Biol. Med.*, 125: 222–227.

Miller, W. S., Demchak, P., Rosenberger, C. R., Dominik, J. W., and Bradshaw, J. L. 1963. Stability and infectivity of airborne yellow fever and Rift Valley fever viruses. *Am. J. Hyg.*, 77: 114–121.

Morley, D. C. 1962. Measles in Nigeria. *Am. J. Diseases Children*, 103(3) : 230–233.

Morris, E. J., Darlow, H. M., Peel, J. F. H., and Wright, W. C. 1961. The quantitative assay of mono-dispersed aerosols of bacteria and bacteriophage by electrastatic precipitation. *J. Hyg.*, 59: 487–496.

Nir, Y. D. 1959. Airborne West Nile virus infection. *Am. J. Trop. Med. Hyg.*, 8: 537–539.

Nir, Y. D., Beemer, A., and Goldwasser, R. A. 1965. West Nile virus infection in mice following exposure to a viral aerosol. *Brit. J. Exptl. Pathol.*, 46: 443–449.

Noble, W. C. 1967. Sampling airborne microbes: Handling the Catch. In *Airborne Microbes*, 17th Symposium Society for General Microbiology (P. H. Gregory and J. L. Monteith, Eds.), pp. 81–101. Camb. Univ. Press, London.

Okano, Y. 1965. Further observations on immunization with live attenuated measles virus by inhalation. *Arch. ges. Virusforsch.*, 16: 294–299.

Personnel of Naval Laboratory Research Unit No. 1. 1944. Experimental human influenza. *Am. J. Med. Sci.*, 207: 306–314.

Ristori, C., Boccardo, H., Borgono, J. M., and Armijo, R. 1962. Medical importance of measles in Chile. *Am. J. Diseases Children*, 103(3) : 236–241.

Rosebury, T., 1947. *Experimental Airborne Infection*. The Williams and Wilkins Co., Baltimore, Md.

Rosebury, T., Ellingson, H. V., and Meiklejohn, G. 1947. Laboratory infection with psittacosis virus treated with penicillin and sulfadiazone; experimental data on mode of infection. *J. Infect. Diseases*, 80: 64–77.

Saslaw, S., Carlisle, H. N., Wolfe, G. L., and Cole, C. R. 1966. Rocky Mountain spotted fever: Clinical and laboratory observations of monkeys after respiratory exposure. *J. Infect. Diseases*, 116: 243–255.

Schulman, J. L., and Kilbourne, E. D. 1962. Airborne transmission of influenza infection in mice. *Nature*, 195: 1129–1130.

Shechmeister, I. L. 1950. Studies on the experimental epidemiology of respiratory infections. III. Certain aspects of the behavior of type A influenza virus as an airborne cloud. *J. Infect. Diseases*, 87: 128–132.

Smith, W., Andrewes, C. H., Laidlaw, P. P. 1933. A virus obtained from influenza patients. *Lancet*, 2: 66–68.

Smorodintseff, A. A., Tushinsky, M. D., Drobyshevskaya, A. I., Korovin, A. A., and Osetroff, A. I. 1937. Investigation on volunteers infected with the influenza virus. *Am. J. Med. Sci.*, 194(2) : 159–170.

Songer, J. R., 1967. Influence of relative humidity on the survival of some airborne viruses. *Appl. Microbiol.*, 15: 35–42.

Spicer, C. C. 1959. Influence of some meteorological factors in the incidence of poliomyelitis. *Brit. J. Prevent. & Social Med.*, **13**: 139–144.

Sulkin, E. S. 1961. Laboratory acquired infections. *Bacteriol. Rev.*, **25**: 203–207.

Taneja, P. N., Ghai, O. P., and Bhakoo, O. N. 1962. Importance of measles to India. *Am. J. Diseases Children*, **103**(3) : 226–233.

Tigertt, W. D., Benenson, A. S., and Gochenour, W. S. 1961. Airborne Q fever. *Bacteriol. Rev.*, **25**: 285–293.

Watkins, H. M. S., Goldberg, L. J., Deig, E. F., and Leif, W. R. 1965. Behavior of Colorado tick fever, vesicular stomatitis, neurovaccinia and encephalomyocarditis viruses in the airborne state. In *A Symposium on Aerobiology, 1963* (R. L. Dimmick, Ed.), pp. 381–388. Nav. Biol. Lab., Nav. Supply Center, Oakland, Calif.

Webb, S. J., Bather, R., and Hodges, R. W. 1963. The effect of relative humidity and inositol on airborne viruses. *Can. J. Microbiol.*, **9**: 87–92.

Webb, S. J., Dumasia, M. D., and Bhorjee, J. Singh. 1965. Bound water, inositol, and the biosynthesis of temperate and virulent bacteriophage by air dried *Escherichia coli*. *Can. J. Microbiol.*, **11**: 141–150.

Wellock, C. E. 1960. Epidemiology of Q fever in the urban East Bay area. *Calif. Health*, **18**: 73–76.

Wells, W. F., and Brown, H. W. 1936. Recovery of influenza virus suspended in air and its destruction by ultraviolet radiation. *Am. J. Hyg.*, **24**(2) : 407–413.

White, R. J., and Madin, S. H. 1964. Pathogenesis of murine hepatitis: Route of infection and susceptibility of the host. *Am. J. Vet. Research*, **25**: 1236–1240.

Wright, D. N., Bailey, G. D., and Hatch, M. T. 1968. Survival of airborne mycoplasma as affected by relative humidity. *J. Bacteriol.*, **95**: 251–254.

13

BIOLOGICAL PROPERTIES OF FUNGAL AEROSOLS

H. B. Levine

NAVAL BIOLOGICAL LABORATORY, SCHOOL OF PUBLIC HEALTH,
UNIVERSITY OF CALIFORNIA, BERKELEY

The fungi, as a group, are uniquely endowed with attributes favoring their establishment and persistence in the aerosolized state (Gregory, 1961). Their spores range in diameter from less than 1 μ to more than 10 μ and, quite characteristically, are resistant over a wide range of temperature, relative humidity, dessication, and light energy. Frequently, the spores are developed on aerial hyphae and may be dislodged by even the most gentle air currents.

The fungi grow abundantly in nature on relatively dry organic and inorganic menstrua. This feature favors a colonial architecture that lacks the cohesive forces imparted by a film of moisture; the colonial elements often develop somewhat discretely and in a fluffy manner, therby contributing to the potential for aerosolization. Septate structures, under conditions of moderate dessication, may disarticulate and disengage from the underlying mycelial growth to form a "dust" that is easily made airborne.

Not surprisingly, then, investigators have found the fungal content of air in different locales to range from a few particles per liter (Pady and Gregory, 1963; Pady et al., 1967, 1968) to more than 10^8 particles per liter. Farmer's lung disease, particularly prevalent in rural Britain, is a consequence of repeated exposure of persons to massive fungal aerosols generated by handling moldy hay (Pepys et al., 1963). Similarly,

340

mycotic bovine abortion has its etiology in hay contaminated by a species of *Aspergillus* and, here too, the fungal aerosols generated from the hay may be so dense as to cloud the surrounding air (Austwick, 1963).

Fortunately, the preponderance of fungal species are unadapted for survival in living organs and tissues; man's experience with them is usually uneventful. Nevertheless, they are antigenic, even if weakly, and allergic sequelae are not uncommon aftermaths of exposure. Some fungi, however, are not obligate saprophytes and those producing respiratory mycoses in man and animals present a problem of considerable medical and economic magnitude. Foremost among the systemic mycoses of respiratory origin in the United States are coccidioidomycosis, histoplasmosis, cryptococcosis, and North American blastomycosis. Additionally, cases of respiratory candidiasis, aspergillosis, actinomycosis, and penicillosis contribute to the problem (Conant *et al.*, 1954). Airborne fungal spores also transmit disease of plants, and their infectiousness is influenced by ambient humidity (Pady *et al.*, 1968).

A consideration of coccidioidomycosis has been selected to exemplify how the aerobiologic and infectivity properties of a pathogenic fungus interact to produce widespread disease. In general, the determinants are similar to those operating with other pathogenic soil fungi, allowing, of course, for certain specific differences in ecology, epidemiology, and particular properties of the organism itself.

Coccidioides immitis grows naturally in the arid soil of our southwestern states, Arizonia, New Mexico, western Texas, and, particularly, in the southern San Joaquin Valley of California. There are other endemic foci in the United States, Mexico, Central, and South America, but the major portion of reported cases stem from infections contracted in the four states mentioned; areas in the vicinities of Bakersfield, Phoenix, and El Paso are noteworthy in this respect. The estimated morbidity is 25,000 cases per year in the United States (Pappagianis, 1961), but the number of exposed persons who contract mild or asymptomatic infections may be twice this number (Fiese, 1958). Eventual recovery, often after prolonged illness, is the rule in symptomatic patients, but disseminated fatal disease frequently is the unhappy fate in persons of Filipino, Negro, and Mexican Indian descent (Fiese, 1958).

Why *Coccidioides* inhabits the regions it does and why it does not spread to adjacent areas is not understood. However, three ecologic factors may relate to the organism's intimate habitat. Specific soil conditions, particularly in respect to elevated salinity (Na^+, Ca^{++}, SO_4^{--}, and Cl^-), encourage the growth of *Coccidioides* and correlate with its spotty distribution (Elconin *et al.*, 1964). Secondly, the high temperature of the endemic regions reduces the indigenous bacterial flora and leaves *Coccidioides* with few natural competitors. Thirdly, the relative lack of water over long periods of time is unfavorable to bacterial growth, but not as harmful to fungal growth and persistence.

During the winter months, the fungus grows in the soil and is often found in areas contaminated by excreta or other decaying organic matter. Apparently, the growth is quite profuse. In the dry summer months the septate mycelium forms spores by contraction of its cytoplasm, development of a tough wall around the cytoplasm, and disarticulation of the cellular members of the mycelial strand. Winds dislodge the spores, called arthrospores, into the air and in the months of June through November, corresponding to the fruit and cotton picking seasons, the numbers of symptomatic cases increase as shown in Figure 13-1 (Smith, 1940). The primary disease is almost exclusively pulmonary.

One might pause at this point to contemplate the nature of the arthrospore for it is this structure that confers upon *Coccidioides* its aerobiologic properties and hence its capacity to produce widespread human disease. Arthrospores of different strains, grown on artificial media, usually range in size from 2 to 8 μ on the long axis, and from 1 to 2 μ on the short axis. In nature under less favorable conditions of nutrition and moisture, they may be somewhat smaller. Their tough walls contain chitin-like polymers and the cell is distinctly hydrophobic. Even in the laboratory, the arthrospores require violent shaking for several hours with glass beads to become thoroughly wetted. The importance of this property in aerobiologic epidemiology should be stressed because the unwetted structure remains light in weight, even in the presence of excessive moisture, and does not become suspended readily in water. As a consequence, it is ever-ready to become airborne whenever energy is imparted to it. Other attributes of the arthrospore contribute to its persistence on the soil and in the air; its contracted cytoplasm is low in moisture and the arthrospore is metabolically dormant or of reduced metabolic activity. In this state it can remain viable for months or years. It can also withstand extremes of temperature and ultraviolet energy. Its small size is compatible with deep penetration into the lungs. Recently Converse *et al.* (1967) produced coccidioidomycosis in dogs and monkeys, housed in cages suspended above ground, in an endemic region of Arizona.

The properties of the arthrospore, listed above which (singly and in combination) favor its propensity to infect by the airborne route, would be of little consequence to human health were it not also for the high virulence of the organism. It is probable that one arthrospore, deposited on pulmonary tissue, would produce an infection (Converse, 1966). This

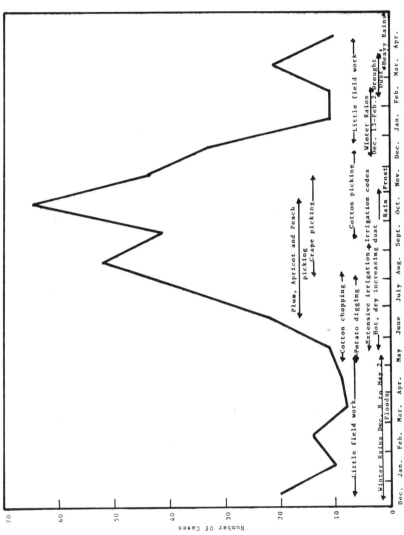

FIGURE 13-1. Seasonal variation in the incidence of primary coccidioidomycosis, Kern and Tulare Countries, California, December, 1937, to April 1939. (Smith, C. E. 1940. *Amer. J. Public Health*, 30: 600–611.) Reprinted with permission of The American Public Health Association, Inc.

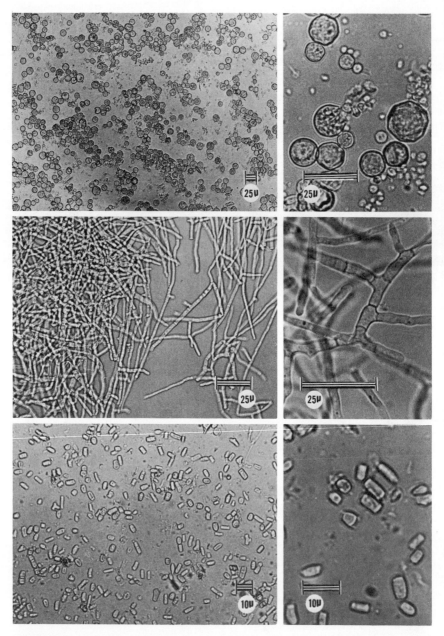

FIGURE 13-2. *Coccidioides immitis;* the spherule-endospore (top), mycelial (center), and arthrospore (bottom), phases of growth. (Levine, Cobb, and Smith. 1960. *Trans. N.Y. Acad. Sci.*, Ser. II **22:** 436–499.) Copyright The New York Academy of Sciences. Reprinted with permission of the New York Academy of Sciences.

feature is related to inherent properties of the arthrospore. Even in immunized animals (Levine *et al.*, 1961; Levine *et al.*, 1962) where an early and augmented leucocytic infiltrate occurs in response to infection, the arthrospore is able to survive and permit the following developments to ensue: The arthrospore in tissue enlarges to a round structure varying in diameter from 15 μ to more than 50 μ. It is now termed a "spherule" and is too large to be engulfed easily by phagocytes. In this relatively safe situation, cleavage of its cytoplasm occurs, resulting in the formation of up to 800 endospores housed within the spherule body. Ultimately the spherule wall ruptures and releases these viable endospores. The spherule-endospore, mycelial, and arthrospore phases of growth of *Coccidioides immitis* are shown in Figure 13-2.

Spherule rupture is associated with necrotizing activity and, not uncommonly, the released endospores become blood-borne and lymph-borne. Metastatic lesions may occur and the host then becomes frankly diseased.

Only one component in nature was, initially, shared by the fungus in the soil and the coccidioidomycotic patient—the air. Through this medium, an unfortunate encounter became almost inevitable. The encounter, which occurs in almost 90% of individuals living near Bakersfield and Phoenix, emphasizes the importance of aerobiology in the respiratory mycoses of man.

REFERENCES

Austwick, P. K. C. 1963. Ecology of *Aspergillus fumigatus* and the Pathogenic Phycomycetes. In *Recent Progress in Microbiology*, VIII, pp. 644–651. University of Toronto Press, Toronto, Canada.

Conant, N. F., Smith, D. T., Baker, R. D., Callaway, J. L., and Martin D. S. 1954. *Manual of Clinical Mycology*. W. B. Saunders Co., Philadelphia, Pa.

Converse, J. 1966. Experimental epidemiology Coccidioidomycosis. I. Epizootiology of naturally exposed monkeys and dogs. Second Symposium on Coccidioidomycosis. Arizona State Public Health Dept., Phoenix, 1965.

Converse, J. L., Reed, R. R., Kuller, H. W., Trautman, R. J., Snyder, E. M., and Rey, J. G. 1967. Experimental Epidemiology of Coccidioidomycosis: 1. Epizootiology of naturally exposed monkeys and dogs. In *Coccidioidomycosis* (L. Ajello, Ed.). Univ. Arizona Press, Tucson, Ariz.

Elconin, F. F., Egeberg, R. O., and Egeberg, M. C. 1964. Significance of soil salinity on the ecology of *Coccidioides immitis*. *J. Bacteriol.*, 87: 500–503.

Fiese, M. J. 1958. *Coccidioidomycosis*. Charles C. Thomas, Springfield, Ill.

Gregory, P. H. 1961. *Microbiology of the Atmosphere*. John Wiley and Sons, Interscience Division, New York.

Levine, H. B., Cobb, J. M., and Smith, C. E. 1961. Immunogenicity of spherule-endospore vaccines of *Coccidioides immitis* for mice. *J. Immunol.*, 87: 218–227.

Levine, H. B., Miller, R. L., and Smith, C. E. 1962. Influence of vaccination on respiratory coccidioidal disease in Cynomolgous monkeys. *J. Immunol.*, 89: 242–251.

Pady, S. M., and Gregory, P. H. 1963. Numbers and viability of airborne hyphal fragments in England. *Trans. Brit. Mycological Soc.*, 46(4) : 609–613.

Pady, S. M., Kramer, C. L., and Clary, R. 1967. Diurnal periodicity in airborne fungi in an orchard. *J. Allergy*, 39(5) : 302–310.

Pady, S. M., Kramer, C. L., and Clary, R. 1968. Periodicity in aeciospore release in *Gymnosporangium juniperi-virginianae*. *Phytopathology*, 58(3) : 329–331.

Pappagianis, D. 1961. Active immunity in coccidioidomycosis: Natural and laboratory features. *Stanford Medical Bulletin*, 19: 35–40.

Pepys, J., Jenkins, P. A., Festenstein, G. N., Lacey, M., Gregory, P. H., and Skinner, F. A. 1963. Farmer's lung: Thermophilic actinomycetes as a source of "farmer's lung hay" antigen. *Lancet*, Sept. 21, 607–611.

Smith, C. E. 1940. Epidemiology of acute coccidioidomycosis with erythema nodosum ("San Joaquin" or "valley fever"). *Amer. J. Public Health*, 30: 600–611.

14

THEORETICAL ASPECTS OF MICROBIAL SURVIVAL

R. L. Dimmick / R. J. Heckly

NAVAL BIOLOGICAL LABORATORY, SCHOOL OF PUBLIC HEALTH,
UNIVERSITY OF CALIFORNIA, BERKELEY

Survival is such an obvious attribute of all living things that the terms "life" and "survival" seem almost synonymous; to live *is* to survive. Many of the aspects of microbial survival we propose to discuss are so simple and obvious that we feel required to apologize to the reader for what he may consider an insult to his intelligence. Nevertheless, when studies of growth, adaptation, mutation, induction, antibiotic resistance, or persistence of bacteria are reported, few authors mention the fact that some of the tested bacteria survived whatever environmental changes the experimental procedures dictated. The act of survival, in and by itself, entered into the process, although it may not have been included as one of the factors influencing the system being studied. Some microorganisms had to survive in order to provide data. Our whole science of microbiology, including its specific application to aerobiology, is based largely on the presence of colonies showing that one or more cells were alive at the time we manipulated them. This *sine qua non* of microbiology is often looked upon as if it operated with the same dependability and simplicity as forces of gravity, and we tend to ignore the decisive role it plays in most of our experiments. Let us briefly review the data-gathering ritual.

We routinely perform assays of viability, and then report data in terms of the number of cells capable of forming colonies. If this number decreases as a result of some manipulation of the cell population, we express the result as a survival ratio variously termed "survival," "percent survival," "persistence," "recovery," "viability," or "percent recovery," and the inverse "loss of viability," "decay," "death," or "death rate." All knowledge concerning apparent "activities" of microbial populations, i.e., infectivity, growth, productivity, genetic potential, and survival, is ultimately based on some type of viability assay. The dynamic relationship of these activities to numbers of viable mircoorganisms is expressed in terms of differences in survival ratios.

We may be interested in a single difference or ratio, but usually we are interested in the way this difference changes with time. For example, most disinfectants do not kill all bacteria instantaneously—nor does exposure to air. When we plot survival ratios as a function of time, we create "survivor curves," or "decay curves," or "death curves," and we compare differences to add a second dimension of meaningful data.

How meaningful these data may be depends on our interpretation of the kinds of events measured by the survivor curve. We would like to be able to say something significant about observed differences between survivor curves. It is often not sufficient to indicate simply that "the bacteria died *more* rapidly this way than that way," or that the applied environment "failed to kill all cells." Rather, we would like to quantitate differences, and to relate the differences to parameters suspected of being involved; e.g., the number of molecules of an additive needed per bacterium to provide protection at a given temperature, or how this number might change with temperature. We are on shaky ground if we fail to understand what the survivor curve tells us.

Throughout the history of microbiology the argument as to whether microbes die according to stochastic (random) processes, or whether deterministic or inherent factors are involved, has been vigorously conducted by both written and verbal word; and the argument continues. The serious reader should examine papers by Rahn (1945), Wood (1956), Vas and Proszt (1957), Lamanna and Mallette (1959), Humphrey and Nickerson (1961), Powers (1962), and Dimmick (1965) that are illustrative of the quandary. In the following we propose to play the role of the Devil's Advocates, arguing that both theories are correct sometimes and incorrect at other times, that the question needs further extended investigation, and that an understanding of the true information content of survivor curves is no less important to applied studies than to fundamental investigations of survival mechanisms.

LOGARITHMIC DEATH

Many microbiologists consider the survivor curve to be an expression of "logarithmic death," based principally on the straight-line relationship obtained from survival data. The use of the logarithm is imperative, because the numbers of microbes we encounter vary over orders of magnitude and the alternative would demand a chart of magnificent dimensions. Possibly the method of plotting adds unconscious bias to the interpretation of results. Regardless, the similarity between these curves and data obtained from other well-recognized chemical reactions makes it tempting for the investigator to interpret his survival data in terms of mono-molecular reaction rates. But what are the assumptions implicit in the mono-molecular theory of decay, i.e., the "target theory"? First, there are sensitive, essential "sites" in each cell being "hit" and destroyed by some "lethal units" (atoms or molecules of high thermal energy, quanta of radiation, reactive end-groups in molecules; all at some assumed energy level higher than the energy level of the sites), and each remaining site has an equal probability of being hit per unit of time. Second, destruction of the site results directly in death of the organism. Note that the time required for this organism to die is not important, since even if the cell is sampled before death occurs, no colony will be formed. Third, the sites in the bacteria are exactly alike and sensitive to the same lethal units and energy levels. Fourth, all bacteria are sufficiently alike, so that interference or shielding by cellular constituents provides the same probability environment for each site. In summary, the "single-hit" argument assumes that all bacteria are exactly alike and that they persist in living only because of select, requisite portions of their total structure. We shall hold strictly to this assumption as long as it is reasonable to do so.

Stated another way, we can write the familiar relationship

$$\frac{N}{N_0} = e^{-Kt}$$

where N_0 = number of cells initially,
N = number of cells at time, t, and
K = a probability constant.

By conditions presently imposed, K is defined as follows: Within the volume occupied by the organisms there is a hypothetical maximum number of positions that can be occupied by both the sensitive site and the lethal unit; that is, a place where a hit will occur. The ratio of the number of positions occupied per time interval by the lethal units (dosage) to the total number of positions per cell is the probability of a hit, so the ratio can be said to be equivalent to percent per unit of time and is easily determined from the slope of the survivor curve.

It is reasonable to assume that the site, if hit, will be destroyed when the energy of the lethal unit is greater than the energy of formation of the site. The energies of the lethal units are likely to be distributed according to the Boltzman factor,

$$(e^{-mc^2/2RT})$$

where m = mass of a unit,
c = velocity of a unit,
R = gas constant, and
T = absolute temperature.

The observed death rate of bacteria, at a constant lethal unit dosage, usually changes as a function of culture age. With the present restrictions, this change could occur only if the energy of activation, or of formation, of all sites also changed equally within the bounds of the energy distribution of lethal units. If the site change was directed toward levels of energy less than the effective dosage energy level, then every site hit would be destroyed regardless of the change. If the change raised site energies above lethal energy levels, then no hit would destroy a site and no cells would die—a kind of *reductio ad absurdum* argument. In effect, when the energies of the lethal units and the sites are within overlapping ranges, a second probability has been added; i.e., the probability of a site being destroyed *if* hit.

The foregoing has really been a reiteration of the theory of first-order reaction kinetics with an important exception: By placing the site within or on a cell, the cell structure could interpose a barrier or shield that might reduce the probability of a hit, or absorb energy from a lethal unit. If so, a general change in cell structure would change the rate of the observed survivor curve. Hence, rate differences noted in the above example of culture age could be caused either by changes in site activation energies or by changes in barrier effectiveness. If the site had changed from being essential to one not essential to the life of the cell, then the bacteria would not die; no decay would be observed.

Although, as we shall see later, within a clone[1] or even a synchronous culture[2] each bacterium differs from others in a multitude of ways, there is one structure assumed to be most uniform in cells of the same species; i.e., the gene. The application of high-energy radiation to bacteria, with cellular constituents thus being practically transparent, should produce logarithmic death as a result of gene damage. This is often the observed situation (Thornley, 1963). Many processes that are highly lethal to bacteria (e.g., autoclaving, open flame, concentrated hypochlorite) probably are stochastic, but the reactions are usually too fast to follow.

[1] All bacteria arising from a single cell, hence all "alike."
[2] All cells divide, or are in the same growth phase at the same time.

However, it is difficult to find examples where true logarithmic death occurred except during early portions of the total decay period. Usually the decay rate seems to decrease, and the process known as "tailing" occurs. Those who cling to the logarithmic decay theory as the *only* possible mechanism argue that deviations such as tailing, or even an apparent increase in numbers of survivors (Jordan and Jacobs, 1949), are caused by aggregation, experimental error, multiple-hit mechanisms, or population differences; the latter belongs to arguments that will be introduced in due time. It is impossible to estimate the extent to which experimental error or aggregates, sometimes acting to protect inner cells or sometimes breaking apart to increase the apparent cell numbers, influence reported data. The critical scientist will examine the data and decide for himself whether the number of reported samples is sufficient to delineate one curve or another. We have, for example, passed suspensions of *Serratia marcescens* through 10-μ membrane filters and found survivor curves resulting from thermal stress experiments decidedly nonlogarithmic and differing only in initial numbers from curves obtained by testing an unfiltered suspension.

Temporal differences caused by errors in mixing can induce unusual behavior. We have found that results were more reproducible when bacteria were injected rapidly by means of a hypodermic syringe below the surface of a test disinfectant than when the disinfectant was added to bacterial suspensions. We suppose the injection method permitted all bacteria to contact the disinfectant at more nearly the same time. In aerobiology experiments, if a chamber is filled by spraying for extended time periods, the contained population will be of different ages with respect to some zero time.

When pipetting with 1-ml pipettes, one might misread 0.8 ml for 0.9 ml, but rarely 0.1 ml for 1 ml. Stated another way, errors of about $\pm20\%$ could be made unconsciously, but regular errors of 1 log or more, which would be required to explain some observed data, seldom occur. Manipulative errors can be located, and corrected, but one still finds most survivor curves to be nonlogarithmic. Technical errors related to aerobiology are discussed elsewhere.

MULTIPLE-HIT DEATH

One may complicate the simple exponential theory by assuming that each bacterium contains more than one sensitive site and that until all are hit, the bacterium will not die. This mechanism would give rise to a survivor curve with an initial "shoulder" as in Figure 14-1, followed by essentially logarithmic decay. A curve of the same shape would be

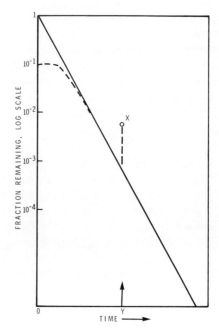

FIGURE 14-1. Hypothetical survivor curve where two or more sites must be hit to produce death, all cells have the same number of sites. At time Y, no living cells could exist in the area of point X, or any place above the line.

found if a sensitive area or volume had to be hit a number of times. Furthermore, the same curve would be found if a "diffusion boundary" existed around a single sensitive site; the time required by a lethal unit to migrate through the boundary would be equivalent to the time needed to hit all lethal units, or to saturate an area. An example might be the time required for a peroxide molecule to migrate through a cell wall (Campbell and Dimmick, 1966). Atwood and Norman (1949), who discuss the problem of multiple-hit curves rather thoroughly, point out the difficulty of determining which mechanism might be operating if one examines only the shape of the survivor curve. They also point out the consequences of assuming that the sensitivities of the sites are dissimilar. In this situation, again, the general curve shape would not indicate directly which mechanism might be involved.

One might think of each cell as having a complement of sites being "bombarded" by lethal units and that within each cell the numbers of sites are declining exponentially. As long as the number of sites remains above a certain minimal level, the cell will live and form a colony; below this level some process occurs that results directly in death. The net result is a time delay—time to destroy all the sites, or time to diffuse through a boundary. These sites could be protein or nucleic acid, hydrogen-bond linkages, as Webb (1965) has suggested; they could be membrane linkages which, if broken, would result in "leakage" of essential components from the cell (or unwanted substances into the cell), or they could be sulfhydryl or carboxyl end groupings essential to enzymatic functions. The boundaries might be membrane thickness, or mitochondrial diameters, or intercellular distances. Regardless of what might be involved, if the cells were all alike, we would expect the curve with a "shoulder" and with no tendency for the curve to "flatten."

If we assume an initial numbers of sites, and a decay rate for those sites, we would obtain the solid line shown in Figure 14-1. Take, for instance, 10 sites per bacterium and note that the initial number of cells is one-tenth the number of sites. At first no bacteria die. Eventually some die—and, finally, the death rate of the bacteria approaches the destruction rate of the sites. At times y, some number of sites remain. The maximum number of cells that can be alive at that time cannot be greater than the number of sites. Therefore, the "area" in which point x is located is "forbidden" insofar as numbers of living cells are concerned; no tailing can occur. Conversely, if the survivor curve for the cells is firmly established, the slope can be extrapolated to "zero" time to yield the average number of sites per cell. Of course, a situation where more than one hit is required to inactivate a site is the same as there being a greater number of sites to be hit.

Multiple-Hit Death with Interference

Suppose each bacterium contained a complement of sites such that the destruction of one caused another site to change in sensitivity. Such a process might occur as a result of the "mesomorphic" nature (Brown and Shaw, 1957) of protein and nucleic acid molecules in "solution." These molecules tend to be oriented with respect to other nearby molecules or charges; a disturbance as a result of a hit might change the configuration of nearby molecules to alter their resistance to lethal units. Correspondingly, if the lethal units represented simply a removal of energy-yielding molecules, such as glucose, the loss could stimulate the utilization of other sugars. The destruction of chemical groupings on a molecule could result in changes in bond angles or distances, and the molecule might not react to the lethal units as it had originally. But this is mere speculation—the over-all process would not change. So far as the individual cell is concerned, the ability to produce a colony would not be lost until all of the sites, regardless of either inherent or acquired sensitivity, had been hit. This process would result in the curve with the shoulder as before. If, as a result of a nearby hit, a site became totally resistant (higher bond energy than the energy in the lethal unit or a new shielding situation) then that cell, as we noted previously, will not die. And, to repeat, if the cells are alike, none will die.

The idea that bacteria are simultaneously very complex (Dean and Hinshelwood, 1963; Weiss, 1963), yet all cells in a given clone or culture could be alike in their complexity, is important enough to discuss in more detail, especially with regard to bacteria in the airborne state. Selective destruction of specific bacterial substances (Webb, 1963) has been assumed to be responsible for "tailing" or flattening of the survivor curve. It should be self-evident that what happens to one bacterium in the air cannot possibly influence what happens to another. Any changes that occur in a single cell relate only to that cell and not to the others. In the case of true homogeneity, after almost all sites have been hit or changed, the number of cells capable of forming colonies will start to decline and this decline will continue. We know of no process whereby dead cells in the air could so change the remaining live ones as to make them more resistant. In fact, once a death process has started in some cells, death processes should be occurring in all other cells by stochastic processes (exponential decay) until all cells are dead. Unless some change in the external environment happens during the course of an experiment—a situation we certainly hope to avoid—there is nothing that will enable some cells, *if they are like all the others*, to become more resistant. Since we *do* find tailing quite often, even with virus, it is evident that we shall be forced into a discussion of the possible heterogeneity of population—whether or not this appeals to our sense of the propriety of applying strictly physical laws to life processes.

DEATH RELATED TO POPULATION HETEROGENEITY

Those who hold that variations from logarithmic decay can be explained on the basis of individual cellular differences do not dispute the obvious physico-chemical activities that eventually lead to death of the cell. Their argument is that differences in cellular structure or function are greater than errors in measurement of physico-chemical effects. That is, difference in individual responsiveness to stress environments are greater than can be attributed to differences arising from pure chance.

The simplest kind of heterogeneity to consider would be that of a difference in the number of sites, all of which are alike, in each microbe. If 10% of the cells had 100 sites and 90% had 10, we could reconstruct the survivor curve shown in Figure 14-2. If 10% of the cells had 10

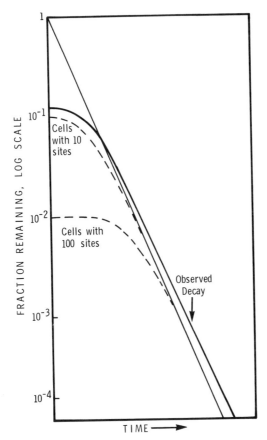

FIGURE 14-2. Hypothetical survivor curve where 10% of the cells contain 100 sites and 90% contain 10 sites, all sites alike. Observed survival curve shown in bold line, site decay by thin line, and separate cell decay by dashed line.

sites and 90% had 100, the curve shown in Figure 14-3 would describe the rate of decay. Note that if one did not know the distribution, one could not easily distinguish the curves from the type shown in Figure 14-1, where cells contained equal numbers of sites. In fact, any distribution of site numbers among the separate units would result in curves that ended in essentially logarithmic decay, as before. Note that the sites decay independently of their distribution within the bacteria. We conclude that this kind of heterogeneity could not account for tailing.

Another kind of heterogeneity to consider is variety in site sensitivity. There are two ways this could occur: (a) a variation of sensitivity,

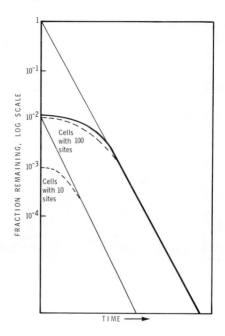

FIGURE 14-3. Hypothetical survivor curve where 10% of the cells contain 10 sites and 90% contain 100 sites, all sites alike. Observed survival curve shown in bold line, site decay by thin line, and separate cell decay by dashed line.

but the variation is the same within each organism; (b) a variation of sensitivity, different within each organism. In either case, three relationships might be visualized, as shown in Figure 14-4. If the average energy of the sites, regardless of distribution, is less than the average energy of the lethal units (A), then arguments used previously would apply; that is, all cells with the same distribution pattern of site energies

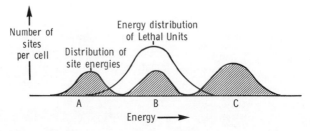

FIGURE 14-4. Possible distribution of site energies compared with energy of lethal units. A, B, C, see text.

would have equal probabilities of losing all of these sites, and the system would act as if neither the site energies nor the lethal units were distributed. When the two are approximately equal (B), a cell is not going to die until the site of highest energy (e.g., 1 per cell) is hit by the lethal unit of highest energy, and the behavior of the system would correspond to a single-site-single-hit model. If the average site energy was greater than the average energy of the lethal units (C), then some essential sites would never be destroyed, and no cell would die. Hence, any distribution of sensitivity, or energy of activation that is identical within each cell or unit, will not produce tailing.

Hedges (1966) described possible survival curves resulting from populations he considered to be homogeneous. One case, which he termed "distributed receptor model," is a final example of how sites might be distributed and could produce tailing. He assumed that the destruction of *any* of several sites would cause death. A cell with 20 sites would have a much greater probability of being hit, and thus die, than a cell with only one site. Stated another way, an entity with a highly complex structure is more likely to malfunction than one with a simple structure. Or, we might suppose that the destruction of a site resulted in the production of a poison, or an abnormal process, that killed the cell. Such populations are not homogeneous; the predicted tailing was the result of the fact that a portion of the cells were less sensitive than the others.

The word sensitivity has crept into the discussion. We would rather use the opposite, resistivity, when speaking of the cell. Let us say for the moment that resistivity is the capacity of a cell to survive in a given environment. The greater the number of sites and the higher the activation energy of these sites, the greater the resistivity. Resistance could be said to be the length of time a cell would survive a given environment. With this definition in mind, we can now say that as long as the resistivity of all cells is alike, the survivor curve of a population of such cells would evince no portion of decreased decay, or tailing.

We are left with the final, irrefutable alternative to explain the commonly observed phenomenon of tailing—that resistivity is distributed within the population of cells. This alternative is not pleasant to accept because it opens a figurative "Pandora's box" of difficulties. We cannot determine the distribution of resistivity without subjecting a population to test, and then we do not have that same population to test in other ways that might relate to resistivity. We are not even sure of the ways in which a population might have become heterogeneous with respect to resistivity. Let us persevere and tackle this "rather sterile philosophy," to quote Wood (1956).

Sources of Cell Heterogeneity

Genetic Diversity

To understand why or how one cell is different from another of the same species, we have to examine the source of cells. Whether cells come from natural surroundings or from carefully controlled cultures in the laboratory, we know that they represent the current endpoint of a continuing process of growth and division that extends backward in time to some eventful beginning.

The presently popular concept is that the genetic thread directs the activities of the cell as well as provides a link to the past and an instructional message for future cells. Knowledge about the structure and function of genetic material is undergoing a figurative explosion, so the reader is left to explore details of this on his own. However, one very important aspect needs to be pointed out here; that is, portions of genetic structure in the form of virus, phage, fragments called "transforming factors," and possibly other portions of the genetic thread, are known to be interchanged between cells to cause affected cells to have attributes related to the source of the genetic fragments. This mechanism provides a kind of eternal message exchange among the population. We will return to this later—right now we want to explore concepts related to whole-cell heterogeneity.

Take, for instance, the concept of division (Dean and Hinshelwood, 1965). The time required for a new cell to divide is known as division time, generation time, or doubling time (Quesnel, 1960). In most instances where division time has been measured, this interval has been found to be distributed in a skewed manner (Koch and Schaechter, 1962; Kubitschek, 1962; Powell and Errington, 1963). We might come to this conclusion without experiment by noting that it is unreasonable to suggest that a cell might divide in 0.1 second but reasonable that it might rest blissfully in some environment to divide once every month. It has been found that, for a given cell, the present generation time is related more to cell size than to previous division time (Schaechter et al., 1962). This means that most populations of cells contain individuals with either different potential moments when they might divide, or the individuals are the result of having been formed after different intervals of growth. This is certainly one way in which cells could be different.

Growth is a complex process (Trucco, 1967; Koch, 1966). Each cell, as it grows, takes nutrients from the environment and releases waste products. As the number of cells in a given environment (the medium held in a flask in the laboratory; water in a pond; or a moist, warm alveolar lining) becomes greater, each cell becomes more competitive. The cell must not only compete for nutrients but must also cope with an increase of waste products greater than its own output. We cannot imagine the cellular machinery remaining static with respect to function, quantitatively or qualitatively, in this situation. Thus there would be differences with respect to time in a given growth environment, and if we suppose that some of these differences relate to survival, we can now account for the observation noted above, that resistance varies with culture age. In fact, cell size, division time, and physiological state resulting from growth processes are probably log-normally distributed (Baker *et al.*, 1964).

Dynamic Diversity

If we admit that cellular machinery does not remain static with respect to environmental changes, we must also imagine regulatory mechanisms—systems that transmit and manufacture needed products in specific amounts at required times, and that destroy and dispose of unwanted material. This is exactly the concept of the cell that molecular biologists and biochemists are currently developing. Certainly there are many points along such complex, interacting paths that act as "essential sites," and the numbers of such sites need not be the same, cell to cell, or time to time.

Then, if we assume the above, we will note that the total process is time-dependent in a manner not entirely predictable. That is, the system responds to changes in environment, or adjusts its complexity, in ways not entirely related to the rapidity of environmental changes. We could admit, for instance, that a given change in environment might elicit different responses, depending on the arrangement and complexity of the cellular machinery at the moment, though we could not explain the how or why of the overall process. Even if we had knowledge of each piece of machinery within the cell, unless we knew how it influenced—and was influenced by—the other pieces, we could arrive at no more than a guess at the total cell response. The theoretical science that treats such complex phenomena is called cybernetics. We recommend a book called *Introduction to Cybernetics* by Ross Ashby (1963) for anyone who wishes to study survival or survival mechanisms in sufficient depth to obtain meaningful answers.

The principles of cybernetics are important enough to have at least a summary discussion (without proof) that will allow some analogies to be examined. The "machine" is considered to be simply a list showing how one set of states is to be transformed to another set of states.

A machine with input is a series of lists (i.e., a table); the input tells which list to use for the given transformation. The output (a set of states) of one table can be the input for another machine; they are said to be coupled, and could be considered to be one machine. Variety (the number of different states) is transmitted from machine to machine, or through machines; the latter is called a transducer. A message (sequence of states) can only be transmitted if there is some variety. If all possible variety is transmitted, there is no control; a transducer that decreases the variety transmitted is a regulator. The output of a machine may also serve as the input; this "feedback" situation can be positive and increase variety, or negative and decrease variety. The former is unstable and can oscillate, the latter is stable and can regulate or control.

The more variety a regulator can accept, the better it regulates. Message is variety and noise is variety; they are synonymous—which is termed which depends on the situation. The table for a machine need not have fixed transforming characteristics (a single trajectory); instead, a transformation may have certain probabilities of going to one state or another. If the probabilities for entering each new state are equal, the trajectory is said to be "Markovian" (essentially a stochastic process). If the probabilities are not equal but depend on the present or past state of the machine, the process is deterministic and is said to depend on the "history" of the trajectory. If the process is deterministic, it has constraints (decrease in variety) as a result of history that prevent some possible events from happening. Living systems operate with constraints hence they are deterministic. The processes are not completely deterministic, however, for the history constraint may, say, "go to 1, 2, or 3 but not to 4, 5, or 6"; whether 1, 2, or 3 occurs may be completely random. The probability involved in the latter is the stochastic element in the process. The decision as to which process is being measured has led to considerable confusion in the literature (Sacher and Trucco, 1962; Bergner, 1962; Heinmetz, 1960).

The concepts embodied above are not especially new; cybernetics only provides a new way to talk about them, and in doing so helps us ask questions with answers we can use. We can describe the living cell as analogous to a number of machines, or parts, with inputs coupled in complex and constantly reoriented ways. The "behavior" of this machine of higher complexity depends on what outputs we measure and how we define what we measure. Mostly, biochemists, biophysicists, and molecular biologists have been studying the structure of the machine. But, to quote Ross Ashby's succinct statement, "That a whole machine should be built of parts of given behavior is not sufficient to determine its behaviour as a whole: only when the details of coupling are added does the whole's behaviour become determinate." Recently, some scientists have turned their attention to a study of couplings within the cell.

Goodwin (1963) has written a monograph concerning temporal functions of the cell. He recognizes that all the couplings may never be defined, so he treats functions of parts as if they were "activities," defines activities as if they were motions of molecules, and has created what amounts to a science of cellular thermodynamics. His theory predicts, as does cybernetics, that parts of the cell's structure and function would oscillate. This means, of course, that certain functions of the whole cell could oscillate (Walter, 1966; Ehret and Trucco, 1967; Winfree, 1967). Recently, papers have appeared showing that some isolated enzyme systems do oscillate (Vaidhyanathan, 1966; Yamazaki et al., 1965; Mustafo, et al., 1966; Morowitz, 1966; Waxman et al., 1966), and that under certain conditions bacterial populations have rhythmic properties (Dimmick, 1965; Heckly et al., 1967).

APPRAISAL BY ANALOGY

Now if we return to the original question of whether the death of bacteria is deterministic or stochastic, the answer would be both—or neither; that is, both processes are involved but the overall behavior could not be predicted on the basis of either one of the processes alone. As a corollary, the survivor curve could not be employed to study a process that, on the molecular level, was assumed to be stochastic and on the cellular level was deterministic. This would be true, not only because of the complexity of processes that lead individual cells to form colonies, but because such activities are not the same in each cell in any population we could produce with present knowledge.

Furthermore, it is evident that a system as complicated as the microbial cell would not possess just two states: growth or nongrowth. Instead, one aspect of the cell's behavior would encompass not only possibility of growth under slightly different environmental conditions for each state, but also the state permitting such growth would vary from moment to moment (Stahl, 1967). Just as a well-regulated temperature bath is said to be functional if it remains within some defined bounds, so the cell forms a colony as long as the internal physiologic state is within the limits set for growth in a given environment. There is nothing to show that if functional properties changed to a state beyond that environmental limit the cell would not grow within some other environmental boundary—for example, at a lower temperature over a longer time period. The converse is also true. To use a familiar analogy, cold water usually required for the cooling system of an automobile can only lead to disaster if applied to an engine that has overheated from lack of water. That is, the state of being hot and dry does not lead through the state of being suddenly wet and cold to the state of proper operation. In the same way, the state of being dry and exposed directly to air might not lead the cell through the state of suddenly being moist and warm in a highly nutrient environment to the state of division; this, despite warmth and moistness, being a required condition for the state of division.

An example of how this process can influence apparent survival, and how it is a temporal phenomenon, was shown as follows. A culture of *Serratia marcescens* was sampled during a 24-hr growth period. Samples were freeze-dried and stored in air. At appropriate intervals over a 7-day period, a portion of each of the above samples was reconstituted and the cells were assayed on two types of agar; one was a minimal medium containing only those substances required for growth (Dimmick, 1965), the other was a medium rich in amino acids, sugars, lipids, etc. (Trypticase Soy Agar, BBL). From the assays we could construct a survivor curve for each representative culture "age" on two kinds of assay medium.

The survivor curves were a variety of shapes, but what was more interesting was the *change* in ratio of the number of colonies, from a given dilution, that formed on one medium compared with the number formed on the other. Before cells were frozen (uninjured cells), the ratio was 1.05 ± 0.05 at the 0.05 level of confidence. Afterwards, ratios as low as 0.1 and as high as 10 were observed, and neither medium was "best" for the whole experiment. The change in the ratio as a function of culture age and storage time is shown in Figure 14-5. Imagine using such data to deduce thermodynamic models!

There are a number of other corollaries that can be stated with the help of simple analogies. One is that mean values cannot be used to determine functional limits; as an example, the amount of material that might "leak" from a population of cells that has just undergone a change of state tells very little about the individual cell's role in control of that leakage (Anderson and Dark, 1967). The analogy would be that the mean temperature of a series of controlled water baths tells us nothing about how well each bath was controlled, and it tells us even less about the principle that might have been involved in the control of the baths. Of course, if all we desire is to describe how the mean temperature of all the baths behaved, we would discover the system to be ideally reproducible. The more baths we added to the system, the less we would know about how the control systems functioned, but the greater the mean reproducibility would be. Essentially the same is true with microbes.

If the external environment were changed to a sufficient extent, some water baths with sluggish control systems might exceed specified limits for some time period, but eventually they would return to the desired

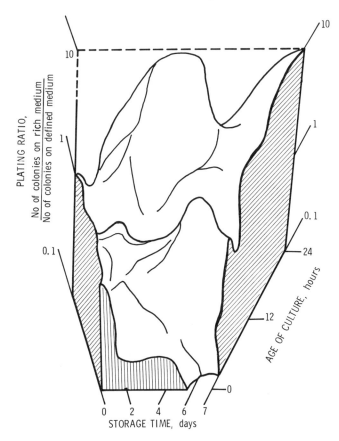

FIGURE 14-5. The change in plating ratio as a function of age of culture and time of storage in air of freeze dried cells of *Serratia marcescens*.

set temperature. In the same way, some cells might exceed some maximal limits with respect to growth in the standard environment unless time enough were allowed for their regulatory systems to return them to within the needed limits. Another way of stating this is to say that not only does the extent of environmental change influence the final observed state, but also the rate of environmental change determines the state in which the mechanism, cell or water bath, will be at a given test time.

Predicted Behavior

In this analogy the water baths have no historical continuity, and they do not influence each other; they respond independently to environmental changes. This is probably not true of microbial populations. Noting what was said previously about exchange of genetic material, and using current concepts of what is meant by a species of bacteria, we could imagine the situation shown in Figure 14-6. The total possible distribution of some attribute within a species might be represented by the large curve. This curve tells us the limits of a characteristic within which a random sample could exist and still be considered to be a part of the species; it is the genotypic potential of the species. In a given environment, only a portion of this total capability would be expressed (phenotypic expression). During the growth of a population in this environment there would be more and more opportunity for information exchange as the cell concentration in the given environment increased; information exchange would occur by means of the exchange of genetic material as well as by the influence of an increase in waste products and utilization of food. It is like saying that the controls on the water baths were loosely cross-connected, so that each was not entirely independent, or that the baths were so close together that the temperature of one influenced the other.

If this characteristic is resistivity, each cell would show a given resistance time when tested; thus the survivor curve would be representative more of the extent and center point of the distribution (actually the cumulative distribution) than it would a specific mechanism operating in each cell. Furthermore, the effect of slight changes in the mean of the population as a result of changes in the growth environment would be observed to be greatest after the majority of the population had died under adverse conditions. This means that the rate and extent of tailing we observed ought to tell us more about possible changes in distribution than the initial portions of the decay curve. Therefore, it is important that we study this often-neglected portion of the total survivor curve. A study of this final portion is also important because

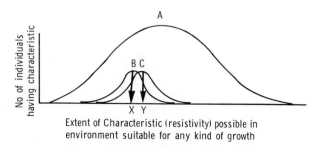

FIGURE 14-6. Expressed environmental characteristics (B, C) of select populations of microbes as related to total potential expression (A). X, Y, measure of mean resistivity.

long-lived cells are ones that contribute most to airborne infectivity, and because if resistant mutants are present in the population, they will be found within this group.

This is not the end of the complexity. Two other very important processes can be postulated by extending the kind of reasoning we have been doing. The first is that a lethal process need not be lethal for the cell in which it starts. When a cell divides, the protoplasm, the genetic mechanism, the cell wall—all that constitutes the being of the cell—are more or less shared with daughter cells. Boundaries that separate or distinguish one cell from another are, in a sense, artificial because all that has really happened is that total bio-mass has increased. Within one cell, a hit could start a process of decaying cellular capabilities that would not reach lethal levels until several generations had passed. The result would be that an injured cell, and some of the offspring, would form a colony, but the colonial size would be less than that of a normal colony. We would expect the extent of the lethal process to be distributed among cells of future generations, so that not all cells in the resulting colony would be "dead"; rather, enough cells would cease to divide so that the growth of the population would be retarded. Abherent growth curves of injured cells were shown by Campbell and Dimmick (1966) and the phenomenon of diminutive or "petite" colonies was discussed by Dimmick (1965).

Cells that form petite colonies have been called mutants or sometimes pseudoauxotrophs (Zamenhof, 1960). When assaying the number of viable cells that have been injured, we do not know whether to consider small colonies part of the "live" count or of the "dead" count. Of course, the sampled cell was capable of dividing, a colony was formed, and the question seems trivial until we consider that our intent is to measure a difference in the number of living cells as a function of time, and to assign a rate process to that difference. If the rate of formation of petite colonies differs from the true death rate, then the question is no longer trivial; we cannot use such rates to infer the influences of external factors on real death mechanisms. And how do such processes influence our estimate of infectiousness based on viability assay? We do not know.

The second process involves internal cellular functions that are inherently identical between cells, but do not operate the same way at the same time. We mentioned this, above, as one way in which population heterogeneity could occur. In a population that was "out of phase," the particular stages of the process in each cell might be expected to be distributed just as generation times are distributed. A sudden change in environment could cause these internal cellular functions to respond in the same way at the same time and the distribution of the phase relationship would change from one relatively wide to one relatively narrow; the cells would be momentarily "synchronized." The synchronous state would not be permanent; the population would oscillate in and out of phase for some period of time and the measured survival capacity would appear to depend on the length of time after the phasing act (Moses and Longberg-Holm, 1966; Heckly et al., 1967).

If we have been less definitive in describing how heterogeneity may occur in a population, and how it might influence results, than we have been with the condition of homogeneity, it is because there is very little worthwhile information available concerning relationships between heterogeneity and survival. Even less is known about the role heterogeneity in pathogenic microbes plays in the measurement of infectivity. We know that infectivity can change as a function of aerosol time (Schlam, 1960). Survivor curves of virus aerosols have decreased decay rates as a function of time, showing evidence of heterogeneity, but we know nothing of why this occurs; we can only speculate that factors operating within cells could cause virus particles to be constructed with differences similar to expected variety in other cellular structures.

Together, these principles, some established and some speculative, add up to the concept that the history or source of the cells plays a decisive role in the subsequent behavior of the population. In natural conditions, or in the laboratory, whether or not we can sample a microbe and cause it to demonstrate its presence by growth depends on the history of the microbe and how we treat the sample. Obviously, what are considered to be ideal growth conditions for the healthy cell may not be so, at least for some time period, for the injured cell.

The best we can do at the present time is (a) to recognize that microbial physiology includes dynamic concepts typical of all living things (King and Paulik, 1967), and (b) to devise new ways of measuring the properties we need to measure in order to reach meaningful conclusions about survival mechanisms. Paul Weiss (1963) has suggested that studies related to concepts of intercellular dynamics be called "molecular ecology"; this name seems to us to be appropriate.

PATHOGENICITY-VIRULENCE-INFECTIOUSNESS

This chapter would not be complete without some remarks about infectivity and its measurement, because the assay of lethal dosages has much in common with measurements of survival. However, infectivity mechanisms are even less understood and more complex—hence, more interesting—than survival. Furthermore, infectivity measurements play a more important role in medical and public health aspects of aerobiology than does mere survival. We shall start by questioning some fundamental assumptions and see where it leads us, if anywhere. These assumptions are that virulence (a) can be measured, (b) is not the same for all microbial species, and (c) can be described by certain probability laws.

Most discussions of how to determine a fifty-percentile infectious dosage (ID_{50}) or lethal dosage (LD_{50}) start with a statement to the effect that if the probability of a single pathogenic cell to infect a host is some value (p), usually a small number, then the probability that n cells will cause an infection is some other value, (p^n). This statement contains a very interesting inconsistency: if the cell is truly pathogenic, then by definition the probability of its infecting a host must be 1.0. A cell cannot be partially pathogenic. What often happens in practice is that we expose a number of hosts to a measured dose of microbes and discover a relationship between dosage and infectivity that follows a maximum likelihood, or log-probit model, and then we argue backward to an ultimate probability function that seems to involve partial pathogenesis of a cell. Sometimes an LD_{50} will be in the range of one cell, and sometimes it will be greater than 10^4; between values less than 20 and, for simplicity, the range of 100 or more, something curious has happened.

For the purpose of reviewing some possibilities, suppose cells are indeed partially pathogenic, and that a dose of 10^3 cells (the cells acting mutually and in unison) was just able to start an infection in half the animals of a group (one-half the group received just under 10^3 cells). Then the additional boost provided if the dose were doubled to 2×10^3 cells should have caused all animals to become infected, because there is no longer reasonable doubt that all animals received more than 10^3 cells. We would probably find, however, that percentage mortality had actually increased from 50% to between 60 and 80%. We must agree that the process was not that of cumulative action, but of some kind of probability function unless we want to explain how some animals still did not receive more than 10^3 cells.

A second possibility is that the measurement of dosage may have been in error. We have pointed out that measured infectivity and measured viability do not always follow the same kinetics during aerosol storage (Schlam, 1960); apparent infectivity often decreases faster than ability to form colonies in agar. So, when animals are exposed to 10^3 "viable" cells that infect only some of the animals, it is reasonable to suggest that the animals did not, indeed, receive 10^3 pathogens, but rather some smaller number of pathogens, and that the remainder of the cells "received" were nonpathogens.

Another possible explanation for high LD_{50} values resides in the definition of "pathogen." Aside from the obvious fact that there must be a host species, there must be a site within the host where pathogenicity can be demonstrated. What do we mean by "be demonstrated"? Well, we simply mean that a pathogenic cell "grows" in the host until an effect is measurable (a lesion, morbidity, mortality, possibly immune response, or that tissue assay shows increased microbial content). A site could be any point in any tissue, or only one specific portion or structure of a given tissue, such as alveoli. The host site, therefore, is only a growth environment or medium. If a microbe placed in the site is able to reproduce continuously, then the microbe was pathogenic; if not, we tend to consider it as having been a nonpathogen. Note the past perfect tense—like the Heisenberg uncertainty principle, we can only speak of what the cell was, not what it is.

This is not mere quibble, for it brings forth a very important question. Can a pathogenic cell (the theoretical one, not an assumed nonpathogenic sister cell) fail to grow in the required site in a host? Just as we said a cell is not necessarily dead because it did not produce a colony under certain conditions at certain times, so a cell that *might* be pathogenic may fail to be so because (a) the "state" of the microbe when it is placed in the site might not be suitable to initiate growth, whereas at another time the microbe might be quite sufficient for the task; (b) sites may differ, not only in a static kind of variety, but the potential of sites to support growth of the microbe could be time dependent (circadian host rhythm; Sollberger, 1965); (c) the site, as a growth medium, might be responsive in a way detrimental to the original microbe or to the offspring. None of these three mechanisms can be described in terms of simple probabilities.

If the rate of response (clearance, immune, etc.) is greater than the growth rate of the pathogen, then the animal wins—if not, he loses. So, what we measure in terms of infectiousness and what we think of as pathogenicity—the cause of the infectiousness—are really two different properties. Infectiousness is measurable, pathogenicity is not.

Another explanation for what is observed is that the dose did not reach the appropriate sites in the measured numbers; i.e., in the case of respiratory exposure, either particles were impacted at points where infection was impossible or animals did not breathe uniformly. This makes another interesting question; in the case of an LD_{50} of, say, 10^3, did half the animals breathe with 1/1000 the efficiency of the other half? It is more reasonable to think that dosage was in error.

The dose is measured by collecting some aerosol in a sampler and performing standard viability assays; the sampler may not collect particles representative of those airborne cells capable of penetrating to the required site. Considerable data are available regarding this aspect (Hatch and Gross, 1964), but as long as particle diameters are in the range of 1 to 3 μ, about 50% of the particles are retained by the alveoli (assuming these are the sites) and this size range is easy to produce with enclosed atomizers.

Furthermore, some microbes (*F. tularensis*) have been shown to have respiratory LD_{50} values approaching that of a single cell, whereas other microbial species, aerosolized in a similar manner and having essentially the same size distribution, have a high LD_{50} (*P. pestis*). This might mean that *F. tularensis* is not site-sensitive and that *P. pestis* is. However, the LD_{50} of *P. pestis* via the intraperitoneal, intramuscular, and intranasal routes also approaches a level of 1 cell, so this species is not exclusively site-sensitive. These arguments strongly suggest that not all cells of a given population of "pathogens" must be pathogenic, nor must the ratio of pathogenic cells to nonpathogens remain constant.

For the purpose of simplifying the overall problem, consider a space that we shall call the breathing volume, defined as the air volume per inspiration $\times$ number of breaths per minute $\times$ minutes of exposure, and consider the volume for the moment as being known and fixed. Place 10 animals in a space of 10 breathing volumes, and in that space place a number of airborne microbes, *all pathogens by definition*. Then we can talk about the probability of a cell being breathed, assuming the cell will be retained by the animal regardless of the site of possible impaction. This is what is meant by "received a dosage." Let us place 1 pathogen in the total breathing volume; only 1 animal can breathe in the cell, so the chance of an animal not being infected is 0.9. If 10 microbes are in the total volume, the probability that an animal will not be infected is 0.9^{10}; 3 animals could be expected to escape infection and 7 to react. With 20 microbes, 9 animals will react, etc.— simple probability statistics. Values derived this way can be plotted on the Goldberg mortality grid (Chapter 5), and will be found to conform to the predicted slope. The LD_{50} (10 animals, 1 cell per animal) is approximately 0.6 cell. As the number of animals is increased, with the same cell ratio, the LD_{50} approaches 0.7 cell, and is a good value regardless of route of infection or microbial species. That is, it doesn't make any difference whether one is assuming the probability of having a cell in a volume of a suspension to be injected, or whether it is the probability of having a cell in 1 breathing volume, the expected LD_{50} is the same.

Of course, the value of 0.7 is a maximum likelihood estimate, and simply means that if one exposed enough groups and averaged the results, the answer would be very close to 0.7 cell. With 10 animals, the 95% confidence interval of a single exposure varies from roughly 0.3 to 1.5 cells, or about two-thirds of a log with respect to dosage values. With 20 animals the interval varies from 0.5 to 1.3, or close to one-half log. The rate of decrease of the expected 95% confidence interval becomes less and less as the number of animals are increased; hence, there is no practicable gain in precision if more than 20 animals are included per test—10 to 16 animals provide maximal information per animal. One cannot expect remarkable precision from these kinds of measurements and should be suspicious of values reported to three significant figures.

The assumption was made that a breathed cell was a "received" cell. The previously mentioned value of 50% retention is reasonable, so a measured dosage of 20 cells is really 1 of about 10, etc. One could, therefore, calculate LD_{50} values based on one-half the measured dose, but we recommend this not be done because there is no firmly established retention values universally applicable to given animal species. This is an area where considerable research remains to be done. Regardless, we can reasonably say that the value is above 10% in most cases, so the effect is not one that should cause an expected LD_{50} value to be 10^3 or greater, if all cells were pathogenic.

What is the situation at this point? Certainly an LD_{50} should be no greater than 20 cells, allowing for maximal probability of reception and retention, and it could be as low as 0.7. We can no longer claim a high accuracy for the measurement, and we can no longer accept the concept that data conforming to the log-probit model at high dosage levels provides an LD_{50} related directly to that dosage. Again, perhaps reluctantly, we conclude that one of two cases is true for high dosage requirements: (1) either not all cells in a given dosage are pathogenic, or (2) not all sites are capable of supporting infection. There is really little difference between the two choices because they are mutually dependent. The cell is not pathogenic if it does not grow in the site, and we cannot determine whether the site or the cell was at fault. The virologist avoids this difficulty by diluting a suspension and describing not how many virus particles are included, but how many LD_{50} units are available; he is satisfied, because a "living" virus particle is by definition an infective particle. We bacteriologists confuse the problem by assuming a viable cell is an infectious one—otherwise we would not list the measured dosage in terms of viable cells.

As in the case of survival of airborne cells, or any survival studies for that matter, these questions would be academic except for one point—the way in which we search for mechanisms. If it is the one cell out of the many that is pathogenic (and we do not know this without testing, thus losing the cell for further test), measurements of average "contents" of suspected "virulence factors" in a cell population has little, if any, meaning unless it might be the case where an additive could change an LD_{50} from some high value to one where all cells appeared to be infective (LD_{50} 1 to 20 cells). Even then, one cannot state with certainty whether the additive changed the cell or the site, or whether the change was either physico-chemical or responsive on the part of the site or the cell.

Certainly we bacteriologists need new and ingenious approaches to these problems; but we shall not find them unless we are able to define, and in some measure understand, the complex and interrelated manner in which living systems respond to their environment and to each other. Biologists who deal with higher forms of life understand this point full well!

SUMMARY

Microbial suspensions are not like chemical reagents that can be taken from the shelf, with the expectation that the label on the bottle will specify contents and behavior. Each cell is a dynamic and responsive mechanism with a multiplicity of momentary, potential characteristics varying within inherent limitations that also change with repeated generations. It is not sufficient that we shrug our shoulders, dismiss the problem as one of biological variation, and continue to draw unwarranted conclusions by statistical methods that reveal nothing about the particular cell that lands in a given spot to initiate an observable reaction. Rather, we need to understand how that cell gained those potentials that caused it to react that way in that place and at that time. Studies of potential infectiousness, and studies of extent of survival, cannot reveal the functional status of internal cellular mechanisms unless we design experiments specifically relating measured functions to observed behavior. Those biochemical functions we might measure, as well as observed behavior, are related to the source or history of the microbial population being tested. Sampling in the field will remain completely unproductive, or at best only a "guesstimate," until we know how source factors relate to behavior because we do not know the source of microbes in natural environments. If our samplers show no colonies (or plaques), we cannot conclude there were no living microbes in the sampled air; if we find some number, we cannot know how many more might have been there, nor can we know how this number might relate to potential infectiousness. We must be able to create desired infectivity levels by manipulating source factors in the laboratory before we can measure potential infectivity in the field. To do otherwise is to practice an art, not pursue a science.

REFERENCES

Anderson, J. D., and Dark, F. A. 1967. Studies on the effects of aerosolization on the rates of efflux of ions from populations of *Escherichia coli* strain B. *J. Gen. Microbiol.*, **46**: 95–105.

Ashby, W. R. 1963. *An Introduction to Cybernetics*. Science editions, John Wiley and Sons, New York.

Atwood, K. C., and Norman, A. 1949. On the interpretation of multi-hit survival curves. *Proc. Natl. Acad. Sci.*, **85**: 696–709.

Baker, G. A., Christy, J., and Baker, B. A. 1964. Analysis of genetic changes in finite populations composed of mixtures of pure lines. *J. Theoret. Biol.*, **7**: 68–85.

Bergner, P. E. 1962. On the stochastic interpretation of cell survival curves. *J. Theoret. Biol.*, **2**: 279–295.

Brown, G. H., and Shaw, Wilfrid. 1957. The mesomorphic state. Liquid crystals. *Chem. Rev.*, **57**: 1030–1147.

Campbell, J. E., and Dimmick, R. L. 1966. Effect of 3% hydrogen peroxide on the viability of *Serratia marcescens*. *J. Bacteriol.*, **91**: 925–929.

Dean, A. C. R., and Hinshelwood, C. 1965. Cel division. *Nature*, **206**(4984): 546–553.

Dean, A. C. R., and Hinshelwood, C. 1963. Integration of cell reactions. *Nature*, **199**(4888): 7–11.

Dimmick, R. L. 1965. Rhythmic response of *Serratia marcescens* to elevated temperature. *J. Bacteriol.*, **89**: 791–798.

Ehret, C. F., and Trucco, E. 1967. Molecular models for the circadian clock. I. The chronon concept. *J. Theoret. Biol.*, **15**: 240–262.

Goodwin, B. C. 1963. *Temporal Organization in Cells. A dynamic theory of cellular processes*. Academic Press, New York.

Hatch, T. F., and Gross, P. 1964. *Pulmonary Deposition and Retention of Inhaled Aerosols*. Academic Press, New York.

Heckly, R. J., Dimmick, R. L., and Guard, N. 1967. Studies on survival of bacteria: Rhythmic response of microorganisms to freeze-drying additives. *Appl. Microbiol.*, **15**: 1235–1239.

Hedges, A. J. 1966. An examination of single-hit and multi-hit hypothesis in relation to the possible kinetics of colicin adsorption. *J. Theoret. Biol.*, **11**: 383–410.

Heinmetz, F. 1960. An analysis of the concept of cellular injury and death. *Int. J. Rad. Biol.*, **2**(4): 341–352.

Humphrey, A. E., and Nickerson, T. R. 1961. Testing thermal death data for significant nonlogarithmic behavior. *Appl. Microbiol.*, **9**: 282–296.

Jordan, R. C., and Jacobs, S. E. 1944. Studies in the dynamics of disinfection. I. New data on the reaction between phenol and *Bacterium coli* using an improved technique together with an analysis of distribution of resistance amongst the cells of the population tested. *J. Hyg.*, **43**: 275–289.

King, C. E., and Paulik, G. J. 1967. Dynamic models and the simulation of ecological systems. *J. Theoret. Biol.*, **16**: 212–228.

Koch, A. L., and Schaechter, M. 1962. A model for statistics of the cell division process. *J. Gen. Microbiol.*, 29: 435–454.

Koch, A. L. 1966. On evidence supporting a deterministic process of bacterial growth. *J. Gen. Microbiol.*, 43: 1–5.

Kubitschek, H. E. 1962. Discrete distributions of generation-rate. *Nature*, 195(4839): 350–351.

Lamanna, C., and Mallette, M. F., 1959. *Basic Bacteriology: Its Biological and Chemical Background.* (2nd Ed.) Williams & Wilkins Co., Baltimore, Md.

Morowitz, H. J. 1966. Physical background of cycles in biological systems. *J. Theoret. Biol.*, 13: 60–62.

Moses, V., and Lonberg-Holm, K. K. 1966. The study of metabolic compartmentalization. *J. Theoret. Biol.*, 10: 336–355.

Mostafo, M. G., Utsomi, K., and Packer, L. 1966. Damped oscillatory control of mitochondrial respiration and volume. *Biochem. and Biophys. Comm.*, 24: 381–385.

Powell, E. O., and Errington, E. P. 1963. Generation times of individual bacteria. *J. Gen. Microbiol.*, 31: 315–327.

Powers, E. L. 1962. Consideration of survival curves and target theory. *Phys. in Biol. and Med.*, 7: 3–28.

Quesnel, L. B. 1960. The behavior of individual organisms in the lag phase and the formation of small populations of *Escherichia coli*. *J. Appl. Bacteriol.*, 23: 99–105.

Rahn, Otto. 1945. Injury and death of bacteria by chemical agents. No. 3 of the "Biodynamics Monographs" (B. J. Luyet, Ed.), *Biodynamica*, Normandy, Missouri.

Sacher, G. A., and Trucco, E. 1962. The stochastic theory of mortality. *Ann. N.Y. Acad. Sci.*, 96: 985–1007.

Schaechter, M., Williamson, J. P., Hood, J. R., and Koch, A. L. 1962. Growth, cell and nuclear divisions in some bacteria. *J. Gen. Microbiol.*, 29: 421–434.

Schlamm, N. A. 1960. Detection of viability in aged or injured *Pasteurella tularensis*. *J. Bacteriol.*, 80: 818–822.

Sollberger, A. 1965. *Biological Rhythm Research.* Elsevier Publishing Co., New York.

Stahl, W. R. 1967. A computer model of self-reproduction. *J. Theoret. Biol.*, 14: 187–205.

Thornley, M. J. 1963. Radiation resistance among bacteria. *J. Appl. Bacteriol.*, 26: 334–345.

Trucco, E. 1967. Collection functions for nonequivalent cell populations. *J. Theoret. Biol.*, 15: 180–189.

Vaidhyanathan, U. S. 1966. Theory of mechano-electric membrane transducers. *J. Theoret. Biol.*, 13: 18–31.

Vas, K., and Proszt, G. 1957. Observations on the heat destruction of spores of *Bacillus cereus*. *J. Appl. Bacteriol.*, 21: 431–441.

Walter, C. 1966. *Enzyme Kinetics: Open and Closed Systems.* Ronald Press, New York.

Waxman, A. D., Collins, A., and Tschudy, D. P. 1966. Oscillations of hepatic δ-aminolevulinic acid synthetase produced in vivo by heme. *Biochem. and Biophys. Res. Comm.*, 24: 675–683.

Webb, S. J. 1965. *Bound Water in Biological Integrity.* Charles C Thomas, Publisher Springfield, Ill.

Webb, S. J. 1965. Radiation, relative humidity and the mechanism of microbial death in aerosols. In *A Symposium on Aerobiology*, 1963. (R. L. Dimmick, Ed.), pp. 369–380. Nav. Biol. Lab., Nav. Supply Center, Oakland, Calif.

Weiss, Paul. 1963. The cell as a unit. *J. Theoret. Biol.*, 5: 389–397.

Winfree, A. T. 1967. Biological rhythms and the behavior of populations of coupled oscillators. *J. Theoret. Biol.*, 16: 15–42.

Wood, T. H. 1956. Lethal effects of high and low temperatures on unicellular organisms. In *Advances in Biological and Medical Physics*, Vol. 4, pp. 119–165. Academic Press, New York.

Yamazaki, I., Yokota, K., and Nakajima, R. 1965. Oscillatory oxidations of reduced pyridine nucleotide by peroxidase. *Biochem. and Biophys. Comm.*, 21: 582–586.

Zamenhof, S. 1960. Effects of heating dry bacteria and spores on their phenotype and genotype. *Proc. Natl. Acad. Sci.*, 46: 101–105.

15

AEROSOL IMMUNIZATION

George A. Hottle

NAVAL BIOLGICAL LABORATORY, SCHOOL OF PUBLIC HEALTH,
UNIVERSITY OF CALIFORNIA, BERKELEY

INTRODUCTION

The techniques for exposure of man and animals to antigenic materials, to develop resistance to disease, have varied only to a limited extent since Jennerian vaccination was first undertaken almost 200 years ago. Initially, the skin was scratched; then a short step led to use of the needle and syringe for injecting foreign materials into one or several parts of the vaccinee's anatomy. With some exceptions, this situation exists today. Finally, oral vaccination has been successfully employed to vaccinate against infantile paralysis.

There are a few indications that changes are being made. Impetus for change has come recently from the need for vaccination of large numbers of individuals and from special requirements surrounding the use of attenuated microorganisms as vaccines. When mass vaccination by injection was necessary, as with influenza vaccine, attenuated measles vaccine, and smallpox vaccine, the jet injector (Hengson et al., 1963) was devised. The technique developed with this apparatus made possible the immunization with measles vaccine of hundreds of thousands of children (Barbotin and Poulain, 1964) in rural areas of Africa. A further improvement in technique of vaccination is exemplified by the oral administration of modified poliovirus to human volunteers by Koprowski and his co-workers (Koprowski et al., 1952). Whole populations have

375

been vaccinated by the alimentary route with the Sabin poliovirus vaccine (Chumakov *et al.*, 1959) when protection against poliomyelitis was sought.

It was only natural that vaccination of animals and people by the aerosol route would be considered for protection against respiratory infections. The introduction of vaccine into animal body by this method has been termed either "vaccination by inhalation," "aerogenic immunization," or "aerosol immunization." If the vaccine is applied by dropping the antigen into the nostrils, the process is known as vaccination by intranasal instillation. There are advantages to each method. With intranasal instillation, less vaccine is needed because the individual dose is administered to each animal. This technique requires very little equipment; however, it is laborious in effort and time, because it is sometimes necessary to anaesthetize the animals in order to have them accept the treatment. With the inhalation method, a large amount of vaccine and specialized equipment are needed, including not only apparatus for nebulizing the vaccine, but also suitable quarters for holding the animals where the aerosolized vaccine will be inhaled in sufficient amounts to produce the desired effect.

The considerations of aerosol immunization include both active and passive immunization. Active immunization will be considered from two points: (*a*) exposure of man or animals to living antigens, and (*b*) exposure of man or animals to nonliving antigens.

Generally, smaller amounts are needed with living antigens because the antigen multiplies within the host and may, in fact, set up an inapparent infection or a mild, transient diseased state. Immunity resulting from such living antigens could be of a solid, long-lasting type because of the marked dissemination and large amount of antigen produced during the infection. On the other hand, if the living antigen remained localized and did not multiply, a limited amount of immnuity or a transient immunity might result.

With nonliving antigens, larger amounts are needed. Since the antigen does not increase after it has been deposited in the vaccinee's tissues, response depends directly on the amount introduced. With these antigens long exposure times may be needed, and the response may be minimal and of short duration. Successive exposures of vaccinated animals are often necessary to build up an effective immunity, i.e., an immunity which will permit animals to survive a challenge with the virulent disease agent.

Passive aerosol immunization has been found of value in preventing infection of animals by respiratory viruses. Besredka (1920) suggested intratracheal injections of immune serums for development of general passive immunity. Although some workers subsequently questioned the superiority of this procedure over intraperitoneal inoculation of serum to confer general immunity, Fox (1936) pointed out the value of the respiratory route for administration of immune serums. In addition to a possible reduction of the hazard of anaphylaxis, as suggested by Besredka, Fox gave the following advantages for introducing immune serums by the inhalatory route in pulmonary diseases: (*a*) a much higher antibody content of the lung than that which follows injection of the immune serum intravenously, and (*b*) prolonged retention of the antibody locally. Lyons *et al.* (1944), working with influenza virus type A PR8), reported (1) that high titer horse serum or its globulin fraction, when administered by intranasal instillation or by inhalation, protected mice against intranasal infection with the virus 4 hours later; and (2) that whole immune serum (or plasma) was superior to any individual globulin fraction in the degree and duration of its protective effect for mice. Other workers (Zellat and Henle, 1941; Taylor, 1941) have found the use of immune serum of value both prophylactically and therapeutically.

To be effective as an immunizing agent by the respiratory route, the antigen must reach the body cells where resistance is built up. Unless it is presented as particles in the size range of 1 to 2 μ (Hatch, 1961), the antigen will not penetrate deeply enough into the pulmonary tissues to modify susceptible cells (Figure 15-1). Living antigens can multiply and be carried to the antibody-forming cells, so penetration may be enhanced. However, with nonliving antigens or with immune serums, particle size is particularly important. This was stressed by Zellat and Henle (1941), who studied the instillation of immune serum to protect mice against infection by influenza virus. In a study of the distribution of inhaled material, Lyons *et al.* (1944) exposed a monkey for 1 hour to an atmosphere of India ink sprayed at a 1:10 dilution. Autopsy of the animal after exposure revealed that some of the inhaled carbon had penetrated into the alveolar spaces of the lungs; also, a considerable amount of ink was seen in the stomach. A comparison was then made by intranasal instillation of a similar amount of a 1:10 dilution of India ink into another monkey. When the tissues of this animal were examined, only a few areas of diffuse blackening were seen in the lung tissue. This was considerably less than that seen in the lungs of the animal exposed to the ink aerosol. Furthermore, the stomach of the second animal contained considerably more of the ink than that of the first animal. This work would seem to establish the superiority of the inhalatory method with respect to distribution and penetration of materials into respiratory tissues.

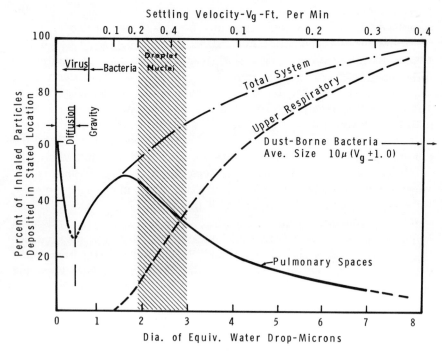

FIGURE 15-1. Total and regional deposition of inhaled particles, in relation to the aerodynamic particle size. (Hatch, T. F. 1961. *Bacteriol. Rev.,* **25**: 238). Reprinted with permission of the American Society for Microbiology.

Vaccination by aerosol particles can function in a variety of ways. It may provide a convenient means for setting up an inapparent systemic infection with an avirulent microorganism, such as *Francisella tularensis* strain LVS. This infection protects against challenge with virulent bacteria and causes serum antibodies to appear (Eigelsbach *et al.*, 1961). Hence, the infection resembles infection by any other route. Aerosol vaccination also provides a unique method for protection against specific respiratory infections. Localized inapparent infection has been established, with increases in serum antibody levels, in 71 to 94% of individuals immunized with an avirulent strain of influenza virus (Zhdanov *et al.*, 1962). That such exposure can provide early protection against specific respiratory pathogens was shown by Schulman and Kilbourne (1963), who vaccinated mice with nonliving influenza virus. Besides producing some degree of immunity, the inactivated virus interfered with infection by live virus administered 1 day after vaccination. Thus, vaccination by the same route along which virulent viruses normally gain access to the host provides special advantages. These advantages have been observed repeatedly in the frequent large-scale use of the oral attenuated Sabin poliovirus vaccine. When this vaccine was given in epidemic times, sharp drops in poliovirus content of stools of children occurred (Sabin et al., 1960). Reductions in excretion of polioviruses were attributed to the well-known phenomenon of virus interference. Finally, aerosol vaccination can introduce nonliving antigens into the body for development of specific immunity (Yamashiroya et al., 1966). The response to the antigen follows closely that obtained when the antigen is injected in the conventional manner, i.e. with syringe and needle. Indeed, Maslov (1959) indicated that when particles of the proper size are used, exposure to an aerosol is equivalent to deep penetration into body tissues, e.g., as in intraperitoneal injection.

When reactions to vaccination are considered, aerosol immunization would be expected to present somewhat the same problems as those encountered with other immunization procedures. The particular vulnerability of the lungs to certain diseases means that special precautions must be observed. Any procedures to be applied to human beings must first be studied very carefully in animals. Only when the animals are found to remain free of disease for long periods of time should treatment of humans be considered. Human studies should be started with very small numbers of vaccinees, and the numbers should not be increased until definitive results are known with small groups representing each sex, age, and racial group.

Special consideration should be given to individuals with known metabolic, pathological, or immunological disorders. Since individuals vary so much, care must be exercised in studying the response of each. When the vaccine consists of viable microorganisms, there exists the hazard of both a pathological response as a result of infection, and an allergic response. A nonviable vaccine presents only the hazard of a dangerous allergic response. With inhalation vaccination, vaccinees must be observed carefully during exposure to the antigen. At the first sign of an untoward reaction, vaccination should be discontinued and supportive therapy initiated.

Animals can be sensitized to microbial products as shown by Larsh (1960), who exposed guinea pigs to aerosols of formalin-killed mycelial fragments of Histoplasma capsulatum. The degree of sensitization to the fungal antigen, histoplasmin, was comparable to that produced by subcutaneous injection. Other work has shown (Gindin et al., 1963) that a second exposure of rabbits to pertussis vaccine aerosols after an interval of 30 days, can lead to an intense reaction in the bronchial lymph nodes and spleen.

Response of animals to aerosol vaccination have induced the same types of antibodies as those seen following parenteral or oral vaccination.

When the vaccine was a living, attenuated microorganism (such as measles virus, *F. tularensis,* or any agent that produced an inapparent systemic infection), the usual antibodies formed by infections are found in the serum; in addition, resistance to infection by virulent disease form was demonstrated. Similarly, when nonliving antigens are administered to animals, the same types of antibodies can be expected. Since the antigenic mass passed into animals by the aerosol route is so limited, there is usually little evidence of antibody production; however, there has been definite evidence of increased resistance to disease by the immunization procedure (Yamashiroya *et al.,* 1966). Nevertheless, Andrew and Wagner (1955) reported that guinea pigs exposed 215 minutes to an aerosol of formalin-killed Q-fever vaccine developed complement-fixing serum antibodies as well as slightly increased resistance to infection by living rickettsiae.

EXPOSURE TO SPECIFIC VACCINES

Although experimental work in human aerosol immunization has been carried on for a number of years, there are no programs of routine use of this technique being carried out in the United States. In the USSR, however, a live influenza vaccine containing type A2 and type B viruses has been in use for vaccination of man since 1961. This vaccine is administered intranasally by spraying with an atomizer. Its use is limited to people over 16 years of age because the reactions are too intense in younger children. A complete immunization consists of three applications of the vaccine at 8- to 10-day intervals.

The techniques of aerosol infection lend themselves readily to aerosol immunization with vaccines containing attenuated, living microorganisms. The mechanism is one of producing infection by the aerosol route. Once infection is established, the host reaction to invasion by foreign substances sets in, and resistance to infection is built up. The problem resolves itself into one of developing an inapparent infection so that immunity will appear without disease. When a vaccine is composed of several strains or types, such as influenza virus vaccine, multiple exposures to the aerosols will be required. Then, if response to one component fails to appear following the first exposure, a second or third opportunity will be provided.

The following instances are cited to show that vaccination by the aerosol route has been investigated with a number of living disease agents. The work is only considered to be related to aerosol immunization when the microorganism was attentuated and thus of very limited disease-producing ability.

Tularemia Vaccine

Vaccination of animals and man with live attentuated *F. tularensis* strain LVS by the aerogenic route was demonstrated by Eigelsbach and Sawyer and their co-workers. Eigelsbach *et al.* (1961) showed that monkeys and guinea pigs vaccinated by aerosols had higher levels of agglutinins, and comparable protection against challenge with virulent *F. tularensis*, than animals vaccinated dermally. This work, as well as new data with human volunteers, was recently discussed by Hornik and Eigelsbach (1966). Some of the results shown in Table 15-1 indicate

Table 15-1 Relationship of route and dose of LVS vaccine to resistance to tularemia aerosol challenge.

| Vaccination | | No. | No. (%) | Percent | |
Dose	Route	challenged[d]	with fever $>100°F$[a]	requiring treatment	Percent protection[b]
10^8	Respiratory	30	18 (60)	0	100
10^6	Respiratory	16	10 (62)	0	100
10^4	Respiratory	56	43 (77)	41	59
—[c]	Dermal	46	29 (63)	46	54
	None	47	44 (94)	89	

[a] Mean incubation period: volunteers vaccinated aerogenically or by acupuncture, approximately 4 days; controls, approximately 3 days.
[b] Uncorrected with respect to control data.
[c] Acupuncture technique.
[d] Aerosol challenge 2 to 18 months after vaccination.
(From: Hornik, R., and Eigelsbach, H. *Bacteriol. Rev.*, **30**: 535, Sept. 1966. Reprinted with permission of the American Society for Microbiology.)

that solid protection was seen in those vaccinated by the respiratory route and challenged with an aerosol dose of 2.5×10^4 organisms, strain SCHU S4. (Hornik and Eigelsbach, 1966.) This protection was greater than that produced by the conventional dermal vaccination.

Sawyer *et al.* (1964) showed that Macaca monkeys could be effectively protected against tularemia and Venezuelan equine encephalitis (VEE) by exposure to a mixed aerosol of live attenuated vaccine strains of microorganisms. The monkeys were exposed directly to aerosols for 10 minutes and received estimated doses of 1,000 to 2,000 viable tularemia bacteria and 200 to 14,000 guinea pig 50%-immunizing doses of VEE virus. There was no fever or other sign of illness following vaccination, and the animals showed increased serum hemagglutination titers for VEE virus and increased serum agglutinins for *F. tularensis,* as well as resistance to challenge by virulent *F. tularensis*. This vaccination procedure was considered to be highly effective, and there was no evidence that the presence of two live antigens, one bacterial and the other viral, in any way inhibited each other.

In a consideration of aerosol vaccination. Buscher and Bellanti (1966) examined the local production of antibody. They found agglutinins to *F. tularensis* in nasal secretions of man following aerosol vaccination as well as following percutaneous vaccination. They concluded that it is not possible to correlate local antibody levels with resistance to infection by *F. tularensis*. The value of aerosol vaccination was seriously questioned by these workers because of the risk involved. They expressed concern that 80% of the people receiving effective doses of vaccine had, as a reaction, overt but mild respiratory tularemia.

Tigertt (1962), in reviewing Soviet work with viable *F. tularensis* vaccine, indicated that, although some workers consider the aerogenic route of vaccination "not without danger, and . . . impossible to recommend . . . for practical use," studies in man are continuing.

BCG Vaccine

The development of resistance to tuberculosis by aerosol vaccination has been studied by Middlebrook (1961). He showed that in guinea pigs, minimal amounts of BCG vaccine given by the aerosol route produced as much resistance to subsequent virulent challenge as a ten-times larger dose given by the intracutaneous route. It was also shown that immunity to infection persisted for at least 2 years. That the species of animal used for this work is important became apparent when the work was extended to include rabbits and mice. Since these animals are less susceptible than guinea pigs to infection with tubercle bacilli, it is not surprising that evidence of aerogenic immunity could not be obtained in either of these species. Middlebrook (1961) also reported some results of aerogenic BCG vaccination of human beings provided by Sol Roy Rosenthal. These data showed that man is similar to guinea pigs in the ease with which he is rendered tuberculin-positive by exposure to not more than 10 guinea pig infective doses of BCG vaccine.

Other Living Bacterial Vaccines

Russian investigators have used dry vaccine preparations for experimental immunization of animals and man. They have published data on vaccines prepared against brucellosis, anthrax, plague, and tularemia. With brucella and anthrax (as reported by Middlebrook, 1961), at least 10 million viable units of vaccine organisms were needed to produce a definite immunological response. These responses indicated that aerogenic vaccination was as effective as the subcutaneous route. When reactions to the brucella vaccine were examined, it was reported (Aleksandrov et al., 1962) that individuals not previously sensitized to brucellosis infection showed no, or very weak, reaction. Those who had been sensitized to brucellosis before vaccination showed more frequent and intense postvaccinal reactions. About half of those reacting exhibited leucocytosis and monocytosis which reached a maximum by the fourth day, remained at that level for the next 5 to 8 days, and disappeared after 1 month. Accompanying this was slight catarrhal inflammation of the mucous membranes of the larynx, trachea, and bronchi, which disappeared after 7 to 10 days. The authors conclude that more extensive and thorough study of this vaccine is needed in sensitized persons.

Virus Vaccines

A good discussion of the influenza virus vaccine used for intranasal vaccination in the USSR is given by Zhdanov (1961). The preparation of the vaccine requires special care, because the growth of influenza viruses on chick embryos is accompanied by a gradual loss of pathogenicity and immunogenicity. After 6 to 8 passages, there occurs a sharp loss of the immunogenic properties of the virus. The pathogenicity is lost earlier, after 2 or 3 passages. Thus the pathogenicity of the virus can be removed without destroying its immunogenicity.

Miller (1966) has reported interesting observations on exposure of white Carneau pigeons to aerosols of VEE virus. As little as 374 mouse 50% lethal doses (mouse LD_{50}) units of virus, presented in 1 minute, produced benign infections in the animals. Viremia levels as high as 10^6 mouse LD_{50}/ml of blood were seen, with disappearance of virus from the blood by the fourth day and appearance of serum-neutralizing antibody 2 weeks later. It was also noted that normal animals were resistant to aerosols containing higher concentrations of virus if inhaled at a rate less than 370 mouse LD_{50}/minute. This resistance may explain why airborne bird-to-bird transmission was found to be a rare occurrence.

In a study of the aerogenic virulence and immunogenicity of the encephalomyocarditis (EMC) viruses for mice, Akers et al. (1966) found that virulence was correlated with large plaque-size. Strains that produced small plaques on L-cell monolayers were avirulent and, in addition, were capable of immunizing the animals. Mice that survived aerosol exposure to about 10,000 plaque-forming units of strains Maus Elberfeld (ME) and Mengo-37A (small plaque-formers) were immune to aerosol challenge by large plaque-forming strains 21 days later.

With the keen interest in use of attenuated measles virus for vaccination of children, there have been studies to examine aerosol administration of the virus. Minamitani et al. (1964) reported that, by a 40-second exposure to nebulized virus, 15 of 19 children responded with complement-fixing (CF) antibody. This was less than all of 32 children who received the virus by the subcutaneous route, but the antibody titers of those who did respond in both groups were the same. The authors noted that reactions in the respiratory group were somewhat less than those in the injected group, possibly because the dose of virus was lower. The high infectivity potential of measles virus for man makes it an ideal microorganism for aerosol vaccination.

Earlier work by Kress et al. (1961) apparently set the pattern for use of measles vaccine by injection rather than by aerosol. These workers exposed children, 5 to 9 years old, to aerosolized measles virus vaccine. Each child received 0.3 to 0.5 ml of nebulized virus into the open mouth while inhaling. The antibody response was good in all vaccinees; they showed serum neutralization titers of 1:400 to 1:600. Although the vaccine produced fewer reactions when administered by the respiratory route than by the parenteral route, marked elevations of body temperature occurred frequently. As a result, the author recommended that the vaccine be given by the parenteral route, with a simultaneous injection of measles gamma globulin (human) at another site in order to reduce the severity of reactions.

Studies with *Coxiella burnettii* (Tigertt et al., 1961) illustrate another method by which aerosol immunization can be carried out. By aerosol exposure of guinea pigs and man to a virulent suspension of rickettsiae, and administration of an antibiotic such as oxytetracycline throughout the incubation period, relapse can be prevented and solid immunity develops. The tricky part of this procedure is to balance the aerosol concentration of microorganisms so that a sufficient antigenic mass is administered, or permitted to grow in the host, and to control further growth of the infecting agent with antibiotics so immune processes can take over. With this system, therapy should be withheld until late in the incubation period of the infection so that sufficient antigen is present to effect an antibody response. Using the above procedure, it was possible to suppress any clinical disease in man after exposure to *Coxiella burnettii* by the respiratory route, and to effect immunization.

Nonliving Antigens

As indicated earlier, response to nonliving antigens is the true test of the value of aerosol immunization. Since these antigens cannot multiply in the host, an effective mass must be provided at the time of vaccination.

Effremova (1963) reported the results of vaccination of people with aerosols of typhoid vaccine. The serum agglutinin response was similar to that obtained in people vaccinated by subcutaneous injection. A second aerosol exposure, 20 days after the first, produced higher serum agglutinin levels. Reactions to vaccination were moderate in adults, less pronounced in children.

In attempting to immunize guinea pigs with atomized tetanus toxoid, Yamashiroya *et al.* (1966) encountered problems with getting enough antigenic stimulation into the animals. With a dose of 15 flocculating units (Lf) of toxoid/animal, a geometric mean serum antibody level of 1:200 (sensitized red blood cell agglutinating titer) was produced 5 weeks after vaccination, and 7 of the 13 animals survived the challenge of 10 minimum lethal doses (MLD) of toxin. With subcutaneously injected animals, the same dose of toxoid produced a mean titer of 1:4,609, and each of 6 animals survived the challenge. When aerosol immunization was carried out as 2 exposures (5 weeks apart), antibody titers like those seen following 1 subcutaneous injection were obtained in the guinea pigs. Data are also included from the work of Muromstev *et al.* (1961) on aerogenic administration of diphtheria toxoid as a revaccination procedure following primary subcutaneous immunization. Guinea pigs that developed a mean titer of 2.6 units of antitoxin/ml of serum after the primary injection of toxoid showed titers of 118 units/ml 2 weeks after the aerosol revaccination. The animals were exposed for 60 minutes to an aerosol prepared from diphtheria toxoid containing 2,100 Lf units/ml. This work also was carried out in children, and aerosol revaccination was found to be effective. It was reported that undue systemic reactions were not observed in children; however, allergic reactions were reported in exposed adults.

Bartlema (1966) reported that mice elicited a much weaker response than guinea pigs to inhaled fluid tetanus toxoid. By adding a bacterial adjuvant (killed *Bordetella pertussis* or a soluble extract of *B. pertussis*) to the inhaled toxoid, he obtained an enhanced antitoxin response. With this adjuvant, the same response was obtained in mice after two exposures to 16 Lf of toxoid by the same respiratory route and after two injections of 0.5 Lf of toxoid. This response showed about 1 unit of antitoxin/ml of serum.

Recently, Waldman *et al.* (1969) vaccinated human volunteers with inactivated influenza virus vaccine. One group of volunteers received aerosolized vaccine, another group was immunized via subcutaneous injection. Identical 1-ml doses were administered to each individual. The aerosol was administered from a DeVilbiss-15 nebulizer, which introduced 0.25 ml into each nostril and 0.5 ml into the posterior portion of the oropharynx during inhalation. Booster inoculations were performed in the same manner two or three weeks later.

Subsequently, an epidemic of A2 influenza occurred. Compared with control groups that received no vaccine, the illness rate among those receiving aerosolized vaccine was reduced by 79%; among those vaccinated subcutaneously the reduction was only 27%.

The 1-ml dose seems larger than required for delivery of a reasonable dose into the respiratory tract. Perhaps a large part of the dose was swallowed rather than being deposited in the bronchi or alveoli. One must admit, however, that the results were impressive.

The report of Waldman *et al.* (1969) also includes an excellent annotated summary of the latest work on aerosol immunization and on antibody in respiratory secretions. They note that influenza virus antibody in these secretions is of the gamma A immunoglobulin class and that the antibody induced in serum by subcutaneous vaccination is of the gamma G class. The authors point out that immunization by the gastrointestinal tract also produces antibodies of the gamma A class. In future studies it would be advisable to include a control group to receive the vaccine by the gastrointestinal route. Perhaps the critical procedure in this work was not introduction of vaccine into the lower respiratory tract, but rather it may have been a combination of the bathing of the oropharynx and then the swallowing of the vaccine that resulted in effective immunization.

Aerosol Vaccination in Veterinary Practice

At present, two vaccines have been licensed for veterinary use by the respiratory route in the United States. These are: (1) Newcastle Disease Vaccine, live virus (B1 type), chick embryo origin; and (2) Avian Bronchitis Vaccine, live virus, chick embryo origin. There is also a divalent vaccine which is a mixture of these two vaccines. These vaccines are administered by aerosol, intranasal, intraocular, or dust methods, and are used widely for the immunization of poultry.

An an illustration of the development of these vaccines, I'd like to describe the first work done with Newcastle Disease Virus Vaccine. Hitchner and Johnson (1948) found an avirulant virus strain, B1, which, when inoculated intranasally into chicks, showed very little reaction and resulted in immunity to challenge with virulent virus 19 days later. They gradually widened the use of the vaccine, first in small isolated flocks, then in larger flocks. Today, five commercial laboratories are licensed by the Veterinary Biologics Division of the Agricultural Research Service for manufacture and sale of these vaccines. There are no other aerogenic vaccines used routinely in animals in the United States.

SUMMARY

There have been numerous research efforts in the application of aerosol immunization to man and to animals. Many of these efforts have been promising. The most attractive results have occurred when immunization against respiratory diseases was investigated; a small amount of antigen was found to have the most noticeable effects. It is also in these areas that practical procedures have been developed in preventive medicine: (a) The use of a bivalent attenuated live virus vaccine for immunization of people in the USSR (this vaccine contains influenza virus types A2 and B and is administered by a nebulizer, to individual vaccinees), and (b) the use of attentuated live virus vaccines for immunization of chickens against respiratory diseases caused by Newcastle disease virus and by avian bronchitis virus. The latter vaccines are available in the United States as monovalent or bivalent vaccines.

One of the problems attendent to aerosol vaccination is that of determining whether the exposed individual really received a proper dose of vaccine. When the vaccine is injected with a syringe and needle, the dose can be accurately controlled; but when it is given as an aerosol, there are several possibilities: (1) The dose may not be accurately measured; (2) a large part of the dose may be swallowed rather than deposited in the lungs (this may be due to the architecture of the nasopharynx or to the particle size of the aerosol), and (3) the vaccinee may not breathe properly to assure that an optimal dose reaches the deep lung tissue where immunity can be developed.

The need for further work is apparent in this field. At present, the emphasis has been on infection by the aerosol route, using attentuated living microorganisms. This has shown results upon which future work can be built. The area where new work is needed is with nonliving antigens. To date, this has not been so promising. New investigations must be approached with imagination. Antigens must be purified so that concentrations can be increased and undesirable components can be removed. Adjuvants must be developed so that the activities of small amounts of antigens are enhanced. Aerosol immunization can fill a practical need in mass vaccination operations and it can also contribute to our knowledge of antibody formation by the utilization of a different organ system for introduction of foreign substances to the host.

REFERENCES

Akers, T. G., Bond, S. B., Papke, C., and Lief, W. R. 1966. Virulence and immunogenicity in mice of airborne encephalomyocarditis viruses and their infectious nucleic acids. *J. Immunol.*, **97**: 379–385.

Aleksandrov, N. I., Gefen, N. Ye., Gapochko, K. G., Garin, N. S., Maslov, A. I., and Mishchenko, V. V. 1962. A clinical study of postvaccinal reactions to aerosol immunization with powdered brucellosis vaccine. *Zhur. Mikrobiol.*, **33:** 31–37.

Andrew, V. W., and Wagner, J. C. 1955. A summary of research studies on the immunology and serology of *Coxiella burnetii*. Unpublished observation.

Barbotin, M., and Poulain, R. 1964. La Campagne de masse de vaccination anti-rougeoleuse en Haute-Volta. *Med. Trop.*, **24:** 405–416.

Bartlema, H. C. 1966. Discussion of aerosol vaccination with tetanus toxoid. *Bacteriol. Rev.*, **30:** 633–635.

Besredka, A. 1920. Infection et vaccination par voie trachèale. *Ann. Inst. Pasteur*, **34:** 361–369.

Buescher, E. L., and Bellanti, J. A. 1966. Respiratory antibody to *Francisella tularensis* in man. *Bacteriol. Rev.*, **30:** 539–541.

Chumakov, M. P., Voroshilova, M. K., Vasilieva, K. A., Bakina, M. N., Dobrova, I. N., Drosdov, S. G., Ashmarina, E. E., Posedlovsky, T. S., Kostina, K. A., Sherman, G. A., Yankevich, O. D., and Uspensky, U. S. 1959. Preliminary report on mass oral immunization of population against poliomyelitis with live virus vaccine from A. B. Sabin's attentuated strains. First International Conference on Live Poliovirus Vaccines. *Scient. Publ. No. 44*, pp. 517–529. Pan American Sanitary Bureau, Washington, D.C.

Efremova, V. N. 1963. Reactogenic and immunological changes in the serum of persons after aerosol immunization with typhoid fever vaccine. *Zhur. Mikorbiol. Epidemiol. Immunol.*, 40(2): 51–57.

Eigelsbach, H. T., Tulis, J. J., Overhold, E. Z., and Griffith, W. R. 1961. Aerogenic Immunization of the monkey and guinea pig with live tularemia vaccine. *Proc. Soc. Exp. Biol. and Med.*, **108:** 732-734.

Fox, J. P. 1936. The permeability of the lungs to antibodies. *J. Immunol.*, **31:** 7–23.

Gindin, A. P., Anosov, I. Ya., and Mayorova, G. F. 1963. Histopathology and histochemistry of the reaction of lymphoid organs to inhalation immunization with pertussis vaccine. *Zhur. Mikrobiol. Epidemiol. i Immunobiol.*, 40(3): 45–49.

Hatch, T. F., 1961, Distribution and deposition of inhaled particles in respiratory tract. *Bacteriol. Rev.* **25:** 237-240.

Hengson, R. A. Davis, H. S., and Rosen, M. 1963. The historical development of jet injection and envisioned used in mass immunization and mass therapy based upon two decades' experience. *Military Med.*, **128:** 516–524.

Hitchner, S. D., and Johnson, E. P. 1948. A virus of low-virulence for immunizing fowls against NDV (Avian Pneumoencephalitis). *Vet. Med.*, **43:** 525–530.

Hornik, R. B., and Eigelsbach, H. T. 1966. Aerogenic immunization of man with live tularemia vaccine. *Bacteriol. Rev.*, **30:** 532–538.

Koprowski, H., Jervis, G. A., and Norton, T. W. 1952. Immune responses in human volunteers upon oral administration of rodent-adapted strain of poliomyelitis virus. *Am. J. Hyg.*, **55:** 108–126.

Kress, S., Schluederberg, A. E., Hornik, R. B., Morse, L. J., Cole, J. L., Slater, E. A., and McCrumb, F. R. 1961. Studies with live, attenuated measles-virus vaccine. *Am. J. Dis. Children.*, **101:** 701–707.

Larsh, H. W. 1960. Natural and experimental epidemiology of histoplasmosis. *Ann. N.Y. Acad. Sci.*, **89:** 78–90.

Lyons, W. R., and Personnel of Naval Laboratory Research Unit No. 1. 1944. The inhalatory route for prophylaxis and treatment of experimental influenza. *Amer. J. Med. Sci.*, **207**: 40–60.

Maslov, A. I. 1959. Concerning the efficacy of the inhalation vaccination. I. The effect of the inhalation method of vaccination on the general immune reconstruction of the body. *Zhur. Mikrobiol. Epidemiol, i Immunobiol.*, **30**(11): 15–18.

Middlebrook, G. 1961. Immunological aspects of airborne infection. Reactions to inhaled antigens. *Bacteriol. Rev.*, **25**: 331–346.

Miller, W. S. 1966. Infection of pigeons by airborne Venezuelan equine encephalitis virus. *Bacteriol. Rev.*, **30**: 589–595.

Minamitani, M., Nakamura, K., Nagahama, H., Fujii, R., Saburi, Y., and Matumoto, M. 1964. Vaccination by respiratory route with live attenuated measles virus, Sugiyama, adapted to bovine renal cells. *Japan. J. Exp. Med.*, **34**: 81–84.

Muromstev, S. N., Borodiyuk, N. A., Menashev, V. P., and Aleshina, R. M. 1961. Revaccination of children with diphtheria toxoid by inhalation. *Zhur. Mikrobiol. Epidemiol i Immunobiol.*, **32**: 589–594.

Sabin, A. B., Alvarez, M. R., Amezquita, J. A., Pelon, W., Michaels, R. H., Spigland, I., Koch, M., Barnes, J., and Rhino, J. 1960. Effects of rapid mass immunization of a population with live, oral poliovirus vaccine under conditions of massive enteric infection with other viruses. Second International Conference on Live Poliovirus Vaccines. *Scient. Publ. No. 50*, pp. 377–385. Pan American Health Organization, Washington, D. C.

Sawyer, W. D., Kuehne, R. W., and Gochenour, W. S. 1964. Simultaneous aerosol immunization of monkeys with live tularemia and live Venezuelan equine encephalomyelitis vaccines. *Military Med.*, **129**: 1040–1043.

Schulman, J. L., and Kilbourne, E. D. 1963. Induction of viral interference in mice by aerosols of inactivated influenza virus. *Proc. Soc. Exp. Biol. and Med.*, **113**: 431–435.

Taylor, R. M. 1941. Passive immunization against experimental infection of mice with influenza A virus: Comparative effect of immune serum administered intranasally and intraabdominally. *J. Immunol.*, **41**: 453–462.

Tigertt, W. D. 1962. Soviet viable *Pasteurella tularensis* vaccines *Bacteriol. Rev.*, **26**: 354–373.

Tigertt, W. D., Benenson, A. S., and Gochenour, W. S. 1961. Airborne Q-fever. *Bacteriol. Rev.*, **25**: 285–293.

Waldman, R. H., Mann, J. J., and Small, P. A. 1969. Immunization against influenza. *J. Am. Med. Assoc.* **207**: 520–524.

Yamashiroya, H. M., Ehrlich, R., and Magis, J. M. 1966. Aerosol immunization of guinea pigs with fluid tetanus toxoid. *J. Bacteriol.*, **91**: 903–904.

Yamashiroya, H. M., Ehrlich, R., and Magis, J. M. 1966. Aerosol vaccination with tetanus toxoid. *Bacteriol. Rev.*, **30**: 624–632.

Zellat, J., and Henle, W. 1941. Further studies in passive protection against the virus of influenza by the intranasal route. *J. Immunol.*, **42**: 239–249.

Zhdanov, V. M. 1961. Recent experience with antiviral vaccines. *Ann. Rev. Microbiol.*, **15**: 297–322.

Zhdanov, V. M., Ritova, V. V., Grefen, N. Ye., Zhukovsky, A. M., Berlyant, M. L., Yevstigneyeva, N. A., Yegorova, N. B., Kreynin, L. S., Leonidora, S. L., Sergeyev, V. M., and Smirnof, M. S. 1962. A comparative study of the intranasal and aerosol methods of vaccination against influenza. *J. Microbiol. Epidemol. i Immunobiol.*, **11**: 63–67.

16

SOME CHARACTERISTICS OF RESPIRATORY INFECTION IN MAN

D. Pappagianis

SCHOOL OF PUBLIC HEALTH, UNIVERSITY OF CALIFORNIA,
BERKELEY AND SCHOOL OF MEDICINE,
UNIVERSITY OF CALIFORNIA, DAVIS

Man is exposed to his environment in numerous ways, some of which are not clearly evident. For example, exposure to those microbiological agents which cause colds or pneumonia may occur in quite an inapparent manner. Invisible though they are to the naked eye, these agents nevertheless can induce abnormal changes apparent in the development of signs and symptoms of respiratory disease. Manifestations as disease do not occur in every instance of contact between man and these agents. The outcome is affected by the nature of the agent, the nature and past history of the human (or other animal) involved, and the nature of the environment in which the two find themselves.

Briefly, with each inspiration, air taken into the respiratory system may contain noxious chemical or biologic agents. These airborne particles may either pass or contact flat, scale-like, surface cells (squamous epithelium), or tall (columnar) cells with fine hair-like projections (cilia) which have a sweeping, rhythmic motion. Various factors, including the sweeping motion of the cilia, may contribute to removal of inspired particles. If the particles gain access to the far reaches of the respiratory system, they may be removed mechanically by the action of phagocytic cells alone, or become biologically inactive through the action of anti-

390

body proteins. If these removal processes fail, abnormal effects may ensue. Thus, infection and disease may result when a microbial agent lights on, or in, a suitable site and proliferates. Some of the elements influencing this outcome are discussed in this chapter.

TYPES OF INFECTIONS INVOLVING
THE RESPIRATORY SYSTEM OF MAN

Infections involving the respiratory system of man can be roughly separated into three groups:

1. Infections acquired via the respiratory route and having pathologic manifestations in the respiratory system, e.g., rhinovirus and adenovirus infections, measles (rubeola), influenza, streptococcosis, pneumococcal pneumonia, tuberculosis, coccidioidomycosis, histoplasmosis, Q fever.

2. Infections acquired via the respiratory system, but with manifestations usually elsewhere, e.g., inhalatory anthrax leading to mediastinitis; viral exanthems, e.g., chickenpox (varicella) (occasionally leads to varicella pneumonia), rubella, measles (rubeola). As Kilbourne (1966) has pointed out, "the site of predominant symptomatology is not necessarily an indication of the site of primary viral invasion."

3. Infection not acquired via the respiratory tract, but having respiratory manifestations, e.g., bronchitis in typhoid fever, pneumonic form ensuing on bubonic plague or on tularemia, lobular pneumonia in visceral leishmaniasis, pneumonitis in ascariasis. [An interesting extension of this is the administration of adenovirus, e.g., type 4, by the *enteric route*, and the induction of resistance to subsequent respiratory challenge as shown by Chanock, and Jackson, and their associates (Jackson, 1967).]

Most respiratory infections are included in category 1 and many are transmissible from man to man. However, some are acquired from nonliving and nonhuman sources, e.g., Q-fever and coccidioidomycosis, and hence do not pose the problem of human-engendered epidemics.

Structural Features in Regard to
Respiratory Infection in Man

Respiratory infections in man have been designated, according to anatomical site, as "upper respiratory infections" (URI) and "lower respiratory infections." It may be difficult to discern precisely where the demarcation lies. Rhinitis (coryza), tonsillitis, pharyngitis, epiglottitis, laryngitis, tracheitis, bronchitis, bronchiolitis (interstitial pneu-

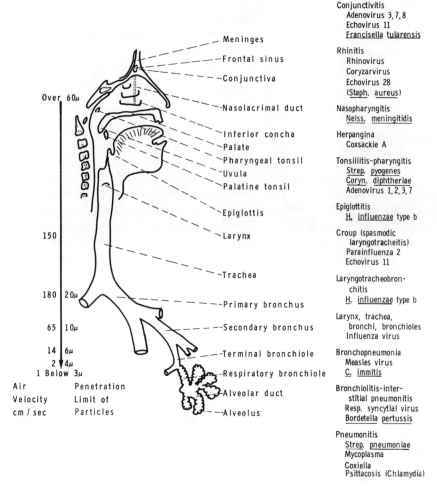

FIGURE 16-1. Characteristics of the respiratory system of man and location of infections. (From Austwick, P. K. C., Figure 3, p. 330, *The Fungus Spore*, edited by M. F. Madelin, 1966. Adapted from Mitchell, R. I. 1960. *Am. Rev. Resp. Dis.*, 82: 630.) Reprinted with permission of Butterworth and Co. (publishers) Ltd., Colston Research Society and National Tuberculosis Association.

monitis), alveolitis or pneumonitis, pleuritis (pleurisy)—all represent the progression and/or localization of pathogens in the respiratory system (Figure 16-1). Hatch (1961) has indicated that consideration of "upper respiratory" would extend to the inclusion of the terminal bronchioles. Others (Wright, 1961; Couch *et al.*, 1966) indicate that "lower respiratory tract" refers to sublaryngeal regions. The larynx appears to offer a logical point of division, inasmuch as there is (with the exception of the true vocal cords) pseudostratified ciliated columnar, secretory epithelium from the lower posterior surface of the epiglottis and upper larynx to the terminal bronchioles. In the latter, the epithelium becomes nonsecretory (no goblet cells) ciliated, simple columnar, or cuboidal. On a comparative basis, we may also reason that the "upper respiratory" area extends to the larynx and "lower respiratory" to *include* larynx and deeper structures; for in the mouse, the "anterior respiratory tract" and "posterior respiratory tract" are separated at the upper (anterior) end of the larynx (Hummel *et al.*, 1966). Briefly, particles breathed into the respiratory system pass or touch the following surfaces:

Vestibule of the nose—stratified squamous epithelium with vibrissae (long, thick hairs), and sebaceous and sweat glands.

Limen or *transitional zone*—stratified squamous with a few mixed serous and mucous glands.

Nasal conchae (bony projections on lateral wall of nasal cavity) and *nasal septum*—ciliated pseudostratified columnar epithelium with numerous goblet cells, prominent basement membrane, and lamina propria rich in mixed serous and mucous glands, large veins (erectile), and infiltrating lymphocytes.

Meatuses (cavities or depressions beneath the conchae)—thinner epithelium with fewer goblet cells and less prominent basement membrane. (The *paranasal sinuses* which are off the main respiratory currents, have a lining similar to that of the nasal meatuses, though the epithelium may be thinner and has only two or three tiers of nuclei.)

Nasopharynx—ciliated pseudostratified columnar epithelium with goblet cells, or stratified, squamous epithelium; the latter may be found where apposition of tissues occurs, e.g., where the soft palate and uvula are in contact with the posterior pharyngeal wall; mucous glands under the stratified squamous surface and mixed serous and mucous glands beneath the pseudostratified columnar.

Eustachian tube—ciliated, pseudostratified columnar with goblet cells and underlying mixed glands at the pharyngeal end.

Pharyngeal tonsil ("adenoids," if hypertrophied)—aggregates of lymphoid tissue in the postero-superior pharyngeal wall, covered by pseudostratified columnar and patches of stratified squamous epithelium, and traversed by ducts of deeper mixed glands.

Ciliated epithelium of the nasal passages and nasopharynx moves the film of mucus from these surfaces, and the secretions of the nasal sinuses toward the oropharynx.

Oropharynx—from level of soft palate and uvula inferiorly, stratified squamous epithelium with a tough, fibroelastic lamina propria.

Epiglottis—upper posterior surface, stratified squamous with underlying mixed glands; middle posterior surface stratified, ciliated columnar epithelium; lower posterior surface, ciliated, pseudostratified columnar with underlying mixed glands.

Larynx—continuation of ciliated pseudostratified columnar with mixed glands in the lamina propria, except at the vocal cords where there is stratified squamous lying on a nonglandular connective tissue, mainly elastic fibers.

Trachea—ciliated, pseudostratified columnar with goblet cells on a distinct basement membrane, and lamina propria with mixed mucous and serous glands and scattered or collected lymphocytes.

Bronchi—in primary and intrapulmonary bronchi, lining resembles that of trachea. As bronchi become smaller, the height of epithelium decreases, and number of goblet cells and seromucous glands diminishes in the thinning lamina propria. As caliber reduces, bronchi become *bronchioles* (about 1 mm in diameter); the last of the air conducting division is the *terminal bronchiole*, about 0.5 mm in diameter with ciliated, simple columnar or cuboidal epithelium, without goblet cells or glands but with underlying smooth muscle.

Respiratory bronchioles—proximally may have ciliated cuboidal epithelium, distally simple cuboidal, with some underlying smooth muscle.

Alveolar duct—walls formed by alveoli but with small bundles of smooth muscle, elastic, reticular, and collagen fibers present.

Alveoli—thin epithelial cells lying on a thin basement membrane; the thin basement membrane of the alveolar capillary endothelium may be so close as to appear merged with that of the epithelium. Two types of cells may be present in alveolar epithelium: (1) alveolar lining cells, called the "membranous pneumonocyte," alternately designated "Type I" or "Type A" cell, and representing about 90% of the alveolar surface; and (2) "granular pneumonocyte" cells, also known as "Type II" or "Type B" cell, which may contribute a surface-active agent coating the alveoli (Kanig, 1963; Felts, 1964; Ryan and Vincent, 1967). The granular pneumonocytes may detach from the alveolar walls and become phagocytic. Alveolar macrophages are considered to be distinct from these, and perhaps have their origin in peribronchial lymphoid tissue (Brooks, 1966) or blood monocytes (see Ryan and Vincent, 1967). The alveolar walls may be punctuated by inter-alveolar holes, the "pores of Kohn," under 10 μ in size (Boatman and Martin, 1963), through which some inter-alveolar communication may occur. It is estimated that there are about 300 million alveoli in the lungs of a human, representing from 30 to 200 m^2 of surface area (von Hayek, 1960).

Movement of the mucociliary "escalator" (Gross, Pfitzer, and Hatch, 1966), which may be at a rate of 15 to 18 mm/min, carries trapped particles in the bronchioles, bronchi, trachea, and larynx toward the pharynx, presumably where secretions and adherent particles, effete mucosal cells, or leucocytes from the upper respiratory system also collect and can be swallowed. This, in part, may be why the normal lower respiratory tract is free of viable bacteria (Kass, Green, and Goldstein, 1966).

It is difficult to make precise comparisons between respiratory intake and retention of particulate materials by man and experimental animals. For example, it is not clear how respiratory infection may be influenced by the upright position of the human, as compared with respiratory infection in quadruped animals. *Size, shape, weight* (density), *electrostatic charge,* and *wettability* (in mucus or alveolar "surfactant," not water) of particles may be important in determining their respiratory distribution. The upright position may be significant, as in the case of injury to the bronchial mucosa by influenza virus, when settling of pneumococci in the deep regions of the pulmonary tree may be more readily accomplished. Mitchell (1960) has indicated the presence of larger ($\geq$ 10 μ) particles lower in the respiratory tract. On the other hand, data presented by Hatch (1961) show that particles between 0.25 μ and 0.5 μ in size incur less "lobular" deposition (i.e., in respiratory bronchiole, alveolar duct, alveoli) than particles 1 to 2 μ in size. At a particle size of 4 μ, 50 to 60% of the particles are retained in the upper respiratory passages. The data (Table 16-1) of Landahl, cited by Druett (1967), do not support this high percentage of retention in the upper respiratory regions, but rather show a greater percentage of 2- and 6-μ particles in the lobular structures. Nevertheless, as appropriately indicated by Hatch (1961), it is probable that some of the larger particles (10 μ) would be deposited deep in the respiratory tract, and some of 1 μ would be retained in the upper tract. Meyers *et al.* (1963) have indeed shown the nasal regions in man to be a significant site of deposition of 1-μ bacterial particles.

Another factor that makes it difficult to translate information from measurements on respiratory intake and retention in experimental animals to that in man is the difference in relative dimensions of the respiratory passages. The contrast is perhaps most striking in comparing mouse with man. This has been presented schematically in Figure 16-2. While the more distal portions of the respiratory tree (e.g., respiratory bronchioles and alveoli) may be comparable, it would seem grossly incorrect to expect similar passage, impingement, sedimentation, or diffusion of particles entering the mouse trachea (approx. 2.5 mm diam) as in the human trachea (approx. 25 mm diam). On the other hand, the relative length from trachea to alveoli may, conversely, favor deeper (e.g., alveolar) deposition in the mouse. Some of these factors appear to have an

Table 16-1 Retention in various regions of the respiratory tract.

Region	300 cc/sec 4-sec cycle 450 cc tidal air — Particle diameter, μ					300 cc/sec 8-sec cycle 900 cc tidal air — Particle diameter, μ					300 cc/sec 12-sec cycle 1,350 cc tidal air — Particle diameter, μ					1,000 cc/sec 4-sec cycle 1,500 cc tidal air — Particle diameter, μ				
	20	6	2	0.6	0.2	20	6	2	0.6	0.2	20	6	2	0.6	0.2	20	6	2	0.6	0.2
Mouth	15	0	0	0	0	14	1	0	0	0	14	1	0	0	0	18	1	0	0	0
Pharynx	8	0	0	0	0	8	1	0	0	0	8	1	0	0	0	10	1	0	0	0
Trachea	10	1	0	0	0	11	1	0	0	0	11	1	0	0	0	19	3	0	0	0
Primary Bronchi	12	2	0	0	0	13	2	0	0	0	13	1	1	0	0	20	5	1	0	0
Secondary Bronchi	19	4	1	0	0	17	4	1	0	0	18	5	1	0	0	21	12	2	0	0
Tertiary Bronchi	17	9	2	0	0	20	9	2	0	1	21	10	2	0	0	9	20	5	0	0
Quaternary Bronchi	6	7	2	1	1	8	7	1	1	1	8	7	1	0	1	1	10	3	1	1
Terminal Bronchioles	6	19	6	4	6	6	24	7	4	6	6	24	8	4	6	1	9	3	2	4
Respiratory Bronchi	0	11	5	3	4	0	10	7	6	6	0	12	11	3	5	0	3	2	2	4
Alveolar Ducts	0	25	25	8	11	0	27	44	17	23	0	27	48	22	25	0	13	26	10	13
Alveolar Sacs	0	5	0	0	0	0	5	4	2	3	0	5	11	9	10	0	18	17	6	7
Total	93	83	41	16	22	97	91	66	30	40	99	94	82	38	47	99	95	59	21	29

(From Druett, H. A., Table 4, p. 178, *Airborne Microbes*, 17th Symp. Soc. Gen. Microbial., edited by P. H. Gregory and J. L. Monteith, 1967. Reprinted with the permission of Cambridge University Press.)

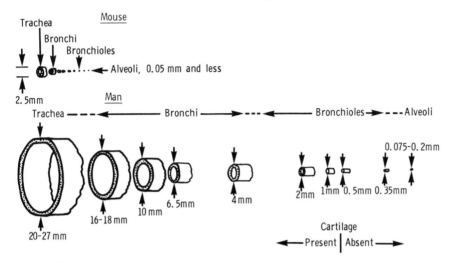

FIGURE 16-2. Schematic comparison of relative diameter and length of the lower respiratory system of mouse and man. * In part, dimensions from Best and Taylor (1955), Boyden (1955), Anson (1966), and von Hayek (1960).

equalizing effect when man is compared with smaller animals, as indicated in Figure 16-3 (Hatch and Gross, 1964).

In a general manner, the location of the lesion and character of the infection are associated with the infecting agent: rhinitis with rhinoviruses; pharyngitis with adenovirus or Group A *Streptococcus pyogenes;* epiglottitis with *Hemophilus influenzae* type b; laryngotracheo-bronchitis

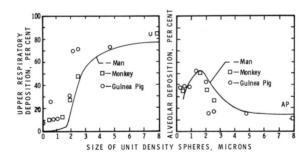

FIGURE 16-3. Deposition versus particle size of inhaled particles in the upper respiratory tract and in the lungs of the guinea pig and monkey compared with man. (From Hatch and Gross, Figure 4.13, p. 67 *Pulmonary Deposition and Retention of Inhaled Aerosols,* Academic Press, 1964. Adapted from Palm, McNerney, and Hatch, 1956. *A.M.A. Archiv. Environ. Health,* 13: (4) Fig. 3, 5, 7, 9, pp. 358-360.) Reprinted with permission of Academic Press, New York, and the American Medical Association, Chicago.

with influenza virus (croup in children with parainfluenza 2 virus); bronchiolitis with respiratory syncytial virus; pneumonitis with *Streptococcus* (*Diplococcus*) *pneumoniae*. Roughly speaking, there seems a preponderance of infections of the upper respiratory passages associated with the smaller (or lighter) viral particles, whereas larger (heavier) agents (bacteria, rickettsiae, Chlamydia,[1] and mycoplasma) appear to induce infections in the lower respiratory tract. It is not clear precisely what influence the dimensions of viruses and bacteria have in the determination of site of infection.

Table 16-2 Frequency of virus growth in cultures of various human organs.

Culture prepared from	Culture inoculated with tissue culture fluid containing:	
	M Rhinovirus	Echovirus 11
Nose	2/3[a]	2/2
Trachea	2/3	2/2
Esophagus	0/3	2/2
Palate	0/1	1/2

[a] Numerator = number of experiments in which virus growth was detected. Denominator = number of experiments performed.
(From Hoorn and Tyrrell, Am. Rev. Resp. Dis. **93** (3) part 2, March, 1966. Reprinted with the permission of The National Tuberculosis Association.)

Allusion has been made to the possible significance of neuraminidase activity of influenza (myxo-) virus in establishing respiratory infection. Neuraminidase has also been demonstrated in pneumococci freshly isolated from human infection (Kelly, Farmer, and Greiff, 1967). However, neuraminic (sialic) acid appears to have rather wide distribution among mammalian cells (Weiss and Mayhew, 1967). Therefore, its presence in respiratory tract cells may not specifically account for the location of pathogenesis by influenza virus and pneumococci.

Doubtless, the specific localization and characteristics of infections may be influenced not only by the physical properties of the particle, but also the nature of the surface on which contact is made, e.g., neuraminic acid receptor sites on the cell periphery.

Hoorn and Tyrrell (1966) have made an interesting approach by studying growth and effects of respiratory viruses on organ cultures of various parts of the respiratory tract. Table 16-2 indicates some of

[1] Psittacosis-lymphogranuloma group.

their results. However, still largely unknown is what precise combination of conditions of temperature, cellular epithelium, moisture, nutrition, CO_2 and O_2, and other microbial "flora," favor the proliferation or persistence of a given organism in a given site. The recovery of the meningococcus from the *nasopharynx* of carriers is a case in point.

Establishment of infection in a given site may also depend on the point of *initial deposition* and on the form of the infectious agent. When Bynoe *et al.* (1961) swabbed rhinovirus on the conjunctiva or *in* the nose, they found that a nasal infection followed; but swabbing virus in the pharynx failed to initiate infection. Tyrrell (1965) cited work which showed that colds were produced when drops of coxsackie A21 virus were placed in the nose. When the virus was administered as a fine aerosol, a respiratory disease resulted, characterized by lower respiratory symptoms, with rhonchi and rales. Perhaps the usual physiologic mechanisms for removing inhaled particles were partially bypassed. These physiologic mechanisms may also be affected by consumption of ethanol, renal acidosis, irritant vapors, or prior viral infection (Kass *et al.*, 1966), or imperfect respiratory cleansing as in "hypostatic" pneumonia. It must also be pointed out that infection of the alveoli may occur after initial infection in the bronchioles, and centrifugal extension from peribronchial lymphatic tissue.

SPREAD (PROPAGATION) OF RESPIRATORY INFECTION

Following proliferation of many of the infectious agents at the appropriate site, their transmission from one person to another may be expected. The conclusion that a respiratory infection is transmitted by the airborne route between an infected human and another "normal" person, is largely inferential. However, we can logically exclude other modes of spread in many instances and this, linked with the primary respiratory nature of these infections, provides presumptive evidence of airborne contagion. The great pandemics of influenza certainly represent examples of this. Although it has been difficult to obtain information on direct man-to-man transmission under controlled experimental conditions (Tyrrell, 1965), evidence for dispersion of infectious organisms from humans with pulmonary disease has been presented by Riley and coworkers (see Riley, 1961). They channeled effluent air, from a ward of patients with advanced tuberculosis, through a chamber housing guinea pigs, and showed relatively frequent infection of the exposed animals. It would be difficult to accomplish similarly controlled exposure of humans with some infectious agents. Williams' (1966) description of airborne transfer of staphylococci, and colonization of the nose, indicates some of the difficulties in distinguishing "contact" spread from airborne spread of microorganisms. This difficulty was also evident in the work of Hamburger *et al.* (1945) who demonstrated a correlation between the density of beta hemolytic streptococci in the nasal passages (not nasopharynx) of patients with streptococcal tonsillitis-pharyngitis and discharge into the nearby environment. These workers also applied serotype studies to demonstrate cross infection of humans in the same hospital ward—presumably by airborne transmission.

Attempts to demonstrate airborne contagion have also been made with some viral agents. Tyrrell (1965) has summarized some of the experiments in which normal subjects were exposed to persons with natural or experimentally induced colds. Infection of the normal subjects was irregular, as only a relatively few subjects developed colds. Tyrrell (1965) estimated that there may be 10,000 "volunteer" (human) infectious doses of rhinovirus in 1 ml of nasal secretion. But in a single sneeze perhaps 0.01 to 0.1 ml is expelled (100 to 1,000 infectious doses). Of this, only about 1% (1 to 10 infectious doses) is finely enough dispersed to remain airborne for more than a minute or two. The photographs of Jennison (1942) indicate with clarity the coarseness of secretions expelled during sneezing, although the photographic technique could not demonstrate the large number of fine particles present. According to Tyrrell (1965), however, a large number of particles produced after a sneeze seem to be formed from the *mouth* even when the sneeze is induced by a cold rather than by experimental stimulation. A *cough* produces a small number of particles from the mouth. There is a greater chance, however, that particles have been formed not from saliva but "from respiratory secretions torn away from the mucous membranes of the lower respiratory tract" (Tyrrell, 1965).

With a tidal volume of approximately 600 ml per breath, the adult human would breathe about 9.0 liters per minute, or 540 liters per hour. Thus, despite a relatively low concentration of infectious virus units, the large volume inspired may allow for sufficient viral particles to establish infection. Large droplets may be retained in the nasal and other upper respiratory passages leading to an upper respiratory tract illness. As previously indicated, Tyrrell has described upper respiratory infections resulting from instillation of drops of cosackie A21 virus, and lower respiratory tract infections from exposure to fine aerosols of this virus. In this connection it should be pointed out that some viruses, e.g., cosackie A21, survive longer in coarse than in fine particles.

Other factors influencing spread are the ambient temperature and humidity, because "viability" and infectivity of viruses and bacteria are affected by desiccation. Influenza virus remains more infective at relatively low (below 40%) relative humidity than at higher moisture level, though this is not true for adeno- and rhinoviruses (Tyrrell, 1965). Interestingly, Lapeyssonie (1963) reported that occurrence of cases of meningococcal meningitis in the "meningitis belt" of Africa was high during the dry season but abruptly decreased when the rainy season began.

Individual variation in the capacity of humans to produce infectious airborne particles is to be expected (Loudun, 1967). One factor which may influence this is the coexistence of bacterial and viral agents in the respiratory tract. "Cloud babies" are particular newborn infants who are highly infectious spreaders of *Staphylococcus aureus* as a result of concomitant infection with respiratory viruses (Eichenwald et al., 1960). On the other hand, in mice experimentally infected with influenza virus and *Bordetella*, transmission of influenza was found to be decreased by concomitant infection with the bacterium (Kilbourne, 1967). Treatment of mice with antibiotics, to reduce their usual respiratory bacterial flora, caused the mice to become better transmitters of influenza virus.

RESISTANCE TO RESPIRATORY INFECTION

Resistance to respiratory infectious disease results from innate and acquired mechanisms (reviewed by Druett, 1967; Green, 1968, and by Rylander, 1968). The mucociliary apparatus discussed earlier constitutes one of the innate mechanisms for expulsion of particles, as does the act of coughing. *Deep breathing* by increasing the velocity of inspired air, may lead to deposition in alveoli of larger particles that normally would have settled in upper parts of the respiratory tract (Dautrebande, 1952). However, deep breathing may prevent atelectasis (local collapse of pulmonary tissue) and thus allow effective normal clearance of secretions.

In the deeper regions of the respiratory tree, phagocytic removal may provide a means of clearing infectious agents. As shown experimentally by Kass et al. (1966), phagocytic cells may kill bacteria effectively even without physically removing them from the lungs. Differing susceptibilities to this killing effect are shown by different species of bacteria. Not only may the inhaled microbial agents be killed, but they may be transported away from the alveolar sites. Phagocytic cells may migrate proximally to the level of the ciliated epithelium where they and their ingested contents may be moved up the escalator as previously discussed. In chronic bronchitis, apparent failure of this mechanism is evident in the recurrent infections. The phagocytic cells (presumably macrophages, since granulocytes are thought not to re-enter the tissues from which they have emigrated) may carry infectious agents into the alveolar or bronchiolar interstitium. Some *direct* penetration, i.e., without the phagocytic cell as an intermediary vehicle, may also occur (Gross, 1964). By either means, penetration into the lymphatic system may occur, directly to the tracheobronchial-mediastinal nodes or via the pleural lymphatics (Engel, 1964).

These clearance mechanisms may be augmented by other factors. Respiratory viral infections may be aborted if *interferon*, a nonspecifically induced protein, is present early after onset of the infectious process. The mechanism involves blockage of viral replication *after* the virus enters the susceptible cell. Specific antiviral immunity may also be exerted through the presence of antiviral antibodies (IgA, IgG) in the mucous secretions of the respiratory tract ("mucoantibody") (Bellanti and Artenstein, 1964; Remington *et al.*, 1964; Jackson, 1967). The significance of IgA-type antibody as a secretory antibody has recently become apparent. The role of specific antibody in favorable response to pneumococcal infection provides an indication of the importance of a humoral response in the respiratory system. Precisely how this antibody reaches the pneumococci in the respiratory tract is not clear. IgG (Bellanti and Artenstein, 1964) and IgM (Jackson, 1967) antibodies may also reach the respiratory epithelial surface under conditions of inflammation, although IgM may be scanty in nasal secretions while abundant in the serum (Buescher and Bellanti, 1966). It is apparent that in those respiratory infections in which a viremic state may occur, e.g., measles, serum antibodies would be expected to have a significant role (see, for example, Krugman *et al.*, 1965).

Although difficult to verify in man, enhancement of specific alveolar (or other respiratory tract) macrophage activity can be expected to follow immunization or prior infection with some respiratory pathogens (Suter and Ramseier, 1964; Nutter and Myrvik, 1966). On the other hand, Tulis *et al.* (1963) reported that focal accumulation of monocytic cells in the pulmonary parenchyma occurred earlier in nonvaccinated monkeys challenged by respiratory route with *Francisella tularensis* than in vaccinated monkeys.

Immunization against respiratory pathogens has proved effective for viral (measles, adenovirus, influenza), bacterial (*Bordetella pertussis, Corynebacterium diphtheriae*, pneumococci, *Mycobacterium tuberculosis*), and mycoplasmal (*Mycoplasma pneumoniae*) agents. Vaccines for these agents have been effective when given by various parenteral routes, or in the case of adenovirus 4 vaccine, by the enteric route. Immunization by the respiratory route is a practical way to immunize chickens against infection with Newcastle disease virus. Middlebrook (1961) has reviewed some of these efforts in man as well as in some laboratory animal hosts. Citing his own results, Middlebrook demonstrated superiority of the airborne route over the subcutaneous route for immunization of guinea pigs with BCG against virulent airborne tubercle bacilli. Russian workers (cited by Middlebrook) also indicated that effective immunization of man against brucellosis could be accomplished by exposure to aerosolized, viable, attenuated brucellae. Indeed, a sizable Russian literature has developed on vaccination by inhalation in tularemia. Hornick and Eigelsbach (1966) adopted this approach with living attenuated *Francisella tularensis* in man, and found a more rapid antibody response and apparently greater immunity to aerogenic challenge than when antigen was given by acupuncture (percutaneously). On the other hand, Buescher and Bellanti (1966) indicated that the dose of viable bacteria (10^6 to 10^8) used in the work of Hornick and Eigelsbach gave sufficient untoward reactions (mild tularemia) to prompt some reservations about the use of aerosol immunization for this disease. Substantial antibody responses were found when the respiratory route was employed for immunization with antigens not customarily encountered in the airborne state, e.g., Venezuelan equine encephalitis virus (Miller, 1966) and tetanus toxoid (Yamashiroya *et al.*, 1966). While these experiments have been limited and may not prove feasible as a means for mass immunization of man, they do indicate the capacity of respiratory surfaces to provide an immunologic portal of entry.

Specific immune mechanisms may not be the only means of favorably responding to infectious agents. The potential use of interferon, and the real use of chemotherapeutic substances provide means of prevention or treatment of viral diseases. Amantadine hydrochloride (aminoadamantanamine-hydrochloride) has shown effectiveness against Influenza A virus (Jackson, 1967). How fully this approach can be utilized remains incompletely explored.

A number of significant findings having to do with human respiratory disease (not just those of infectious origin) may derive from studies in experimental aerobiology. As has been indicated in some of the foregoing, it is difficult to transpose information directly from experimental animals to man. However, experimental studies may provide important clues to effects of atmospheric pollutants on respiratory tissues and pulmonary function. The average dose or number of infectious particles needed to establish respiratory infection can be estimated and thus provide support for estimating infectious doses in man. The latter can, of course, be approached directly with some diseases that can be effectively treated or that are not particularly dastardly toward man. It is also within the realm of experimental aerobiology to determine what peculiar circumstances promote meningococcal infection in some groups of humans but not in others. Thus, used in a restrained manner, experimental aerobiology can offer both inferential and direct information bearing on the various respiratory challenges to man.

REFERENCES

Anson, B., (Ed.). 1966. *Morris' Human Anatomy* (12th ed.). McGraw-Hill Book Co., New York.

Austwick, P. K. C. 1966. The role of spores in the allergies and mycoses of man and animals, in *The Fungus Spore* (M. F. Madelin, Ed.). Proc. of the 18th Symposium of the Colston Research Society. Butterworth and Co. Publ. Ltd., London.

Bellanti, J. A., and Artenstein, M. S. 1964. Mechanisms of immunity and resistance to virus infections. *Pediat. Clin. North Am.*, 11: 549–561.

Best, C. H., and Taylor, N. B. 1955. *The Physiological Basis of Medical Practice* (6th ed.), p. 345. The Williams and Wilkins Co., Baltimore, Md.

Boatman, E. S., and Martin, H. B. 1963. Electron microscopy of the alveolar pores of Kohn. *Am. Rev. Respirat. Diseases*, 88: 779–784.

Boyden, E. A. 1955. *Segmental Anatomy of the Lungs*. McGraw-Hill Book Co., New York.

Brooks, R. E. 1966. Concerning the nomenclature of the cellular elements in respiratory tissue. *Am. Rev. Respirat. Diseases*, 94: 112–113.

Buescher, E. L., and Bellanti, J. A. 1966. Respiratory antibody to *Francisella tularensis* in man. *Bacteriol. Rev.*, 30: 539–541.

Bynoe, M. L., Hobson, D., Horner, J., Kipps, A., Schild, C. C., and Tyrrell, D. A. J. 1961. Inoculation of human volunteers with a strain of virus isolated from a common cold. *Lancet*, 1(No. 7188): 1194–1196.

Couch, R. B., Cate, T. R., Douglas, R. G., Jr., Gerone, P. J., and Knight, V. 1966. Effect of route of inoculation on experimental respiratory viral disease in volunteers and evidence for airborne transmission. *Bacteriol. Rev.*, 30: 517–529.

Dautrebande, L. 1952. Physiological and pharmacological characteristics of liquid aerosols. *Physiol. Rev.*, 32: 214–256.

Druett, H. A. 1967. The inhalation and retention of particles in the human respiratory system, in *Airborne Microbes* (P. H. Gregory and J. L. Monteith, Eds.). 17th Symposium of the Society for General Microbiology. The University Press, Cambridge, England.

Eichenwald, H., Kotsevalov, O., and Fasso, L. 1966. The "Cloud Baby": an example of bacterial-viral interaction. *Am. J. Disease Children*, 100: 161–173.

Engel, S. 1964. Comparative anatomy and pulmonary air cleansing mechanisms in man and certain experimental animals. *Health Phys.*, 10: 967–971.

Felts, J. 1964. Biochemistry of the lung. *Health Phys.*, 10: 973–979.

Green, G. M. 1968. Pulmonary clearance of infectious agents. *Ann. Rev. Med.*, 19: 315–336.

Gross, P. 1964. The processes involved in the biologic aspects of pulmonary deposition, clearance and retention of insoluble aerosols. *Health Phys.*, 10: 995–1002.

Gross, P., Pfitzer, E., and Hatch, T. 1966. Alveolar clearance: its relation to lesions of the respiratory bronchiole. *Am. Rev. Respirat. Diseases*, 94: 10–19.

Hamburger, M., Green, M. J., and Hamburger, V. G. 1945a. The problem of the "dangerous carrier" of hemolytic streptococci. I. Number of hemolytic streptococci expelled by carriers with positive and negative nose cultures. *J. Infect. Diseases,* **77**: 68–71.

Hamburger, M., Green, M. J., and Hamburger, V. G. 1945b. The problem of the "dangerous carrier" of hemolytic streptococci. II. Spread of infection by individuals with strongly positive nose cultures who expelled large numbers of hemolytic streptococci. *J. Infect. Diseases,* **77**: 96–108.

Hatch, T. 1961. Distribution and deposition of inhaled particles in respiratory tract. *Bacteriol. Rev.,* **25**: 237–240.

Hatch, T., and Gross, P. 1964. *Pulmonary Deposition and Retention of Inhaled Aerosols,* Fig. 4.13, p. 67. Academic Press, New York.

Hoorn, B., and Tyrrell, D. A. J. 1966. Effects of some viruses on ciliated cells. *Am. Rev. Respirat. Diseases,* **93** (part 2): 156–161.

Hornick, R., and Eigelsbach, H. 1966. Aerogenic immunization of man with live tularemia vaccine. *Bacteriol. Rev.,* **30**: 532–537.

Hummel, K. P., Richardson, F. L., and Fekete, E. 1966. "Anatomy," Chap. 13, in *Biology of the Laboratory Mouse* (2nd ed.) (Earl L. Green, Ed.). McGraw-Hill Book Co., New York.

Jackson, G. G. 1967. Control of acute viral respiratory disease. *Arch. Environ. Health,* **14**: 759–767.

Jennison, M. W. 1942. Atomizing of nose and mouth secretions into the air as revealed by high-speed photography. In *Aerobiology,* Publication #17 of the American Association for the Advancement of Science, Washington D.C.

Kanig, J. L. 1963. Pharamaceutical aerosols. *J. Pharm. Sci.,* **62**: 513–535.

Kass, E. H., Green, G. M., and Goldstein, E. 1966. Mechanisms of antibacterial action in the respiratory system. *Bacteriol. Rev.,* **30**: 488–496.

Kelly, R. T., Farmer, S., and Greiff, D. 1967. Neuraminidase activities of clinical isolates of *Diplococcus pneumoniae. J. Bacteriol.,* **94**: 272–273.

Kilbourne, E. D. 1966 Discussion. *Bacteriol. Rev.,* **30**: 530–531.

Kilbourne, E. D. 1967. Respiratory disease outbreaks. *Arch. Environ. Health,* **14**: 768–775.

Krugman, S., Giles, J., Friedman, H., and Stone, S. 1965. Studies on immunity to measles. *J. Pediat.,* **66**: 471–488.

Lapeyssonie, L. 1963. La Méningite cérébro-spinal en Afrique. *Bull. World Health Organ.,* **28**: (Suppl.).

Loudon. 1967. cited by Meade in discussion section of paper by Kilbourne, E. D. 1967 q. v.

Meyers, C. E., Hresko, J., Zippin, C., Coppoletta, J., and Wolochow, H. 1963. Recovery of aerosolized bacteria from humans. *Arch. Environ. Health,* **6**: 643–648.

Middlebrook, G. 1961. Immunological aspects of airborne infection: reactions to inhaled antigens. *Bacteriol. Rev.,* **25**: 331–346.

Miller, W. S. 1966. Infection of pigeons by airborne Venezuelan equine encephalitis virus. *Bacteriol. Rev.,* **30**: 589–595.

Mitchell, R. I. 1960. Retention of aerosol particles in the respiratory tract. *Am. Rev. Respirat. Diseases,* **82**: 627–629.

Nutter, J., and Myrvik, Q. 1966. In vitro interactions between rabbit alveolar macrophages and *Pasteurella tularensis. J. Bacteriol.,* **92**: 645–651.

Remington, J. S., Vosti, K. L., Lietze, A., and Zimmerman, A. 1964. Serum proteins and antibody activity in human nasal secretions. *J. Clin. Invest.,* **43**(No. 8): 1613–1624.

Riley, R. L. 1961. Airborne pulmonary tuberculosis. *Bacteriol. Rev.*, 25: 243–248.

Ryan, S., and Vincent, T. 1967. Chapter 6, p. 117, in *Ultrastructural Aspects of Disease* (D. W. King, Ed.). Hoeber Medical Division, Harper and Row, New York.

Rylander, R. 1968. Pulmonary defense mechanisms to airborne bacteria. *Acta, Physiol. Scand. Suppl.* 306.

Suter, E., and Ramseier, H. 1964. Cellular reactions in infection. *Advan. Immunol.*, 4: 117–173.

Tulis, J. J., Eigelsbach, H. T., and Kerpsack, R. W. 1963. Host-parasite relationship in monkeys vaccinated against tularemia and subsequently challenged. *Bacteriol. Proc.*, M-98, p. 79.

Tyrrell, D. A. J. 1965. *Common Colds and Related Diseases*, The Williams and Wilkins Co., Baltimore, Md.

von Hayek, H. 1960. *The Human Lung*, Hafner Publishing Co., New York.

Weiss, L., and Mayhew, E. 1967. The cell periphery. *New Engl. J. Med.*, 276: 1345–1362.

Williams, R. E. O. 1966. Epidemiology of airborne staphylococcal infection. *Bacteriol. Rev.*, 30: 660–672.

Wright, G. 1961. Structure and function of respiratory tract in relation to infection. *Bacteriol. Rev.*, 25: 219–227.

Yamashiroya, H., Ehrlich, R., and Magis, J. 1966. Aerosol vaccination with tetanus toxoid. *Bacteriol. Rev.*, 30: 624–632.

17

AEROBIOLOGY AND HOSPITAL SEPSIS

Edward L. Fincher

SCHOOL OF BIOLOGY, GEORGIA INSTITUTE OF TECHNOLOGY,
ATLANTA, GEORGIA

INTRODUCTION

The hospital as a center of patient treatment will harbor many potential and actual sources of infectious microorganisms among patients and staff personnel. Direct contact between primary infection sources, carrier, and susceptible host is the most important mechanism in the spread of common infections. However, transmission does occur by other routes. Indirect environmental routes, some incompletely understood, must be considered as inseparable components of an extremely complex interaction of infection sources, transmission patterns, and susceptible hosts.

The necessity of special engineering design in surgical areas, infant and newborn nurseries, and other critical care areas of the hospital is recognition of the most important factor of host susceptibility in cross-infection. Healthy individuals living under normal conditions are in frequent contact with large numbers of microorganisms in the environment without incurring infections. Excepting those diseases considered naturally or potentially transmitted via the airborne route, *e.g.*, measles, smallpox, Q-fever, tuberculosis, pneumonic plague, anthrax, brucellosis, psittacosis, influenza, adenoviral URTI, streptococcal and staphylococcal infections, and allergic reaction to airborne allergens (pollen etc.), contact with the microflora of intramural air does not usually present an infection hazard to individuals in a normally susceptible population.

However, the patient undergoing protracted and extensive surgery or treatment for major burns, the newborn infant, and patients with primary disease states affecting the normal immuno-defensive system of the body present abnormal states of risk to cross infection from multiple sources. Certain drug therapies, including antibiotics and immuno-suppressants, can have the secondary effect of producing an imbalance in the "normal" microflora of the patient and of significantly depressing normal levels of the immuno-defense system. These disadvantages are secondary to the principal and beneficial effects of the prevention or treatment of primary disease states, but they serve nonetheless to emphasize the point that the population in a general hospital presents a distribution of susceptibilities to infection extending from patients with "minimum normality" to patients of high susceptibility.

A balanced evaluation of the role of air in the total environmental hygiene index of the hospital must proceed from the standpoint of patient susceptibility. Despite the lack of convincing quantitative evidence that standard counts of airborne microbial contaminants correlates with rates of infection, present requirements for low air counts in operating rooms and newborn nurseries are based on the fact that in the absence of an infectious agent no infection will occur. The logical extension of this reasoning would be to eventually produce microbe-free patients—highly effective but hardly practical. Nevertheless, delivery to critical areas of air with a low viable particle content, and ventilation and air control within critical areas, is within engineering practice; hence, air as an environmental element should be controlled.

The possible significance of microorganisms in the air must be considered in terms of (a) areas where exposure of highly susceptible patients occur, and (b) a measure of an intramural hygiene control index. Importance of the latter in air contamination control is indicated by the rapidity of dissemination of airborne bacteria when marker organisms or inert tracers were released in patients' wards (Rubbo et al., 1962), laundry chutes (Hurst et al., 1958; Michaelsen, 1964), and in a basement (Walter, 1958; Wells, 1935) of the hospital. These results demonstrate that contamination from a focal source can be distributed vertically through several floors and laterally into corridors, and they emphasize the uniqueness of air as the common element in intimate and dynamic contact with the majority of hospital areas.

SOURCES OF AIRBORNE CONTAMINATION

Patients and Staff Personnel

Saphrophytes

Viable microbial particles found in the intramural air of the hospital arise from a variety of sources. Many of these microorganisms are saphrophytic, and are representative of soil microflora brought into the building as contaminants on clothing and shoes. Green *et al.* (1962) reported a qualitative analysis of airborne microbial contaminants in several hospital locations. The frequency distribution of various species was dependent on where in the hospital the air sample was taken. Yeasts and actinomycetes occurred relatively infrequently; molds were found more frequent in animal-care areas, autopsy rooms, food-preparation areas, and some patient areas. The number of gram-negative rods varied from place to place and time to time; gram-positive rods also varied, but their average numbers tended to be higher and less variable than gram-negative rods. Gram-positive cocci were found as often as gram-positive bacilli, and were the dominant bacterial type reported in areas where extramural contamination was lowest and conditions of environmental hygiene highest.

Under especial conditions, controlled to minimize extraneous air contaminants, the principal source of airborne bacteria (gram-positive cocci and diphtheroid organisms) are the personnel themselves (Speers *et al.*, 1965); the anaerobic or microaerophilic *Corynebacterium acnes* appears to be the most numerous of the skin microflora (Rosebury, 1962).

The apparent numbers of airborne microbial particles in hospital air depend to a significant extent on where the air sample is taken, as well as the time of day—factors that reflect the level of personnel activity. Activity and concentration of airborne contaminants have been shown repeatedly to be closely related, but have not been correlated sufficiently to permit prediction of specific concentrations, numbers of people, and activity levels.

Neither have total quantitative counts of airborne cells proven useful in accurately predicting the probability that one level of contamination is significantly different from another in its relationship to direct cross infection. The presence of microorganisms in the air is in itself of indeterminate consequence when viewed as a general fact of the intramural hospital environment. This is not, however, a negation of air as a significant component of hospital environments that demand a high general index of hygiene.

Gram-positive Cocci

Intramural sources of airborne contamination are cardinal to the significance of the types and numbers of microorganisms found in the air. Bacteria recovered from the environment in patient areas are related to bacteria directly recoverable from lesions (Selwyn *et al.*, 1964), particularly from body surfaces. Such observations emphasize that patients and staff personnel, as carriers or active foci of infection, are the principal source of airborne microbial contamination.

Some healthy individuals, as well as patients with subclinical and clinical infection who produce abnormally large numbers of airborne microbial particles, have been termed "dispersers" (Noble, 1962). The individuality of patients as dispersers of staphylococcal particles is indicated by the finding that 8 of 3,675 patients admitted to a ward during a period of four years possessed a markedly greater ability to disperse staphylococci into the air than did "normal" patients. Ability to disperse large numbers of staphylococcal particles was not related to clinical disease; however, dispersal about the ward appeared to be mediated by the bedclothes, for when these were disturbed large numbers of staphylococci were disseminated into the air (Noble, 1962). Direct transport by droplets, or droplet nuclei from nasal staphylococcal carriers, was not indicated, but rather the indirect transmission sequence of contamination of skin, clothing, and bedding by nasal secretion and release of the staphylococcal particles into the air (Hare *et al.*, 1956) was apparent.

In the last two decades considerable interest has been directed to sources and factors influencing the transmission of *Staphylococcus aureus*. A high percentage of normal individuals harbor this organism, the anterior nares being a common site of colonization. Individual asymptomatic carriers of this organism are not considered especially hazardous as an infection source (Burke *et al.*, 1961), but do present a potential of transmission to others. A recent paper by Sciple *et al.* (1967) showed the extreme variability in numbers of organisms shed by individuals. Three carriers of *S. aureus* were tested, but only one individual was shown to shed this organism. Although their test procedure did not measure aerosol concentration, but rather total numbers deposited in the "room," 10^8 organisms were recovered after 60 minutes, of which 5% were *S. aureus* of the same type as found in the man's nares. Staff individuals carrying a recognized epidemic strain of staphylococcus with high communicability and virulence should be removed from service functions in high-risk patient care areas.

The environment, including air, is readily contaminated by heavily colonized nasal carriers of hemolytic streptococci. It seems evident that such individuals, actually a small percentage of patients with positive nose cultures, are more likely to spread infection than those who expel only small numbers of streptococci.

Seven incidences of hospital cross infections were traced to certain patients whose nose cultures were positive for the infecting type of streptococcus, and who expelled large numbers of these organisms; the number of cross infections in hospital wards by different types of streptococci was proportional to the number of each type recovered from the air. However, the precise route of transfer from carrier to host was not established (Hamburger *et al.*, 1944, 1945). A definite causal relationship between microorganisms in the air and subsequent infection cannot be stated. With streptococcal infection, for example, exposure of the organism to desiccation in the airborne state resulted in attentuation of infectivity and failure to induce infection by direct insufflation; these observations are pertinent to the significance of the airborne streptococcus (Rammelkamp *et al.*, 1958).

White (1961) found that heavy nasal colonization correlated with the frequency of recovery of staphylococci from skin, environment, and air in the vicinity of the carrier. Aerial dissemination from carriers among hospitalized patients was most pronounced among patients having more than 100,000 per nasal culture; over 20 staphylococcal particles/ft³ of air were recovered from 35% of air samples taken around carrier-patients, whereas only 8% of air samples were positive for staphylococci when taken near patients who were either noncarriers or carriers of less than 100,000 colonies per nasal culture. The hazard presented by carriers of small numbers of staphylococci appeared to be no greater than the hazard of patients who are not carriers. Seventy percent of the staphylococci species isolated from the air samples taken near nasal carriers were the same phage type as those isolated directly from the carriers. A similar association has been reported to exist between wound sources and environmental isolates of staphylococci (Selwyn *et al.*, 1964).

Nasal administration of antibiotics has been shown to decrease both the number of nasal carriers of staphylococci and the number of staphylococci isolated from patients who remained carriers (Varga *et al.*, 1961), and the method is considered by Solberg (1965) to be a valuable measure for preventing dissemination of staphylococci by nasal carriers.

A consistent correlation has been shown between case histories where nasal swabbing showed 1,000 or more indicator organisms had persisted for several days to two weeks, and the subsequent spread of infection by one naturally and two artificially induced carriers of tetracycline-resistant *S. aureus*. Treatment with tetracycline increased the numbers of the indicator strain of staphylococci in the nose to a level of 1,000 per swab, and increased the number of the organisms in air near the carrier, resulting in the transmission of indicator staphylococci to nearby susceptible persons. These effects were not found when a carrier was treated with penicillin. Air was considered to be a possible route through which the indicator $S.$ $aureus$ was transmitted from one person to another (Ehrenkraz, 1964). Bersten et $al.$ (1960) observed an increased incidence of staphylococci infection among ward patients receiving tetracycline. There was no evidence, however, that drug-resistant strains were intrinsically more transmissible than drug-susceptible strains. He believed the important factor increasing transmissibility of the organisms was drug-induced interference with the usual interspecies relation among the nasopharyngeal flora. Solberg (1965) reported dispersal of $S.$ $aureus$ increased with increasing nasal counts, but found a better correlation between skin (fingers and hands) counts and air counts than between nasal and air counts; usually the presence of a high skin contamination accompanied high nasal counts.

The number of staphylococci in the nares, and the frequency of isolation of staphylococci from the air, were found to interrelate with patient activity (Varga et $al.,$ 1961). Duguid et $al.$ (1948) found that the number of viable particles liberated by individuals held in an experimental chamber varied from about 1,000/minute during slight activity to about 10,000/minute during vigorous activity. Nasal carriers of $S.$ $aureus$ contaminated the surrounding air more regularly and to a greater degree by the act of liberating dust from clothing than by sneezing. Generation of airborne microbial particles from skin surfaces is most often caused by the mechanical action of movement, and the abrasive effect of clothing on skin surfaces (Duguid et $al.,$ 1948; Hare et $al.,$ 1956). Generation of airborne particles from skin surfaces is not, however, dependent solely on this mechanism since it has been shown that small differences are found between numbers of bacteria from individuals in street clothing and the same persons unclothed. It is interesting that individuals appear to shed more bacteria immediately after a shower bath than before (Speers et $al.,$ 1965).

The association of antibiotic-resistant strains of staphylococci with the dissemination of the organism suggests a factor in addition to that of massive nasal colonization. White et $al.$ (1964) reported that nasal carriers of large numbers of penicillin-resistant staphylococci disseminated organisms onto the skin and into the air more frequently than similar nasal carriers of penicillin-sensitive strains. However, changes in the penicillin-resistant phage types of staphylococci in the nose of another group of nasal carriers did not alter the frequency of dissemination.

Increased dissemination of resistant staphylococci by such carriers could not be explained solely on the basis of increased numbers. Rather, extent carriers of large numbers of penicillin-resistant staphylococci disseminators had previously received antimicrobial agents during hospitalization, and with the nearly simultaneous acquisition of resistant staphylococci.

Another type of disseminator of staphylococci is the "cloud baby" reported by Eichenwald et al. (1960). The majority of babies possessed low indices of infectivity while a small number, the "cloud baby," were highly infectious to others. The "cloud baby" did not present overt signs of disease, but nonetheless contaminated the surrounding air, principally from his respiratory tract. The factor responsible for producing the staphylococcal clouds surrounding the baby appears to consist of a number of respiratory viruses in a bacterial-viral interaction.

The perineal area may be colonized by a sufficiently large number of S. aureus cells to classify certain individuals as perineal carriers— carriers who may play an important role in communicating staphylococcal disease (Brodie et al., 1956). The perineal-carrier state may persist for months in a small percentage of male individuals. The organism is able to multiply in the perineal skin and is not a fecal or nasal contaminant because Ridley (1959) found a phage type of perineal staphylococcus different from those recovered from these two other sources.

The numbers of bacteria dispersed by both nasal and perineal carriers is variable and, on the basis of equal numbers of cells per area, the variability is the same in both instances. In a study of dispersers among men and women, Bethune et al. (1965) found that a high percentage of men disseminated more organisms than women, and that dispersal was more profuse below the waist in both men and women.

Numbers of staphylococci on skin, and degree of aerial dissemination with bedmaking, has been shown to increase with increasing perineal counts. Perineal carriers dispersed far greater numbers of staphylococci into the air than nasal carriers and represent a greater problem in the control of infection than the contaminant source would suggest (Solberg, 1965).

In an 11-week study, McKee et al. (1966) reported 11 cases of hospital-acquired Group A beta-hemolytic streptococcal infection caused by a tetracycline-resistant M-nontypable T-9 strain. Epidemiological evidence strongly suggested a medical attendant as the source. Multiple cultures taken from his nose, throat, and skin were negative, but streptococci could be detected in the air samples when the medical attendant entered the room and increasing numbers of airborne streptococci were obtained in sequential samples of the air as his activity increased. Eight

cases of streptococcal puerperal endometritis and three cases of streptococcal post-operative wound infection occurred during this epidemic.

Patients are a principal source of cross infection. Staphylococcal outbreaks in a medical and surgical ward have been described as due to
the admission of a patient with superficial staphylococci infection (Barber
et al., 1959); also heavy and persistent contamination of the environment can be caused by open infection (Burke et al., 1961). Staphylococcal
"broadcasts" into the environment, including the air, could be attributed to patients with infected open skin lesions, who are particularly
apt to disperse staphylococcal particles (Shooter, 1958). Hare et al.
(1961) found little environmental contamination by S. aureus when organisms were derived from skin lesions covered with a dressing. However,
extensive and heavy contamination occurred when the source of the
organism was an infection secondary to a dermatological infection, pneumonia, or entercolitis—situations where it was impossible to cover the
source of the organisms with an occlusive dressing.

Most airborne S. aureus cells in hospital wards are derived from relatively few patients. Patients with superficial infections, such as eczema
and burns, are particularly prolific disseminators of S. aureus. Thomas
et al. (1961) found that a patient with exfoliative dermatitis, who was
enclosed in a plastic cubicle, shed 314 S. aureus particles per square
foot per minute, which amounted to 69.3% of the total organisms liberated. In comparison, 20 healthy people in ordinary clothing shed 5.2
S. aureus (1.9% of total) particles at the highest count. The patient
was also implicated as a source of airborne staphylococci because the
ratio of antibiotic-resistant strains of staphylococci to the nonresistant
strains increased when the total count increased. Dermatological patients
infected with S. aureus showed wide variation in the extent to which
these organisms were dispersed; the extent seemed to depend on the
amount of skin that had undergone such change that S. aureus could
multiply on it (Cooke et al., 1963).

The potential of subclinical skin diseases as a hazard has been stressed
by Selwyn et al. (1965). Heavy subclinical infections were found to
be common in a large group of patients with psoriasis. In studies on
dispersal of organisms in a cubicle, such patients were generally shown
to produce very high levels of airborne contamination. The degree of
dispersal could not be correlated with clinical assessment of the lesion
or the nasal carrier status of the patient. Relatively few S. aureus organisms were shed by nasal carriers whose skin lesions were not infected.
In the absence of specific antibacterial measures, consistently high levels
of S. aureus were found throughout the environment. Attempts to control
sources of contamination by nasal disinfection techniques produced no

change in the level of environmental contamination or in incidence of cross infection. In contrast, disinfection of skin lesions markedly reduced the numbers of pathogens in the environment and virtually abolished both cross infection and development of nasal carriage. Solberg (1965) found disinfection of skin with hexachloraphene to be a valuable measure that prevented dispersal of staphylococci from patients with staphylococcal skin lesions. Patients with extensive pyodermias yielded far greater skin and air counts than individuals with minor skin infections.

An infant with streptococcal infection of the skin superimposed on an infantile eczema was found by Loosli *et al.* (1950) to have been the source of infections occurring in 25 infants and in 18 adults, including attending personnel and visitors. The infant contaminated the ward environment (air, dust, and bedclothes) with Group A, type 33, streptococci to a high degree. Aseptic techniques to avoid contact and droplet infection between nurse, doctor, or attendant and patient were carefully employed. However, during the period when the wards were contaminated, secondary infections (throat, skin, and wounds) occurred in other infants and streptococcal pharyngitis infections in the adults. The route of spread of the organisms was from the skin to the bedclothes, then to the air and dust, then to other infants, and finally to attending personnel and visitors. Antibiotic treatments of the patients promptly brought the infection under control and simultaneously the ward environment became essentially free of hemolytic streptococci.

Gram-negative Enteric Bacilli

The etiological agents of gastro-intestinal infections are usually transported via the medium of infected food or water as carrier components in the fecal-oral route of transmission. In a typical outbreak of salmonellosis, for example, bacteria are commonly derived from the consumption of infected food. The infection rate of a majority of epidemics is characteristically explosive in nature, suggesting the ingestion of large numbers of the organisms in a single item of food contaminated with a salmonella, for example.

Explosive food-borne outbreaks of salmonellosis can occur in hospitals, but outbreaks may occur in a different sequence with infection rates more characteristic of cross infection than massive ingestion. Once a source is introduced, infection continues to occur over a period of weeks or months and is apparently not related to a common source of food. The hospital environment is suggested as a source of cross-infection, bacteria being transmitted by direct contact involving personnel, patients, fomites, or possibly by the airborne route. Although general patient groups appear to be liable to cross-infection with salmonella (Datta *et al.*, 1960; Close *et al.*, 1960), published reports suggest that cross-infection with salmonella occurs predominantly in nurseries and wards where infants are kept (Leeder, 1956; Szanton, 1957; Murray *et al.*, 1958; Abramson, 1947; Jellard *et al.*, 1959; Bate *et al.*, 1950; Edgar *et al.*, 1963; MacKerras *et al.*, 1949; Rubbo, 1948; Watt *et al.*, 1958; Epstein *et al.*, 1951; Rubenstein *et al.*, 1955).

The potential airborne spread of salmonellosis was suggested by findings of Varela *et al.* (1942), who recovered salmonella from the upper respiratory tract. The spread of salmonellosis as an airborne infection was suggested by two examples reported by Neter (1950), where the causative organisms *Salmonella oranienburg* and *Salmonella cholerasuis* appeared to be present in the nasopharynx and throat. *S. oranienburg* was the predominant organism in the nasopharynx of a baby that served as a source of cross infection among premature infants. The tentative conclusion was made that an additional avenue of exit of salmonellae from patients clinically diagnosed as having enteric disease was the upper respiratory tract; airborne transmission of the infection was thus possible. Laurell (1952) indicated that coliform organisms present in the upper respiratory tract of children behave in the same manner as other bacteria at this site. In the investigation of carriers, the coliform organisms were found to be transmitted to a greater extent when they were present in both the nose and throat, although spread could occur when they were present in the throat alone. When carrier-spreaders exist, the infecting agent can usually be recovered from dust in the environment. Furthermore, a possible selection of types of coliform organisms most likely to invade (colonize) the mucosa of the upper respiratory tract was suggested by the consistent uniformity of serotypes recovered from patients in one ward.

Datta *et al.* (1960) recovered *S. typhimurium*, phage type 27, from the sputum of 5 out of 14 patients on a chest ward; none of the patients excreted the organisms via feces. These cases were part of an outbreak in a large hospital where 102 cases of enteritis occurred over some 20 weeks' time. Although samples of the air in wards taken alongside beds did not yield the causative organism, the isolation of the organism from dust and from patients' sputum was taken as an indication that aerial spread of the infection can occur.

During an epidemic of 309 cases of infection with *S. enteritidis* over a period of 9 months, Szmuness (1966) found the organism in the upper respiratory tract of 27.5% of those examined; throat carriers were found in all the hospital wards involved in the epidemic. Positive cultures from the upper respiratory tract were found in 78.2% of infants up to 6 months of age; localization of the organism in this area was considered a secondary phenomenon as a result of the presence of the organism in the intestine. The chief transmission route was probably the insufficiently disinfected hands of personnel. Evidently some cases were caused by airborne infection, because epidemiological techniques demonstrated that: (1) the infection spread rapidly, (2) the intensity and extent of the epidemic in many children's wards exceeded the possibilities of the fecal-oral route, (3) the occurrence of a single infection in a ward caused a wave of new infections in one to three days among children from different rooms who had no direct or indirect contact, and (4) hygienic measures that had been successfully applied in other episodes of intestinal infections were not successful. The ineffectiveness of such measures in other episodes of salmonellosis involving cross-infection has been reported by others (Datta *et al.*, 1960; Neter, 1950). Two outbreaks of *S. montevideo* and *S. barilly* among newborn infants were suggested to have been caused in both instances by contamination of the delivery room by air exhausted from a resuscitator apparatus. The causative organisms in each outbreak were recovered from the water trap fluid of the machine (Rubenstein *et al.*, 1955).

Frequent isolation of *S. derby* from dust suggests that this source of air contamination plays a role in cross infection in surseries. The outbreak described by Rubbo (1948) was not characterized by explosive suddenness, but involved many cases scattered throughout the hospital over a period of 8 months; 78.7% of the 47 cases were hospital-acquired infections. The significance of such infections is emphasized by the finding that the majority of cases of *S. derby* infection were superimposed on other debilitating illnesses resulting in a mortality of 1.2% among the infected cases.

Similar findings have been reported from 7 outbreaks of *S. typhimurium*, phage type 2, gastro-enteritis that occurred over a period of 11 months in the infants' ward of a children's hospital. The infection was neither spread by human carriers nor from a central source of food, but the causative organism was recovered from dust samples taken from floor cleaning machines. The recovery of *S. typhimurium*, phage type 2, from a vacuum cleaner dust bag 10 months after the last case had been removed from the ward (Bate *et al.*, 1958) led to the suggestion that dust has a protective effect. Increased survival of dust-borne, gram-negative organisms associated with epidemic gastro-intestinal infections in infants has been reported for *Bacterium coli* (Rogers, 1951), *S. enteritidis* (Szmuness, 1966), *S. typhimurium*, phage type 27 (Datta *et al.*, 1960).

Heavy contamination of the environment in infants' wards and the high carriage rate for antibiotic-resistant gram-negative bacilli among the newborn suggest that these are principle sources of cross function (Shallard, *et al.*, 1965). The majority of newborn babies were shown to acquire antibiotic-resistant gram-negative organisms soon after entering the hospital. One week after admission, 90% of the gram-negative organisms isolated from nasal swabs were resistant to antibiotics. Certain gram-negative bacilli were responsible for a series of small epidemics in the ward when they appeared in infants' throat, nose, rectal, and umbilical swabs; these bacilli occurred sometimes in infections of babies, but were not isolated from nose, throat, or rectal swabs from nurses (Shallard, 1965, 1966).

Clemmer *et al.*, (1960), who studied experimental airborne salmonellosis in chicks, found that the oral route of inoculation frequently resulted in hematogenous spread of organisms to the lungs as well as other tissues. Concurrent enteric and hematogenous infection often resulted from exposure of chicks to aerosols of salmonella. In some cases, challenge of 1- to 2-day-old chicks with less than 20 viable, airborne cells of salmonella initiated respiratory infection; with certain exceptions, inhaled organisms multiplied in the lungs over a period of several days, but infections were usually self-limited. In chicks inoculated *per os* by capsule, the enteric infection frequently resulted in hematogenous spread of organisms to the lungs as well as other tissues. Fatalities following exposure to aerosols did not exceed those expected from *per os* inoculations with a majority of salmonella strains; however, two strains of *S. pullorum* multiplied in the lungs and mortality was at least twice that observed among chicks infected *per os*.

Darlow *et al.* (1961) demonstrated a marked difference in susceptibility of strains of mice challenged with aerosols of *S. typhimurium* (Strain VI). Inocula of 4.56 to 6.63×10^4 organisms were lethal for one mouse strain whereas 68 to 100 organisms were required for the other. There appeared to be no difference caused by the strain of infecting microbe, regardless of whether the route of inoculation was intra-peritoneal, *per os,* or by inhalation. The disease produced by inhalation was characterized by a specific primary pneumonia. With some minor differences, aerosols of *S. typhosa* were shown in studies of Tully *et al.* (1963), to produce the same clinical picture observed in animals given an oral challenge. A prolonged incubation period with the absence of typhoid bacilli in the stools after the sustained bacteremia caused by respiratory infection, and a delayed fever response, may suggest several differences between pathogenesis of orally induced and respiratory-induced infections of *S. typhosa*. The conjunctiva has been suggested as a portal of entry for the salmonella organism in infections of guinea pigs (Moore, 1957).

Problems of Sampling

Selection of bateriological techniques used in the recovery of micro-organisms from the environment is important. Airborne organisms have impaired viability as a result of physical influences of temperature, humidity, rate of drying, and extended exposure to oxygen. Because formation of a macro-colony of bacterial cells is the most practical method of demonstrating viability of cells recovered from the air, procedures enchancing viability should be based on this requirement. For example, recovery of gram-negative organisms from the environment, particularly salmonella, should be guided by such findings as those of Datta et al. (1960), who reported that the majority of salmonella isolates from fecal specimens were made from enrichment media and not from direct plating procedures. In their investigation of a general hospital outbreak of infections with *Salmonella typhimurium*, phage type 27, among laundry workers and domestic staff, the only ones who contracted the infection were those who collected the used sheets. Sampling of sheets shaken above exposed petri plates containing a selective medium were negative. However, when sheets used by asymptomatic excretors were sampled by liquid extraction and the samples cultured in selenium broth, *S. typhimurium*, phage type 27, was found.

Bacterial contaminants found in the environment represent a residual surviving population, reduced in numbers by the lethal influences of temperature, humidity, drying, and light. Many remaining cells have an impaired growth capacity and may not reproduce unless they are initially exposed to an enriched culture medium. Failure to recover such cells on initial isolation by the use of selective media often occurs and can be explained by the additional influence of selective, inhibitory substances incorporated in the growth medium—substances not inhibitory to the unstressed cell. On this basis, attempts to recover microbial samples from the environment should be done under growth conditions conducive to the reversal of impaired cell functions.

Control of Shedding by Personnel

Control of dispersal of bacteria by enclosing patients in disposable paper instead of cotton sheets has been suggested by Bethune et al. (1965). Bacteria arising from the nasopharynx constitute a small fraction of the total shed into the environment (Bernard et al., 1965). A significant reduction of dissemination of skin bacteria can be effected either by application of 70% ethyl alcohol or lanolin to the skin, or by use of suitable clothing made of tightly woven fabric (Speers et al., 1965;

Bernard *et al.*, 1965). Studies on the effect of bathing to reduce the dispersal of microbial particles from the skin surfaces of healthy subjects revealed that the number of airborne bacteria increased significantly after shower bathing (Speers *et al.*, 1965; Bethune *et al.*, 1965). The effect was most pronounced between 10 and 45 minutes after showering and usually ceased within 1 to 2 hours. Nevertheless, the recommendation that surgeons, nurses, and out-going patients should not have a shower immediately before going into the operating room (Bethune *et al.*, 1965) has been considered inappropriate as an overall recommendation (Bowie *et al.*, 1965).

Nature and Size of Airborne Particles

Microscopic analysis of dust from patients' wards reveals many flake-like particles that resemble skin scales, but a smaller number of fibre-like particles are often seen. Skin particles, measured by a cascade impactor, have a mean equivalent diameter of approximately 8 μ; however, no estimate has yet been made of the mean equivalent diameter of skin-scales carrying viable bacteria (Davies *et al.*, 1962). Measurement of airborne viable microbial particles obtained from a dermatological ward was reported to be from 4 to more than 50 μ in diameter (Selwyn, 1965). The size distribution of airborne particles carrying *S. aureus* has been reported to depend on the diagnosed disease of the skin producing these particles. In psoriasis, 78% of the viable particles were greater than 8 μ; in eczema, 39% were found in the range 10 to 18 μ (Selwyn *et al.*, 1965).

A difference in particle size associated with different types of bacteria has been reported for airborne bacterial particles present in a nursery for the newborn. Mean equivalent diameters of viable particles carrying *S. aureus* were found to be 6, 18.5, and 24 μ; airborne particles carrying Group G streptococci were 5, 7, and 7.5 μ (Hughes, 1963). The "cloud-baby" reported by Eichenwald *et al.* (1960) disseminated airborne particles that were almost entirely less than 5 μ in equivalent diameter.

A study of the size of airborne particles disseminated by a staphylococcal carrier who was the source of airborne infection during surgery was made by Walter *et al.* (1963). After establishing controlled conditions of dressing and undressing in a cubicle, they obtained Andersen samples during the last 5 minutes of confinement. They found a count median diameter (CMD) of 5.3 μ; shaking the carrier's woolen suit jacket, trousers, and lab coat under the same conditions produced a cloud with a CMD of 8.2 μ. The lower particle CMD indicated by these studies may be related to the time sequence of taking air samples during the last 5 minutes of a 15-minute period, with the result that a segment of the bacterial cloud having the largest particles was removed by gravitational settling.

Noble *et al.* (1963) reported equivalent diameters of airborne particles on hospital wards to be in the range of 13.3 to 15.7 μ with 25% of the particles less than 8 μ and 25% greater than 20 μ in equivalent diameter. They also reported equivalent diameters of airborne fungi to be in the range of 3 to 13 μ with 25% smaller than 2 μ and 25% larger than 15 μ. Many fungi appear to be present in the air as single spores. The summary of their results indicated that organisms associated with human disease or carriage were usually found on particles in the range of 4 to 20 μ equivalent diameter.

Air-Conditioning and Ventilating Equipment

In air-conditioning systems using water spray for humidification, parts where moisture and dissolved nutrients exist may provide growth conditions that are potential sources of contaminating microorganisms. The extent of involvement of aerosols formed from contaminated humidifying spray-water in contamination of intramural air is not known, but there seems no doubt that this potential source of infection should be minimized (Blowers *et al.*, 1962).

Conditions simulating those found on direct-expansion air conditioning coils at 14°C, and in condensate from the coil warmed intermittently to room temperature, were found by Cole *et al.* (1964) to cause a decrease in numbers of *E. coli, S. aureus,* beta-streptococci, and *Bacillus sp.* Bacteria placed on cooling fins disappeared rapidly, and the conclusion was made that direct-expansion air-conditioning coils of the window-type unit are unlikely sources of airborne contamination.

Many organisms isolated from the condensate of cooling coils or humidification units have not been associated with infection. *Pseudomonas aeroginosa* is one of the notable exceptions. This organism, which can adapt to a wide range of growth conditions, has been reported to be a heavy contaminant of water associated with cooling systems, and has been recovered from air passing through the system (Andersen, 1959). In hospitals, such air is most often implicated in infections of major surgery, burns, the newborn, and debilitating disease. Cells may survive for several weeks in a protected particle such as eschar from burns (Hurst *et al.*, 1966). Air in wards without cases of pseudomonas infection contain only an occasional organism; more frequent recovery has resulted from sampling areas adjacent to patients with respiratory disease whose sputum contained the organism (Gould, 1963).

One survey of air-conditioning units in a hospital showed that approximately 24% of the units sampled contained *Staphylococcus aureus,* and approximately 63% of the units located in the nurseries were contaminated with this organism (Shaffer *et al.*, 1962). Although such sources of airborne contamination are comparatively minor, adequate maintenance should insure that they do not become sources of gross contamination that might be significant in producing infection in patients of high susceptibility. It should be emphasized that the presence of microbial contaminants in the environment is not direct evidence of an infection hazard. Attenuation of virulence as a result of drying has been shown for *S. aureus* (Hinton *et al.*, 1960) and is a factor in the relatively low overall frequency of infections where environmental microbial contaminants of infectious potential exist.

Patient-Care Equipment

Gram-positive cocci are a major group of bacteria causing cross infection in hospitals; in particular, coagulase positive *Staphylococcus auereus* has been a major concern in hospital epidemics; it remains a significant cause of nosocomial infection. Development of new antimicrobial drugs has greatly enhanced the therapeutic control of the gram-positive infections. Improvement in the treatment of these infections has been accompanied by a rising frequency of infections caused by gram-negative bacilli. Many of these organisms are considered to be of low or negligible pathogenicity, and are frequently found as natural contaminants in water or where conditions of high moisture content prevail—the so-called "water-bugs."

During the past decade several authors have directed attention to the emerging importance of infections caused by gram-negative bacilli (Finland *et al.*, 1959; Sandusky, 1961; Rogers, 1959; Selwyn *et al.*, 1964). In a study of *Enterobacteriaceae* and *Pseudomonas aeruginosa* in the respiratory flora of hospital and nonhospital related populations, Benham *et al.* (1960) found that the carrier rate for antibiotic-resistant *S. aureus* among surgeons decreased during the period 1952–58, whereas gram-negative organisms increased. Respiratory transfer of the patients' flora to the surgeon was considered a mode of acquisition. As carrier rates among surgeons changed in relation to kinds of organisms, a similar change was reflected in organisms causing infection in clean surgical wounds.

Thomas *et al.* (1961) from a retrospective analysis of laboratory records for 1954 and 1959, reported an increase in incidence of coliform organisms, *Proteus*, and yeasts isolated from sputum and throat swabs from hospital patients. Evidence was also found of antibiotic-resistant "hospital strains" of gram-negative bacilli. Colonization of the upper respiratory tract with gram-negative bacilli after surgery is frequent and occurs rapidly. However, appearance of these organisms in the trachea and sputum after an operation seems to indicate colonization rather than infection (Redmond et al., 1967). Emergence of gram-negative microflora occurs in the nasopharynx, frequently in association with the administration of antibiotics, and is not necessarily evidence of clinical disease.

Gram-negative, bacillary lung infections appear to be a result of opportunistic invasion in many patients with underlying disease, and are of the bronchopneumonia variety. These organisms are an important cause of disease when they implant in the respiratory tract (Lepper, 1963). *Klebsiella sp.* were frequently found by Weiss et al. (1954) in the sputum of patients with chronic pulmonary disease, yet did not seem to play a role in the patients' disease. In an outbreak of 5 cases of fatal pneumonia on a single hospital ward the causative agent (*Klebsiella pneumoniae*, an organism of unusual antibiotic-resistance) was cultured from an aerosol solution used in nebulizers of the intermittent positive pressure breathing (IPPB) equipment of these patients (Mertz et al., 1967). The risk of patient infection by inhalation of contaminated aerosols is emphasized by the results of experimental studies that have shown a marked reduction in the size of an inoculum as a function of particle sizes in the aerosol required to produce infection.

Patient equipment required for "assisted" breathing, for maintaining high-humidity breathing air, and administration of medications by inhalation of liquid aerosols may be a potential source of infectious aerosols. Effectiveness of design in certain therapy machines depends on the formation of aerosols containing particles of sizes permitting optimal penetration of the respiratory tract (Mercer et al., 1965; Kelsch, et al., 1965).

In a survey of 52 IPPB machines in several hospitals, Reinarz et al. (1965) found 45% of equipment generated aerosols with particles in the 1.4- to 3.5-μ diameter range, containing greater than 330 viable particles per liter of air. More than 80% of the units generated aerosols with greater numbers of bacteria than ambient air. The major site of generation of bacterial aerosols was the reservoir nebulizer; machines not equipped with reservoir nebulizers did not generate aerosols with sufficient bacterial contamination to be considered a greater risk of infection than room air. Other workers have reported high bacterial contamination of inhalation therapy humidifiers with gram-negative bacilli (Knudsin et al., 1962; MacPherson, 1958).

The pathogenicity of *Pseudomonas aeruginosa* in the newborn, especially in the premature baby, has been pointed out by several authors (Neter et al., 1955; Hoffman et al., 1955; Jacobs, 1964), and neonatal infections with *P. aeruginosa* have been reported in association with contaminated resuscitation equipment (Bassett *et al.*, 1965) and incubators (Barrie, 1965). Attention has also been directed to a high degree of gross contamination found in the water of humidifying devices in use in the newborn nursery (Sever, 1959).

Significance of such contamination in the formation of infectious aerosols is indicated by incidents (caused by *P. aeruginosa* in the nebulizer solution) of infections among premature infants. In one instance, all infected infants had been exposed to an aerosol from the nebulizer, whereas all well ones had not (Anon., 1959).

Delivery-room resuscitators contaminated with *P. aeruginosa* via a water-sink aerator were assumed to have been the source of aerosols infecting 22 infants (2 fatalities) in studies by Fierer *et al.* (1967). Cross infection among other infants not exposed to the resuscitators was thought to have occurred via contaminated hands of personnel in the nursery.

In some patients, impaired respiratory function and difficulty of breathing can be improved by the performance of a tracheostomy—an external opening made into the trachea and causing a by-pass of the nasopharynx. The patient respires directly through his tracheostoma, and airborne infection of the bronchial tree may occur directly (Gotsman *et al.*, 1964). These authors reported infections due to antibiotic-resistant staphylococci and *P. aeruginosa,* the nature of the infecting organisms clearly indicating the hospital environment as the source. However, no determination of the frequency of airborne infections via the tracheostoma can be made from these results.

A number of patients, all with severe lung disease, were reported by Phillips *et al.* (1965) to have become infected via a contaminated ventilator that pumped the organism as an aerosol directly into the patient's trachea. Two other patients on the same ward became infected, but the low ward contamination and negative air samples for *P. aeruginosa* did not indicate cross infection from these sources.

Edmundson *et al.* (1966) examined aerosols produced by nebulizer equipment and found the following genera of gram-negative bacilli (percentage in parentheses): *Herellea* (53), *Pseudomonas* (40), *Alcaligenes* (33), *Achromobacter* (23), *Flavobacterium* (23), *Paracolobactrum, Escherichia,* and *Mima* (13). The organism *Flavobacterium meningosepticum* has been reported by several workers from infections in infants (Cabrera *et al.*, 1961; George *et al.*, 1961).

Equipment employed in critical areas, such as surgery, may produce potentially hazardous aerosols. Ranger *et al.* (1958) reported dissemination of airborne microorganisms by a surgical pump. Certain types of pumps used in aspirating contaminated body-fluids from patients in surgery have been found to discharge extensive contamination into the operating room air (Blowers *et al.*, 1955).

Dental procedures involving rotary, vibrating, grinding instrumentation, and the use of air and water syringes and suction devices that might exhaust into the room, are potential sources of airborne contamination. Inclusion of these sources in the medical-care environment is indicated by the fact that over 40% of hospitals in the United States have dental facilities, the percentage increasing with hospital size (Amer. Hosp. Assn., 1967).

Burton *et al.* (1963) listed infectious diseases that might be considered in the category of occupational risks for dentists in the course of their direct contact with patients' mouths. Belting *et al.* (1964) has shown that when a dental rotor was used in tooth repair in patients with pulmonary tuberculosis and sputum positive for *Mycobacterium tuberculosis*, positive cultures were obtained several feet from patients' mouths. In a preliminary study to compare aerosols generated from air-turbine and conventional cutting instruments, Madden *et al.* (1963) found that 85% of the total count was represented by particles of 3.5 μ or less. The implication for pulmonary deposition of microbial aerosols of such particle sizes is direct. A later report (Hausler *et al.*, 1966) emphasized the importance of the water-flow rate to the cutting site and, as have others (Miller *et al.*, 1963; Brown, 1965; Larato *et al.*, 1966), verified the production of bacterial aerosols by ultrahigh-speed cutting instruments.

No relationship between bacterial aerosols generated by dental procedures and transmission of infection, either to the dental operator or to individuals in the immediate area, has been established. Sufficient evidence is available, however, to include consideration of this source as potentially signifcant in air contamination of the hospital area.

EPIDEMILOGY OF AIRBORNE CROSS INFECTION

An unresolved problem in hospital sepsis is the significance of airborne cross infection in terms of its frequency of occurrence, its function as a direct and indirect route of contamination transport, and the determination of its predictable patterns in cross-infection dynamics. Epidemiological evidence is required to evaluate the significance of microbial contaminants in the environment and their relationship to infection. Presence of contaminants in environmental areas may be only the expression of an active infection process and not the source of the infecting microbial agent.

Sources of microbial contaminants in the hospital environment, uniqueness of the host susceptibility represented by the patient population, and selected episodes where the microbial aerosol can be directly implicated in cross infection emphasize the potential circumstances and opportunities for infection that are peculiar to the hospital environment. Assumption of this viewpoint requires a broad interpretation of the role of airborne contamination in critical patient areas.

The possible role of airborne cross infection is difficult to assess because most information has been largely retrospective from studies of epidemics of infection in the hospital. Where airborne transmission might be suspected, its real role remains conjectural in the absence of data contemporaneous to pre-outbreak and infection outbreak events. The dynamic behavior of air, produced by ventilation and movement of personnel, and fluctuating contamination levels create a continuous variety of exposure conditions. Where air monitoring is available for determining viable particle assay, as well as qualitative analysis, samples represent contaminating events as temporal episodes that are averaged with respect to time. The results are then used to reconstruct conditions that are assumed to represent those to which the host (patient) was exposed.

Air is only one of many sources of contamination in contact with the patient. Other transmission routes (mainly direct contact) can be controlled to a high degree, but they are, nonetheless, variables not generally amenable to a control that would completely eliminate them from a multi-factoral study. Control of airborne contaminants as an experimental variable, with reduction of contamination levels as a goal, is one direct approach to evaluation of the role of aerosols in nosocomial infection.

Bacterial infection of surgical wounds has been a complicating postoperative sequela during the history of surgery. Major prevention has been most singularly effective as a result of development of aseptic techniques and the practice of barrier interception between operating staff and patients. The contaminant-laden air remains an indirectly controlled factor. If infections acquired during surgery are the result of wound contamination by airborne bacteria, then a reduction in infection rates should follow removal of these contaminants. This evidence can be sought in instances when air contamination control was attempted and where apparent airborne transmission occurred.

One method of reducing the level of airborne contamination is to use ultraviolet (UV) irradiation of the air. The bactericidal action of UV is well established (Rentschler et al., 1941; Wells, 1940; Lidwell, 1946), and practical intensities of UV under standardized test conditions have been shown to reduce the numbers of bacteria in the air to a level equivalent to that produced by 100 or more air changes per hour (Wells, 1944, 1945).

Hart (1941) found that infections continued to occur despite rigorous attention to the elimination of contamination sources from personnel and equipment and concluded that airborne pathogenic bacteria were the cause of "unexplained infections" in clean wounds. A review (Hart, 1960) of 29 years' experience with UV irradiation of operating room air showed an unquestionable reduction in infections of clean general surgical, orthopedic, and neurosurgical wounds. From records of 4,382 operations a comparison of post-operative infection (including reopened incisions) rates showed that a 11.6% rate without radiation (1,782 operations) was reduced to 0.62% with radiation (2,600 operations). An infection rate of 0.24% with radiation was obtained in operative wounds (2,460) with primary incision. Others have reported approximately 65 to 80% reductions in clean surgical operations where UV irradiation was used; Overholt et al. (1940) showed a reduction from 13.8% to 2.7%, and Woodhall et al. (1949) reported infections in clean neurosurgical operations were reduced from 1.1% to 0.4%.

A less marked reduction in infection rates over a 2-year period was reported from a cooperative study evaluating the efficacy of UV radiation in the operating rooms of five university hospitals. The incidence of infection in refined-clean wounds was reduced from 3.8% in controls to 2.9% in irradiated wounds, a reduction that was significant at the 5% level. Refined-clean wounds are the least susceptible to contaminants from sources other than air. If irradiation influences infection, these wounds would most likely show it—ergo, the airborne source of infection. The absence of a measurable reduction in infections in other wound classes indicates that contamination sources other than air are largely responsible for infections in these wounds (NAS-NRC, 1964).

Shooter et al. (1956) reported an immediate reduction of average counts of bacteria in operating-room air, estimated to be from 30 to 40/ft³ to 4 to 10/ft³, was obtained by reversing the flow of air from the passage outside the operating room to the operating room itself. No alteration in operating room technique was made, so observations were the result of the ventilation change only. The reversal of air-flow and positive pressure ventilation was followed by a reduction in surgical wound infection rate from 9% to 1%. Although there were large total numbers of bacteria in the operating room air, relatively few staphylococci were recovered.

In an investigation of infection in a thoracic surgical unit, Blowers et al. (1955) found inefficient ventilation, the use of blankets, and personnel activity contributed to the high concentration of S. aureus and other bacteria in air. Most infections were considered to have been caused by airborne organisms present at the time of operation. Correction of these deficiencies was followed by a reduction in infection rate from 10.9 to 3.9%.

Instances of surgical infection caused by airborne *S. aureus* were reported by Walter *et al.* (1963) who observed that 2 of 169 operations became infected. A disseminating nasopharyngeal staphylococcal carrier was present in the periphery of the operating room during each of the operations and, although suitably gowned and masked, the carrier was shown to have contributed to the contamination of the instrument table, surgical masks of various members of the surgical team, and the general environment. As many as 3.7 ± 2.7 organisms per cubic foot were recovered from the air, and the assay showed *S. aureus* present during 11.2% of the operations. Particles shed by the carrier were less than 10 μ CMD. The two patients who became infected were healthy individuals undergoing elective surgery. Thus, absence of high susceptibility to infection of the affected patients, low numbers of bacteria in the air, and the presence of a "disperser" in a critical area emphasize the significance of a disseminating source and the equivocal significance of total counts of airborne bacteria. The size of the particle also suggests that reliance on settling plates might be inadequate as a method of sampling.

Kinmouth *et al.* (1958) did not observe a reduction in infection rates (av. 10%) when a change in ventilation and personnel protective clothing decreased air counts of bacterial particles from ca. $22/ft^3$ to ca. $8/ft^3$. However, a failure in the operation of the positive-pressure ventilation system was accompanied by an increase in infection rate to an estimated 35%. This deficiency apparently existed for only a short time, but it is quite suggestive of the importance of maintaining low levels of air contaminants in surgery. One surgical infection attributed to reversion to high counts (30 to $40/ft^3$) of bacteria in the air was reported by Shooter *et al.* (1956) to have occurred during reversal of the positive-pressure ventilation system of the operating room.

Determination, both quantitative and qualitative, of bacterial contaminants in the air is required to predict the infective potential for open surgical wounds. Additional information required is whether these airborne bacterial particles gain entrance to the wound, and, once in the wound, whether infection follows. Partial answers were provided by Burke (1963) from a study of bacteria recovered from surgical wounds before closure and a study of phage-types of staphylococci in the air in contact with the wound. Forty-six of 50 wounds contained coagulase positive staphylococci with an average of 14 colony-forming units per wound. A maximal number was reached in the wound within 1 to 2 hours and remained relatively constant in wounds made during surgery lasting up to 4 hours. Total staphylococcal count in the wound gradually increased up to an average of 24.2 in wounds up to 2 to 3 hours operating time and a slight decrease during 3 to 4 hours procedures. Sixty-eight percent of the wounds studied showed the same strain of staphylococcus in the wound as in the air in contact with it. Two infections occurred, one of which was caused by a strain of staphylococcus recovered only from the air. Staphylococcal strains were found in the air and in the wounds in all 50 operations, thus emphasizing the role of host susceptibility factors and virulence of staphylococci in the production of infection.

The significance of airborne transmission of bacterial contaminants is also suggested by circumstances in which infection is not a direct sequela but where a potentially infectious microorganism is transmitted from patient to patient. Nasal colonization of infants by staphylococci in the nursery is an example of this type of transmission. Mortimer *et al.* (1966) studied comparative rates at which infants acquired staphylococci in the anterior nares and umbilicus, and found that 10% acquired an index staphylococcus via the airborne route as compared to 14% by physical contact with nurses who carefully washed their own hands.

The effectiveness of ventilating air in transporting bacterial particles within intramural environments has been clearly demonstrated. The significance of this source of contamination can be realized only by considering the multiple and complex factors that are unique to the medical-care environment and to the represented host population. Direct contact is undoubtedly the most significant route of transmitting infection in this situation. Evidence for airborne transmission is most plentiful from studies on staphylococcal wound infection and indicates that air is significant in this regard, but the role is subordinate to other, more common routes. Under conditions of close environmental control, such as in well-run surgeries, a low level of airborne contamination can be maintained. Bourdillon *et al.* (1946) suggested a maximum count of 10 bacterial particles/ft^3 of air for major surgical operations on tissues with normal resistance to infection and 0.1 to 2 bacterial particles/ft^3 of air for special conditions. Evidence from published reports indicates that air contamination of this magnitude and lower are presently encountered in operating rooms (Baldwin, 1965; Walter *et al.*, 1963; Fincher, 1966; Howe *et al.*, 1963).

One epidemiological problem of providing comparative data involves the number of observations (cases) required to determine a statistical difference when a further reduction in counts is made. The frequency of cross infection via the airborne route is believed to be low under conditions of high environmental control. The case numbers needed to demonstrate a statistically significant reduction in a given rate by an improvement in the air component of control is considerable, as pointed out by Lidwell (1963).

Levels suggested by Bourdillon *et al.* (1946) for special conditions would appear to be valid in meeting the requirements of a patient with high susceptibility to infection. However, an ability to assess the susceptibility of patients would be of major significance in setting practical and effective levels of environmental control adequate to remove this source of cross infection. This applies to airborne as well as to other environmental routes of transmission.

REFERENCES

Abramson, H. 1947. Infection with *Salmonella typhimurium* in the newborn. Epidemiologic and clinical considerations. *J. Dis. Child.*, **74**: 576–586.

American Hospital Association—Guide Issue. 1967. *J. Am. Hosp. Assoc.*, **41**(part 2): 470.

Anderson, K. 1959. *Pseudomonas pyocyanea* disseminated from an air-cooling apparatus. *Med. J. Aust.*, **1**: 529.

Anon. 1959. Outbreak due to *Pseudomonas aeruginosa. Morbidity and Mortality, Weekly Report*, PHS, U.S. Dept. of Health, Education and Welfare, **18**(17): 2.

Baldwin, M., Weatherby, R. J., and MacDonald. F. D. S. 1965. Microbial characteristics in a neurosurgical environment. *Hospitals*, **39**: 71–78.

Barber, Mary, and Dutton, A. A. C. 1958. Antibiotic-resistant staphylococcal outbreaks in a medical and a surgical ward. *Lancet*, **2**: 64–68.

Barrie, Dinah. 1965. Incubator-borne *Pseudomonas pyocyanea* infection in a newborn nursery. *Arch. Dis. Child.*, **40**, 555–559.

Bassett, D. C. J., Thompson, S. A. S., and Page, B. 1965. Neonatal infections with *Pseudomonas aeruginosa* associated with contaminated resuscitation equipment. *Lancet*, **1**: 781–784.

Bate, J. G., and James, Ursula. 1958. *Salmonella typhimurium* infection dust-borne in a children's ward. *Lancet*, **2**: 713–715.

Belting, C. M., Haberfelde, G. C., and Juhl, L. K. 1964. Spread of organisms from dental air rotor. *J. Am. Dent. Assoc.*, **68**: 648–651.

Benham, R. S., Havens, Isabelle, and Landy, J. J. 1960. Respiratory flora of hospital-related populations. *J. Inf. Dis.*, **107**: 1–10.

Bernard, H. R., Speers, R., Jr., O'Grady, F., and Shooter, R. A. 1965. Reduction of dissemination of skin bacteria by modification of operating-room clothing and by utlraviolet irradiation. *Lancet*, **2**: 458–461.

Bernsten, C. A., and McDermott, W. 1960. Increased transmissibility of staphylococci to patients receiving an antimicrobial drug. *New Engl. J. Med.*, **262**: 637–642.

Bethune, D. W., Blowers, R., Parker, M., and Pask, E. A. 1965. Dispersal of *Staphylococcus aureus* by patients and surgical staff. *Lancet*, **1**: 480–483.

Blowers, R., Mason, G. A., Wallace, K. R., and Walton, M. 1955. Control of wound infection in a thoracic surgery unit. Lancet, **1**: 786–794.

Blowers, R., Lidwell, O. M., and Williams, R. E. O. 1962. Infection in operating theatres in relation to air conditioning equipment. *J. Inst. Heat. & Vent. Eng.* (London), pp. 244–245, October.

Bourdillon, R. B., and Colebrook, L. 1946. Air hygiene in dressing-rooms for burns or major wounds. *Lancet,* 1: 601–605.

Bowie, J. H., Tonkin, R. W., and Robson, J. S. 1965. Shower-baths and the control of hospital infection. *Lancet,* 1: 909–910.

Brodie, J., Kerr, M. R., and Somerville, T. 1956. The hospital staphylococcus; a comparison of nasal and faecal carrier status. *Lancet,* 1: 19–20.

Brown, R. V. 1965. Bacterial aerosols generated by ultra high-speed cutting instruments. *J. Dent. Child.,* 32: 112–117.

Burke, J. F., and Corrigan, E. A. 1961. Staphylococcal epidemiology on a surgical ward. Fluctuations in ward staphylococcal content, its effect on hospitalized patients and the extent of endemic hospital strains. *New Engl. J. Med.,* 264: 321–326.

Burke, J. F. 1963. Identification of the sources of staphylococci contaminating the surgical wound during operations. *Ann. Surg.,* 158: 898–904.

Burton, W. E., and Miller, R. L. 1965. The role of aerobiology in dentistry. In *A Symposium on Aerobiology, 1963* (R. L. Dimmick, Ed.), pp. 87–94. Nav. Biol. Lab., Nav. Supply Center, Oakland, Calif.

Cabrera, H. A., and Davis, G. H. 1961. Epidemic meningitis of the newborn caused by flavobacteria; epidemiology and bacteriology. *Am. J. Dis. Child.,* 101: 289–295.

Clemmer, Dorothy I., Hickey, J. L. S., Bridges, Joan F., Schliessmann, D. J., and Shaffer, M. F. 1960. Bacteriologic studies of experimental air-borne salmonellosis in chicks. *J. Inf. Dis.* 106: 197–210.

Close, A. S., Smith, M. B., Koch, Marie L., and Ellison, E. H. 1960. An analysis of ten cases of salmonella infection on a general surgical service. *Am. Med. Assoc. Arch. Surg.,* 80: 972–976.

Cole, W. R., Bernard, H. R., and Dunn, B. 1964. Growth of bacteria on direct expansion air-conditioning coils. *Surgery,* 55: 436–439.

Cooke, E. Mary, and Buck, H. W. 1963. Self-contamination of dermatological patients with *Staphylococcus aureus. Brit. J. Derm.,* 75: 21–25.

Darlow, H. M., Bale, W. R., and Carter, G. B. 1961. Infection of mice by the respiratory route with *Salmonella typhimurium. J. Hyg.,* 59: 303–308.

Datta, Naomi, and Pridie, R. B. 1960. An outbreak of infection with *Salmonella typhimurium* in a general hospital. *J. Hyg.,* 58: 229–241.

Davies, R. R., and Noble, W. C. 1962. Dispersal of bacteria on desquamated skin. *Lancet,* 2: 1295–1297.

Duguid, J. P., and Wallace, A. T. 1948. Air infection with dust liberated from clothing. *Lancet,* 2: 845–849.

Edgar, W. M., and Lacey, B. W. 1963. Infection with *Salmonella heidelberg,* an outbreak presumably not foodborne. *Lancet,* 1: 161–163.

Edmondson, E. B., Reinarz, J. A., Pierce, A. K., and Sanford, J. P. 1966. Nebulization equipment. A potential source of infection in gram-negative pneumonias. *Am. J. Dis. Child.,* 111: 357–360.

Ehrenkranz, N. J. 1964. Person-to-person transmission of *Staphylococcus aureus.* Quantitative characterization of nasal carriers spreading infection. *New Engl. J. Med.,* 271: 225–230.

Eichenwald, H. F., Kotsevalov, O., and Fasso, L. A. 1960. The "cloud baby": an example of bacterial viral interaction. *Am. J. Dis. Child.,* 100: 161–173.

Epstein, H. C., Hochwald, A., and Ashe, Rosemary. 1951. Salmonella infections of the newborn infant. *J. Pediatrics,* 38: 723–731.

Fierer, J., Taylor, P. M., and Gezon, H. M. 1967. *Pseudomonas aeruginosa* epidemic traced to delivery-room resuscitators. *New Engl. J. Med.,* **276:** 991–996.

Fincher, E. L. 1966. Air sampling application, methods, recommendations. In *Control of Infections in Hospitals.* Continuing Education Series No. 138, pp. 200–209. The University of Michigan, School of Public Health, Ann Arbor, Mich.

Finland, M., Jones, W. F., Jr., and Barnes, Mildred W. 1959. Occurrence of serious bacterial infections since introduction of antibacterial agents. *J. Am. Med. Assoc.,* **170:** 2188–2197.

George, M. M., Cochran, G. P., and Wheeler, W. E. 1961. Epidemic meningitis of the newborn caused by flavobacteria. *Am. J. Dis. Child.,* **101:** 296–304.

Gotsman, M. S., and Whitby, J. L. 1964. Respiratory infection following tracheostomy. *Thorax,* **19:** 89–96.

Gould, J. C. 1963. *Pseudomonas pyocyanea* infections in hospitals. In *Infection in Hospitals* (R. E. O. Williams and R. A. Shooter, Eds.). pp. 119–130. F. A. Davis Co., Philadelphia, Pa.

Greene, V. W., Vesley, D., Bond, R. G., and Michaelsen, G. S. 1962. Microbiological contamination of hospital air. II. Qualitative studies. *Appl. Microbiol.,* **10:** 567–571.

Hamburger, M., Jr., Puck, T. T., Hamburger, Virginia G., and Johnson, Margaret A. 1944. Studies on the transmission of hemolytic streptococcus infections. III. Hemolytic streptococci in the air, floor dust, and bed clothing of hospitals wards and their relation to cross-infection. *J. Inf. Dis.,* **75:** 79–94.

Hamburger, M., Jr., Green, Margaret Johnson, and Hamburger, Virginia G. 1945. Strep. II. Spread of infection by individuals with strongly positive nose cultures who expelled large numbers of hemolytic streptococci. *J. Inf. Dis.,* **77:** 96–108.

Hare, R., and Thomas, C. G. A. 1956. Transmission of *Straphylococcus aureus. Brit. Med. J.,* **2:** 840–844.

Hare, R., and Cooke, E. M. 1961. Self-contamination of patients with staphylococcal infections. *Brit. Med. J.,* **2:** 333–336.

Hart, D., and Upchurch, S. E. 1941. "Unexplained infections" in clean operative wounds: The importance of the air as a medium for the transmission of pathogenic bacteria and bactericidal radiation as a method of control. *Ann. Surg.* **114:** 936–959.

Hart, D. 1960. Bactericidal ultraviolet radiation in the operating room. Twenty-nine-year study for control of infection. *J. Am. Med. Assoc.,* **172:** 1019–1028.

Hausler, W. J., and Madden, R. M. 1966. Microbiologic comparisons of dental handpieces. 2. Aerosol decay and dispersion. *J. Dent. Res.,* **45:** 52–58.

Hinton, N. A., Maltman, J. R., and Orr, J. H. 1960. The effect of desiccation on the ability of *Staphylococcus pyogenes* to produce disease in mice. *Am. J. Hyg.,* **72:** 343–350.

Hoffman, M. A., and Finberg, L. 1955. *Pseudomonas* infections in infants associated with high-humidity environments. *J. Pediatrics,* **46:** 626–630.

Howe, C., and Marston, Alice T. 1963. Qualitative and quantitative bacteriologic studies on hospital air as related to postoperative wound sepsis. *J. Lab. & Clin. Med.,* **61:** 808–819.

Hughes, M. H. 1963. Dispersal of bacteria on desquamated skin. *Lancet,* **1:** 109.

Hurst, Valerie, Grossman, M., and Ingram, F. R. 1958. Hospital laundry and refuse chutes as source of staphylococcic cross-infection. *J. Am. Med. Assoc.,* **167:** 1223–1229.

Hurst, Valerie, and Sutter, Vera L. 1966. Survival of *Pseudomonas aeruginosa* in the hospital environment. *J. Inf. Dis.,* **116:** 151–154.

Jacobs, J. 1964. The investigation of an outbreak of *Pseudomonas pyocyanea* infection in a pediatric unit. *Post. Grad. Med. J.*, **40**: 590–594.

Jellard, C. H., Jolly, H., and Brown, R. N. 1959. An outbreak of *Salmonella bovis-morbificans* infection in a children's ward. *Lancet*, **1**: 390–392.

Kelsch, R. C., Barr, M., Jr., and DeMuth, G. R. 1965. Mist therapy in lower respiratory tract infection: a controlled study. *Am. J. Dis. Child.*, **109**: 495–499.

Kinmouth, J. B., Hare, R., Tracy, G. D., Thomas, C. G. A., Marsh, J. D., and Jantet, G. J. 1958. Studies of theatre ventilation and surgical wound infection. *Brit. Med. J.*, **2**: 407–411.

Kundsin, Ruth B., and Walter, C. W. 1962. Asepsis for inhalation therapy. *Anaesthesiology*, **23**: 507–512.

Larato, D. C., Ruskin, P. F., Martin, A., and Delanko, R. 1966. Effect of a dental air turbine drill on the bacterial counts in air. *J. Prost. Dent.*, **16**: 758–765.

Laurell, G. 1952. Airborne infections. IX. Coliform organisms in the upper respiratory tract of children, with particular reference to their mode of spreading in a children's hospital. *Acta Path. et Microbiol. Scand.*, **31**: 112–123.

Leeder, S. F. 1956. An epidemic of *Salmonella panama* infections in infants. *Ann. N.Y. Acad. Sci.*, **66**: 54–60.

Lepper, M. H. 1963. Opportunistic gram-negative rod pulmonary infections. *Dis. Chest.*, **44**: 18–26.

Lidwell, O. M. 1946. Bactericidal effects of the partial irradiation of a room with ultra-violet light. *J. Hyg.*, **44**: 333–341.

Lidwell, O. M. 1963. Methods of investigation and analysis of results. In *Infection in Hospitals* (R. E. O. Williams and R. A. Shooter, Eds.) pp. 43–46. F. A. Davis Co., Philadelphia, Pa.

Loosli, C. G., Smith, M. H. D., Cline, J., and Nelson, L. 1950. The transmission of hemolytic streptococcal infections in infant wards with special reference to "skin dispersers." *J. Lab. & Clin. Med.*, **36**: 342–359.

Mackerras, I. M., and Mackerras, M. J. 1949. An epidemic of infantile gastroenteritis in Queensland caused by *Salmonella bovis-morbificans* (Basenau). *J. Hyg.*, **47**: 116–181.

MacPherson, C. R. 1948. Oxygen therapy—an unexpected source of hospital infection? *J. Am. Med. Assoc.*, **167**: 1083–1086.

Madden, R. M., and Hausler, W. J., Jr. 1963. Microbiological comparison of dental handpieces. I. Preliminary report. *J. Dent. Res.*, **42**: 1146–1151.

McKee, W. M., DiCaprio, J. M., Roberts, C. E., Jr., and Sherris, J. C. 1966. Anal carriage as the probable source of a streptococcal epidemic. *Lancet*, **2**: 1007–1009.

Mercer, T. T., Goddard, R. F., and Flores, R. L. 1965. Output characteristics of several commercial nebulizers. *Ann. Allergy*, **23**: 314–326.

Mertz, J. J., Scharer, L., and McClement, J. H. 1967. A hospital outbreak of *Klebsiella pneumoniae*, from inhalation therapy with contaminated aerosol solutions. *Am. Rev. Resp. Dis.*, **95**: 454–460.

Michaelsen, G. S. 1964. Waste handling. In *Proceedings of the National Conference on Institutionally Acquired Infections*. pp. 65–69. USPHS Publication No. 1188, U.S. Govt. Printing Off. Washington, D.C.

Miller, R. L., Burton, W. E., and Spore, R. W. 1965. Aerosols produced by dental instrumentation. In *A Symposium on Aerobiology, 1963* (R. L. Dimmick, Ed.), pp. 97–120. Nav. Biol. Lab., Nav. Supply Center, Oakland, Calif.

Moore, B., 1957. Observations pointing to the conjunctiva as the portal of entry in salmonella infection of guinea-pigs. *J. Hyg.*, **55**: 414–433.

Mortimer, E. A., Wolinsky, E, Gonzaga, A. J., and Rammelkamp, C. H. 1966. Role of airborne transmission in staphylococcal infections. *Brit. Med. J.,* 1: 319–322.

Murray, J. W., and Walker, J. H. C. 1958. An outbreak of enteritis (*Salmonella heidelberg*) in a maternity unit. *Med. Officer,* 100: 221–223.

National Academy of Science—National Research Council. Report of an Ad Hoc Committee. 1964. Postoperative Wound Infections: The influence of ultraviolet irradiation of the operating rooms and of various other factors: *Ann. Surg.,* 160(2): 1–192 (Supplement).

Neter, E. 1950. Observations of the transmission of salmonellosis in man. *Am. J. Publ. Hlth.,* 40: 929–933.

Neter, E., and Weintraub, D. H. 1955. An epidemiological study on *Pseudomonas aeruginosa* (*Bacillus pyocyaneus*) in premature infants in the presence and absence of infection. *J. Pediatrics,* 46: 280–287.

Noble, W. C. 1962. The dispersal of staphylococci in hospital wards. *J. Clin. Pathol.,* 15: 552–558.

Noble, W. C., Lidwell, O. M., and Kingston, D. 1963. The size distribution of airborne particles carrying micro-organisms. *J. Hyg.,* 61: 385–391.

Overholt, R. H., and Betts, R. H. 1940. A comparative report on infection of thoracoplasty wounds; experiences with ultraviolet irradiation of operating room air. *J. Thorac. Surg.,* 9: 520–529.

Phillips, I., and Spencer, G. 1965. *Pseudomonas aeruginosa* cross-infection due to contaminated respiratory apparatus. *Lancet,* 2: 1325–1327.

Rammelkamp, C. H., Jr., Morris, A. J., Catanzaro, F. J., Wannamaker, L. W., Chamovitz, R., and Marple, E. C. 1958. Transmission of Group A streptococci. III. Effect of drying on the infectivity of the organism for man. *J. Hyg.,* 56: 280–287.

Ranger, I., and O'Grady, F. 1958. Dissemination of microorganisms by a surgical pump. *Lancet,* 2: 299–300.

Redman, L. R., and Lockey, Eunice. 1967. Colonization of the upper respiratory tract with gram-negative bacilli after operation, endotracheal incubation and prophylactic antibiotic therapy. *Anaesthesia,* 22: 220–227.

Reinarz, J. A., Pierce, A. K., Mays, Benita B., and Sanford, J. P. 1965. The potential role of inhalation therapy equipment in nosocomial pulmonary infection. *J. Clin. Invest.,* 44: 831–839.

Rentschler, H., Nagy, R., and Mouromseff, G. 1941. Bactericidal effect of ultraviolet radiation. *J. Bacteriol.,* 41: 745–774.

Ridley, M. 1959. Perineal carriage of *Staph. aureus. Brit. Med. J.,* 1: 270–273.

Rogers, D. E. 1959. The changing pattern of life-threatening microbial disease. *New Engl. J. Med.,* 261: 677–683.

Rogers, K. B. 1951. The spread of infantile gastro-enteritis in a cubicled ward. *J. Hyg.,* 49: 140–151.

Rosebury, T. 1962. In *Microorganisms Indigenous to Man.* p. 315. Blakiston Division, McGraw-Hill Co., New York.

Rubbo, S. D. 1948. Cross infections in hospitals due to *Salmonella derby. J. Hyg.,* 46: 158–163.

Rubbo, S. D., Stratford, B. C., and Dixson, Shirley. 1962. Spread of a marker organism in a hospital ward. *Brit. Med. J.,* 2: 282–287.

Rubenstein, A. D., and Fowler, R. N. 1955. Salmonellosis of the newborn with transmission by delivery room resuscitators. *Am. J. Publ. Hlth.,* 45: 1109–1114.

Sandusky, W. R. 1961. *Pseudomonas* infections: Sources and cultural data in a general hospital with particular reference to surgical infections. *Ann. Surg.*, **153**: 996–1005.

Sciple, G. W., Riemensnider, D. K., and Schleyer, C. A. 1967. Recovery of microorganisms shed by humans into sterilized environments. *Appl. Microbiol.*, **15**: 1388–1392.

Selwyn, S., MacCabe, A. F., and Gould, J. C. 1964. Hospital infection in perspective: The importance of the gram-negative bacilli. *Scot. Med. J.*, **9**: 409–417.

Selwyn, S. 1965. The mechanism and prevention of cross-infection in dermatological wards. *J. Hyg.*, **63**: 59–71.

Selwyn, S., and Chambers, D. 1965. Dispersal of bacteria from skin lesions: a hospital hazard. *Brit. J. Derm.*, **77**: 349–356.

Sever, J. L. 1959. Possible role of humidifying equipment in spread of infections from the newborn nursery. *Pediatrics*, **24**: 50–53.

Shaffer, J. G., and McDade, J. J. 1962. Air-borne *Staphylococcus aureus*. *Arch Env. Hlth.*, **5**: 547–551.

Shallard, M. A., and Williams, A. L. 1965. A study of the carriage of gram-negative bacilli by new-born babies in hospitals. *Med. J. Aust.*, **1**: 540–542.

Shallard, M. A., and Williams, A. L. 1966. Studies on gram-negative bacilli in a ward for new-born babies. *Med. J. Aust.*, **2**: 455–459.

Shooter, R. A., Taylor, G. W., Ellis, G., and Ross, J. P. 1956. Postoperative wound infections. *Surg. Gyn. & Obst.*, **103**: 257–263.

Shooter, R. A., Smith, M. A., Griffiths, J. D., Brown, Mary E. A., Williams, R. E. O., Rippon, Joan E., and Jevons, M. Patricia. 1958. Spread of staphylococci in a surgical ward. *Brit. Med. J.*, **1**: 607–613.

Smuness, W. 1966. The microbiological and epidemiological properties of infections caused by *Salmonella enteritidis*. *J. Hyg.*, **64**: 9–21.

Solberg, C. O. 1965. A study of carriers of *Staphylococcus aureus* with special regard to quantitative bacterial estimations. *Acta. Med. Scand.*, **178**, Supplement **436**: 1–96.

Speers, R., Jr., Bernard, H., O'Grady, F., and Shooter, R. A. 1965. Increased dispersal of skin bacteria into the air after shower-baths. *Lancet*, **1**: 478–480.

Szanton, V. L. 1957. Epidemic salmonellosis. A 30-month study of 80 cases of *Salmonella oranienburg* infection. *Pediatrics*, **20**: 794–808.

Thomas, C. G. A., and Griffiths, P. D. 1961. Air borne staphylococci and the control of hospital cross-infection. *Guy's Hosp. Rept.*, **110**: 76–86.

Thomas, C. G. A., Griffiths, P. D., and Huntsman, R. G. 1961. Coliform organisms and yeasts in the respiratory tract. *Guy's Hosp. Rept.*, **110**: 87–95.

Tully, J., Gaines, S., and Tigertt, W. D. 1963. Studies on infection and immunity in experimental typhoid fever. V. Respiratory challenge of chimpanzees with *Salmonella typhosa*. *J. Inf. Dis.*, **113**: 131–138.

Varela, G., and Olarte, J. 1942. Investigation de Salmonelas en las amigdales. *Rev. Inst. Salub. Enf. Trop. Mexico*, **3**: 289–292.

Varga, D. T., and White, A. 1961. Suppression of nasal skin and aerial staphylococci by nasal application of methicillin. *J. Clin. Invest.*, **40**: 2209–2214.

Walter, C. W. 1958. Environmental sepsis. *Mod. Hosp.* **91**: 69–78.

Walter, C. W., Kundsin, Ruth B., and Brubaker, Mary M. 1963. The incidence of airborne wound infection during operation. *J. Am. Med. Assoc.*, **186**: 908–913.

Watt, J., Wegman, M. E., Brown, O. W., Schliessman, D. J., Maupin, Elizabeth, and Hemphill, Emmarie C. 1958. Salmonellosis in a premature nursery unaccompanied by diarrheal disease. *Pediatrics*, **22**: 689–705.

Weiss, W., Eisenberg, G. M., Alexander, J. D., Jr., and Flippin, H. F. 1954. Klebsiella pulmonary disease. *Am. J. Med. Sci.,* **228:** 148–155.

Wells, W. F. 1935. Air-borne infection and sanitary air control. *J. Indust. Hyg. and Toxicol.,* **17:** 253–257.

Wells, W. F. 1940. Bactericidal irradiation of air. I. Physical factors. *J. Franklin Inst.,* **230:** 347–372.

Wells, W. F. 1944. Ray length in sanitary ventilation by bactericidal irradiation of air. *J. Franklin Inst.,* **238:** 185–193.

Wells, W. F. 1945. Circulation in sanitary ventilation by bactericidal irradiation of air. *J. Franklin Inst.,* **240:** 379–396.

White, A. 1961. Relation between quantitative nasal cultures and dissemination of staphylococci. *J. Lab. and Clin. Med.,* **58:** 273–277.

White, A. C., Smith, J., and Varga, D. T. 1964. Dissemination of staphylococci. *Arch. Intern. Med.,* **114:** 651–656.

Woodhall, B., Neill, R. G. and Dratz, H. M. 1949. Ultraviolet radiation as an adjunct in the control of post-operative neurosurgical infection; clinical experience 1938–1948. *Ann. Surg.,* **129:** 820–825.

18

DENTAL AEROSOLS

M. A. Mazzarella / D. D. Flynn

NAVAL MEDICAL RESEARCH UNIT NO. 1
UNIVERSITY OF CALIFORNIA, BERKELEY

RELATIONSHIPS BETWEEN AEROSOLS AND DENTAL PRACTICE

The phenomenon of airborne infection, or the act of transmitting disease from person to person by airborne, microbe-laden particles, has been recognized since the days of Lister and Pasteur. Langmuir (1961) discusses (1) the underlying principles of generation of microbial aerosols both from artificial and natural sources, (2) techniques of sampling microbial aerosols, (3) the great variation in the capacity of infected persons to contaminate their environment and to serve as dangerous carriers or spreaders of disease, (4) the extraordinary capacity of certain routine laboratory procedures to set up fine-particle aerosols and thus infect laboratory workers through inhalation, (5) the sites in the respiratory tract that may serve as portals of entry of infection and the great variation of the dosage required to infect at these different sites, and (6) the crucial importance of particle size in determining penetration and retention of inhaled particulates. Of extreme importance to the dentist is the fact that such particles can be expelled by dental procedures as well as the physiologic acts of breathing, talking, coughing, and sneezing.

In 1934, Wells stated that human oral microorganisms are disseminated in two ways: (1) directly, by expelling large droplets from the mouth and nose; and (2) indirectly, by "droplet nuclei" formed from small droplets that evaporate and remain bouyant in the air. He considered the first as a form of contact transmission, the second as true, airborne microbial transmission. He reasoned that each droplet size range has a different etiological significance—droplets initially over 100 microns (μ) in diameter (0.1 mm), which did not strike another person directly, quickly fell to the floor, and droplets with diameters smaller than 100 μ evaporated to form "droplet nuclei" which could remain airborne for many hours or days. Duguid (1945) interpreted this to mean larger droplets (the ones we see) caused dust-borne infections, whereas smaller droplets (the ones we don't see) produce true airborne infections. We now know that it is size, not source, that is important with respect to respiratory penetration.

Wells' "direct dissemination" is akin to the visible ballistic splatter of large particles easily seen in the dental operatory when the air turbine handpiece, water and air syringes, or rotating cups, brushes, and discs, are operated. The smaller particles are illustrative of true aerosols that are dispersed in the same way as larger droplets, but that result in the formation of invisible particles that remain floating in the environment for extended periods of time ("indirect dissemination"). Such invisible aerosols may consist of any combination of oral secretions, viable bacteria, enamel and dentin dust, and metallic particles from dental restorations.

The similarity of certain dental instruments and the mouth to laboratory apparatus used to disperse microbial aerosols is interesting (Figure 18-1). For example, the spinning disc (May, 1949) is directly comparable to sandpaper or sulci discs, rubber prophylaxis paste cups, or other circular devices that are rotated against dental structures and that disseminate oral detritus of varied composition and size. Dental hand instruments such as hoes, hatchets, scalers, and the electro-mechanical scaler, that operates at a near-ultrasonic vibratory speed at the working tip, simulate the flicking action of the vibrating reed (Rayner and Hurtig, 1954). The most intriguing similarity is that of the mouth and nasopharynx to the Wells (1941) refluxing atomizer, another widely used device for the laboratory production of bacterial or viral aerosols. The cuspidor used for patient expectorations, now fortunately going into disuse, can also be likened to the Wells atomizer because of the constant, fine mist created by the water swirling around the inner periphery of the bowl.

From the illustration in Figure 18-2 it is obvious that the breathing space of the dentist and his auxiliary personnel can be contaminated with saliva, bacteria, tooth and metallic debris, and sometimes hemorrhagic elements resulting from soft tissue manipulation. There seems

SANDPAPER DISC

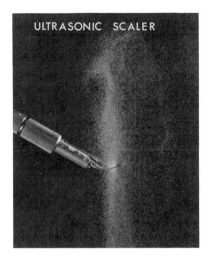

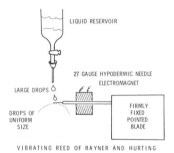

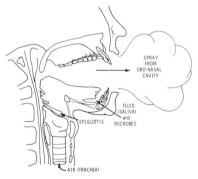

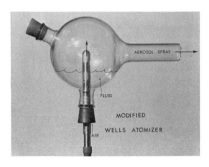

FIGURE 18-1. Similarity of certain dental instruments and the mouth to laboratory apparatus used to generate aerosols. Vibrating Reed (Rayner and Hurtig, 1954, Science, *120:* Fig. 1, p. 672). Reprinted with permission of the American Association for the Advancement of Science, Copyright 1954.

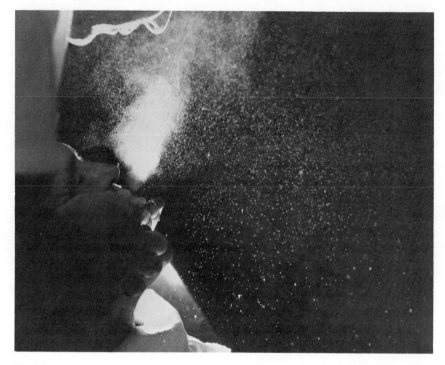

FIGURE 18-2. An aerosol generated during use of an air turbine handpiece. The illustration was not staged, but was backlighted and photographed during actual dental procedures.

little doubt that aerosols are easily generated, observed, and propagated within the dental operatory.

AEROSOLS GENERATED BY COMMON NASO-ORAL ACTIVITIES

Duguid (1945) delineated the sites of origin and enumerated the number of microbial droplets formed during expiratory activities such as breathing, talking, laughing, coughing, and sneezing, and studied the effects of varying the intensities of these activities. He seeded *Bacillus prodigiosis* into the oral cavity of patients, and used the number of viable cells recovered from the vicinity of the mouth to indicate the extent of dissemination of particles. Although he sampled at short distances (3 to 6 inches) for short periods of time (60 to 90 seconds), he recovered numerous organisms in both large and small droplet form; the recovery number increased in proportion to increase in vigor of oral physical activity of the subject. Duguid concluded that nearly all droplets originate from the front of the mouth as opposed to the nose or throat, and therefore assumed that the problem of airborne infection depended only on the pathogens present in the oral secretions.

In 1947, duBuy, Arnold, and Olson, using only those lactobacilli present in the patients' mouths, attempted to determine whether oral microbes were spread by a direct airborne route, or indirectly by "suspension and resuspension in air." When they studied dissemination of lactobacilli by the act of speaking, no viable bacteria were recovered from the atmosphere. Nebulized lactobacilli, however, lived for only a short time in the air and did not travel far from the site of origin. Viability varied with composition of the suspending fluid from which the microorganisms were aerosolized. No viable bacteria were detected (using selective growth media) if air samples were dispersed from fluids containing less than 10^6 or 10^7 microorganisms/ml.

These results were less dramatic than Duguid's, probably because duBuy *et al.* (1947) sampled at greater distances from the mouth with longer periods elapsing between samples, and used a less nutritious sampling medium than Duguid. Data offered by duBuy *et al.* included only a small fraction of the actual microbial aerosol dispersed by dental instruments. Recent investigators have used more selective media, sampled at greater distances from sites of bacterial dispersion, and sampled for longer time intervals after dispersion has started. In addition, their primary concern has been the dispersion of material resulting from dental instrumentation rather than with dispersal arising from physiological acts such as talking, coughing, or sneezing.

AEROSOLS OF DENTAL ORIGIN

The propagation of aerosols generated by dental procedures and containing particles less than 5 μ in diameter had not been investigated before 1947. Such aerosols are capable of invading the environment around a patient's oral cavity, as well as the air of an entire dental suite (Burton and Miller, 1965; Miller *et al.*, 1965). One of the first dentally oriented articles which questioned dental office air hygiene was that of Hoffman (1958). He concluded that the dentist carries out a number of procedures which involve real risks of aerial transmission of disease.

Two major sources of air contamination resulting from the operation of dental instruments and equipment are: (1) air-water-oil spray from the air turbine handpiece, and (2) oral microbes, tooth fragments, metallic particles wrenched from existing dental restorations, and sometimes hemorrhagic components arising from soft tissue manipulations— disseminated by either high-speed or conventional handpieces (Grundy,

1967). All sources will ultimately yield a microbial dispersion. It has been pointed out by Miller *et al.* (1965), based on work performed at the Naval Biological Laboratory and in a private dental office, that the amount and size of airborne particles varies according to the dental procedure (Table 18-1). A brief summary of pertinent research follows.

Table 18-1 Summary of concentration and particle size distribution of colony-forming units (CFU) in aerosols generated in mouths of patients during selected phases of dental treatment.

Operation	Field condition	CFU/ft³ Low	CFU/ft³ High	Median particle diameter	Percentage of particles less than 5 μ diameter
None (environmental air)	None	7.3	29.4	5.9	40
Polishing teeth using prophy cup and pumice	Wet	35.0	1,488.0	4.9	45
Drying of teeth using air syringe	Wet	98.0	944.0	1.3	90
Cutting of Preparation in Tooth, Using:					
Air turbine with air	Dry	12.5	42.5	7.0	35
spray	Wet	16.2	85.3	6.7	33
Air turbine and water spray 557 bur	Wet	40.0	212.0	4.8	52
Contra-angle 557 bur	Variable wet and dry	23.0	194.0	5.5	42
"Finishing" restoration; contra-angle with $\frac{5}{8}''$	Dry	32.0	86.0	5.1	40
cuttle disk	Wet	4,679.0	6,537.0	2.7	80
Polishing restoration; straight hand piece	Dry	18.0	226.0	4.2	55
with Robinson bristle brush and pumice	Wet	1.1×10^4	1.4×10^5	6.5	30

(From Miller, Burton, and Spore, 1965, Aerosols Produced by Dental Instrumentation. In *A Symposium on Aerobiology, 1963*. (R. L. Dimmick, Ed.), p. 100. Nav. Biol. Lab., Naval Supply Center, Oakland, Calif.)

Kazantzis (1963) assayed the calcium and oil droplet content of air samples taken from the vicinity of a patient's mouth during the operation of an air-turbine handpiece with various water-flow and oil-drip feed rates. A fine mist of oil droplets was visible in a beam of light close to the turbine head. Oil droplet counts varied from a few to 180/ml;

their average diameter was 1.0 μ. Solid particles varied from a few to 170/ml, the average diameter being 1.5 μ. A concentration of 0.62 mg of oil and 0.047 mg of calcium per cubic meter of air was found close to the patient's face—and 0.3 mg of oil per cubic meter of air at a 3-ft distance. For comparison, the average daily oil droplet concentration in the atmosphere of a machine shop where semi-automatic lathes operated was 0.5 to 1.0 mg oil per cubic meter of air. Kazantzis stated that "the introduction of oil into the lung is by no means innocuous," and he further discussed possible pulmonary sequelae of inhaled mineral oil droplets. He urged the use of minimal amounts of purified vegetable oils free from impurities (e.g., fatty acids) as a lubricant for the air turbine handpiece. However, he did not consider the inhalation of solid particles produced by the air-turbine handpiece to be pathogenic in the concentrations he encountered.

Using an oil-flow rate of 33 drops/min and water flow rate of 2.6 ml/min, Kazantzis found that the quantity of respirable dust produced by the high speed drill was small, especially when it was used with a water jet. Samples taken while a conventional handpiece was in operation contained a very small concentration of solid particles; when the cavity preparation was blown clean by an air blast, a much higher concentration of particles was found than at any time during the use of the air turbine handpiece.

Kazantzis recovered both *Staphylococcus albus* and *Streptococcus viridans* on blood agar plates during operation of either the air turbine or conventional handpieces. Greater concentrations of oral airborne microorganisms and debris were recovered during use of the air turbine as compared to the conventional handpiece. However, he reasoned that because of the shortened working time with the turbine, there was "no increased risk to the operator of bacterial contamination associated with the use of the high speed drill."

Pistocco and Bowers (1962) inoculated *Serratia marcescens* into the water supply of an air turbine, operated the handpiece in the open, and recovered tracer organisms up to 6 feet from the handpiece; the maximum concentration was at 18 inches.

When Miller *et al.* (1965) seeded a denture with *S. marcescens,* they not only recovered tracer organisms from the oral cavity of the dental laboratory technician who polished the denture, but found that the air-conditioning system had distributed the organism from the laboratory throughout the entire dental suite, including the waiting room.

Madden and Hausler (1963) seeded Class II cavity preparations with a known inoculum of *S. marcescens* and then operated both the air-turbine and conventional handpiece. The air turbine handpiece was powered by air pressure at 30 psi, the water-flow rate was adjusted to 2.0 ml/min and the oil-flow rate to 15 drops/min. The conventional handpiece was operated dry. They found that the number of airborne bacteria decreased as distance from the cutting surface to the point of sampling increased. At sampling distances similar to those used by Kazantzis, who found little difference in the concentration of "aerial contaminants" whether generated by a high- or conventional-speed handpiece, Madden and Hausler detected much greater concentrations of the tracer organism when the air-turbine handpiece was used. Of the particles in the air sample taken while the air turbine was running, 84.5% were less than 5 μ diameter; 47.0% of the particles from the air sample generated by the conventional handpiece were of such size. They concluded that the operation of the air turbine handpiece created a potential hazard for the patient and practitioner.

In parallel experiments performed in a controlled environment room, Hausler and Madden (1964) found that air turbine handpieces generate an aerosol 2 to 20 times the magnitude of a conventional handpiece, and that 60 to 65% of the particles in the aerosol were small enough to penetrate the alveolar spaces. The number of smaller-sized particles increased as the volume of water spray from the air-turbine handpiece increased.

The dispersion of S. marcescens during operation of the air turbine or conventional handpieces without air-water spray, in a controlled environment room were compared (Madden and Hausler, 1966). S. marcescens was not recovered after operation of the conventional handpiece and only 11% of the particles generated by the air turbine were less than 5 μ diameter. The investigators hypothesized that the difference between this and prior studies was the water spray used previously, and "that the single most important factor in aerosol generation by the air turbine is the water flow rate." They also found the number of viable bacterial particles decreased rapidly as the tooth-to-sampler distance was increased. However, the relative number of particles smaller than 5 μ was found to increase with distance from the operative site. This observation is compatible with Wells' and Duguid's theories on droplet nuclei formation and suspension in air.

Miller et al. (1965) reported that a wet field consistently produced aerosols of higher microbial content than did the same procedure performed in a field dried by cotton rolls or a rubber dam. Stevens' (1963) experiments showed that, whereas the rubber dam lowered the bacterial count, concomitant use of an oral evacuator gave varied results.

Brown (1965) claimed that use of a rubber dam and an efficient oral evacuation system would reduce the concentration of bacterial aerosols produced by the air turbine to a level as low as that produced by the conventional handpiece. He found the greatest concentration of airborne bacteria during Class II cavity preparations, and he related the quantity of bacteria disseminated into the air to the patient's oral hygiene, the amount of carious dentin cut, and to the length of the cutting time.

Shay and Clendenin (1963) attempted to evaluate possible dangers to patients when dental prophylaxes were performed by dentists who were known carriers of coagulase-positive staphylococci of known phage patterns. They studied 10 different dentist-carriers who performed prophylaxes through a total of 19 clinical contacts. As the dentists worked, air samples were taken with the Decker slit sampler. In two cases identifiable staphylococci were found to be transmitted to the environment of the dental operatory. One microbial isolate originated from one of the dentist-carriers; the other originated from a patient-carrier. This patient harbored coagulase-positive staphylococci of a distinct phage pattern during his entire dental treatment period; his dentist carried staphylococci of another typeable pattern which was not recovered from the air.

To determine if oral pathogens could be dissminated into the air by the spray for a high speed handpiece, an air turbine was run for 60 sec at full speed (30 psi without cutting) over the occlusal surface of the left mandibular first molar of 5 patients who had active pulmonary tuberculosis and positive sputums (Belting, Haberfelde, and Juhl, 1964). For each of the 5 patients, 3 experimental "runs" were made: (1) air-water spray on, (2) air-water spray off, and (3) no air turbine, but with the mouthspray attachment substituted and operated at 10 psi. For each experiment, three pairs of petri dishes (one of each pair containing phosphate buffer solution for culturing *Mycobacterium tuberculosis* and the other containing blood agar) were placed: (1) 6 inches in front of the patient at chin level, (2) on the bracket table 2 feet in front of the patient, and (3) on the instrument cabinet 4 feet to the right front of the patient. They found *Mycobacterium tuberculosis* was disseminated into the area of the operatory in which the dentist and his assistant would be working. Forty-two of the 45 blood agar plates contained colonies of various oral microorganisms in addition to 2 identified as beta-hemolytic streptococci; results are summarized in Table 18-2.

Belting, Haberfelde, and Juhl concluded that "unless protective measures are taken, the dentist and his assistant are exposed to a serious health hazard when operating with an air rotor on patients having pathogens in their oral flora." There can be no argument with this viewpoint.

Table 18-2   Colonies of *Mycobacterium tuberculosis* obtained after 15-week incubation from aerosol produced by dental air rotor and mouthwash spray operated in mouths of tuberculosis patients for one minute.

Patient	Air rotor-water on 6 in.	2 ft	4 ft	Air-rotor-water off 6 in.	2 ft	4 ft	Mouthwash spray 6 in.	2 ft	4 ft	Total
1	0	+ 2	0	+15	+ 3	+17	+ 1	0	+2	40
2	0	0	+ 2	0	0	0	0	+1	+1	4
3	0	0	+ 1	0	+25	+ 9	+ 3	+2	+1	41
4	+2	+15	+ 2	+ 1	0	+ 2	0	+5	+1	28
5	+1	0	+ 5	+ 2	+ 1	0	+25	+1	0	35
Total	3	17	10	18	29	28	29	9	5	148

(From Belting, Haberfelde, and Juhl, *J. Amer. Dental Assoc.*, **68**: 650, May 1964. Reprinted with permission of the American Dental Association.)

RECENT DEVELOPMENTS WITH CONTROLLED ENVIRONMENTS

The application of aerobiologic principles to certain problems of the dental profession is rapidly opening the door to a new and exciting realm of dental research. Unlike controlled laboratory procedures, variables abound in most of the clinical studies that are reported in the dental aerosol literature. For example, some investigators report their findings on dental aerosols generated by the air-turbine handpiece with an air-water-oil spray, yet the water- and oil-flow rates, and the air pressure used to drive the turbines, vary according to the investigator; some air turbines use no oil. Clinical environments differ widely in bacterial background as does the equipment used to sample the environments.

Madden and Hausler (1966) realizing the difficulty in trying to establish baseline data on bacterial backgrounds in an open clinic, conducted their later studies in a controlled-environment room. To further isolate the clinical patient from the surrounding environment, a human aerosol test chamber (HATCh) has been designed and constructed for use in an applied research program of the Materials and Technology Branch of the U.S. Public Health Service's Dental Health Center in San Francisco (Micik *et al.*, 1967). The test chamber, large enough to house a patient, is fitted with a viewing window and rubber gloves to permit dental treatment from a position outside the test chamber. Sterile air is introduced into the chamber at a flow rate of 4 ft³/min and withdrawn through 4 samplers (2 Andersen sieve samplers and 2 slit samplers) containing heart infusion agar. To further discriminate against the environmental microbial background and to delineate human microbial dissemination more precisely, the chamber was placed in an 8 × 10 ft "clean room" (Micik et al., 1969). Using these chambers, the composition of dental aerosols from patients undergoing treatment can be studied without interference from the microbial background of the external environment. Inhibitors that previously had to be added to the sampling media to suppress undesirable background bacterial growth can also be eliminated.

Some results obtained in the HATCh (corroborating earlier work) are shown in Table 18-3. Oral physical activities produced few airborne microbes with the exception of coughing, sneezing, and toothbrushing, all of which produce significant, and potentially dangerous, numbers

Table 18-3 Common naso-oral activities or dental procedures. Characteristics of bacterial aerosols generated from 11 patients in human aerosol test chamber (HATCh).

Activities or procedures	No. of tests	Rate of production			Percent $\leqq 5\,\mu$
		Colony-forming units per minute			
		Range		Median	
		Low	High		
Breathe oral and nasal quiet and heavy	53	0	9	2	...
Whisper, speak, shout, yawn, gargle, eat, whistle, clear throat.	99	0	72	10	...
Hiss, cough	30	2	1,200	33	30
Sneeze	17	6	2,700	770	50
Brush teeth	16	49	>70,000	2,100	54
Examination scaling	16	2	61	8	...
Prophylaxis (pumice cup)	13	8	312	78	43
High-speed instrumentation (Air-turbine Handpiece) with air coolant	23	0	1,900	12	80
with water coolant	10	24	3,400	500	95
Use of triplex syringe water stream	9	6	30	16	...
air	16	20	4,100	120	65
spray	9	40	>20,000	2,000	94
Polishing restoration (bristle brush)	12	28	>80,000	3,300	55

(Data from Micik, et al. 1967).

of microbes in the air. Similar numbers were produced by the rotating prophylaxis cup and the bristle brush whereas very high levels were produced by the air-water jet spray and the air turbine handpiece. Commencement of dental treatment is accompanied by an immediate increase in airborne microbial counts, just as Miller *et al.* (1965) and Madden and Hausler (1966) demonstrated. This increase probably arises from the dissemination of oral microbes rather than arising from dust residues in the office spaces.

In recent work (Mazzarella, unpublished data), it has been possible to recover exclusive, oral inhabitants such as *Streptococcus salivarus, Streptococcus mitis,* and a PPLO (pleuropneumonia-like organisms) tentatively identified as *Mycoplasma salivarum,* from aerosols disseminated into the dental operatory. In spite of the fact that meningococci were sensitive to the sampling procedures employed, the organism was recovered from the air in close proximity to known carriers receiving dental treatment. The hazard to dental personnel becomes great with an increased oral carriage rate of a dangerous microbe or the occurrance of an epidemic.

Mazzarella and Hancock (1968) employed experimental laboratory techniques simulating actual clinical use of the ultrasonic scaler in patient treatment, and found that within 10 minutes tracer organisms had spread throughout the operatory, and could be assayed in the Andersen sampler for as long as 40 minutes. Bacterial particles less than 5 μ diameter were recovered in numbers greater than 5,000/ft^3 of air. This obviously represents a potential hazard not only to the dentist but also to subsequent patients—especially if one considers that 1 pathogen/ft^3 of air is considered to be able to cause disease in a susceptible host.

Finally, if size-analysis of particles disseminated from sources of potential pathogens (e.g., the oral cavity, the skin, instruments) is done superficially, and if mean sizes are reported without considering possible variance, the true hazard may be grossly underestimated. If, in an aerosol of a mean particle size, for example, of 20 μ, there exists only 1 airborne pathogen capable of alveolar penetration, and it is breathed by a person, then to that person the infectiousness of the aerosol was 100%. It is of little comfort to the patient debilitated by pneumonia to assure him that he has suffered an extremely rare accident which everyone regrets deeply.

THE FATE OF INHALED PARTICLES

In any discussion of hazards associated with aerosols generated in the dental operatory, we must consider the capacity of airborne particles to by-pass the defenses of the upper respiratory tract and penetrate to the lung alveoli. Early observations of Lister, Hildebrandt, and Owens that demonstrate the remarkable cleansing capacity of the upper respiratory tract, and consequent protection against invasion of the lungs by inhaled dustborne bacteria, have been summarized by Hatch and Gross (1964).

Briefly, the largest particles are filtered out by hairs within the nares; others excite the cough reflex and are expelled with mucus. The mechanical action of the ciliated epithelium, lining a major portion of the respiratory tract, escalates particles on a mucus blanket to the posterior nasopharynx, thence to the oropharynx for elimination by swallowing. Finally, particles are also ingested by macrophages of the respiratory tract.

Bloomfield (1919–1921) studied the pathogenesis of infections via the upper respiratory portals of entry by swabbing bacterial cultures onto the tongue and nasal mucosa, and into the tonsillar crypts. Although he observed an immediate decrease in recoverable viable bacteria, some tracer organisms were recovered during the next 24 hours. In no cases did he recover viable tracer bacteria after 24 hours. Bloomfield concluded that disappearance of "seeded" organisms was due to antibacterial effects of saliva and other oral secretions, plus mechanical flushing procedures of the mouth and nose, and that bacteria in general, when introduced into the upper air passages are not likely to colonize and live there indefinitely. It thus appeared that the oronasal system is able to trap and eliminate the larger particles in the air we breathe.

Wells et al. (1948) studied "droplet nuclei" infection of rabbits exposed to aerosols of tubercle bacilli. The average number of tubercles in lungs of rabbits exposed to small particles was 16 times greater than in rabbits that breathed large particles. They concluded that the larger particles could be strained out and eliminated by the upper respiratory tract.

Whereas particles greater than 5 μ in size usually are eliminated by respiratory defences before they penetrate past the larger bronchioles, maximal alveolar retention occurs with particles from 0.5 to 5 μ diameter. Hatch and Gross (1964) state that (1) penetration into the pulmonary air spaces is nearly zero for particles of 10 μ, but reaches a maximum at and below 1 μ; and (2) percentage deposition of particles which have penetrated to the air spaces is maximum for particles between 1 and 2 μ. Note, however, that the probability for deposition of particles 1 μ and less decreases until particle sizes of 0.5 to 0.25 μ are reached, when the probability for deposition again rises. It should be understood that those particles in the 1- to 2-μ range which do reach the alveolar spaces settle out by gravity, whereas particles 0.5 μ and less are precipitated by forces of diffusion, the diffusive forces increasing as particle size further diminishes.

As a result of filtration by the upper respiratory regions, one can see that all inhaled particles from a dental aerosol are not deposited in the lungs. However, particles that do penetrate to the air sacs become physiologically important. Lungs can hold both living and inanimate particles in the air spaces until they are either phagocytized or acted upon by absorptive processes that introduce them to the general circulation. In this manner, the stage is set for infections, allergic, and immunologic reactions, not only in the lung, but throughout the entire body (Bensch et al., 1967; Nungester, 1951). Inhaled microbes or antigenic substances of less than 0.5 μ diameter, if not immediately absorbed into the circulatory system, may remain suspended and be washed out more readily than wet ones.

Palm et al. (1956) speculated that greater deposition of bacterial particles occurs than deposition of inorganic or inanimate particles, and that bacterial particles might increase in size within the moist respiratory air to cause increased retention. This speculation was, in fact, confirmed by Meyers et al. (1961), who showed that "seeded" bacteria could be recovered from the respiratory tract for as long as 120 minutes after a 5-minute exposure to a bacterial aerosol of 5,000 cells. They believed that more organisms were deposited than could be explained from findings with inorganic materials of similar size.

Ostrom et al. (1958) reported recovery of 0.1% of the inhaled dose if subjects were exposed to more than 1,000 cells and if samples were taken immediately after exposure. When the dose was increased to 2,000 cells, microorganisms were recovered from at least 50% of the persons sampled. The microbial concentrations disseminated by common dental procedures can be seen from Tables 18-1 and 18-3 to far exceed concentrations used in these experiments.

Since recovery of inhaled bacteria was greater in nasal mucus samples than in oropharyngeal rinse or oral swab samples, Meyers et al. (1963) postulated a "nasal depot" for organisms originally located in other portions of the upper respiratory tract. Wright (1961), however, has estimated the cyclic clearing and renewal of the nasal mucus blanket to repeat every 10 to 20 minutes, and Bloomfield (1919–1921) postulated that normal respiratory mucosa and secretions were highly bactericidal to foreign microorganisms. Jay et al. (1933) failed to implant B. acidophilus into the mouths of caries-free children (who normally did not carry this organism), using acidophilus milk prepared with strains isolated from the saliva of caries-rampant patients. In contrast, Ames and Nungester (1949) observed no clearance from nasal turbinates of animals

4 hours after exposure to *Bacillus globigii,* and found that only 25 to 50% of the cells initially recoverable had disappeared from the naso-pharynx within this time period. Meyers *et al.* (1963) recovered viable tracer organisms from the nasal mucus for 180 minutes after inhalation. Nungester (1951) feels that particles may be deposited in the nasal passages during both inspiration and expiration. Whereas foreign mi-crobes are eventually eliminated, they can certainly persist in the re-spiratory passages for periods too long for complacency on the part of the dental profession.

Since concentration and dosage of inhaled microbial particles are also related to infection of the host, two questions then arise: How large an inhalatory dose is required? And how often can a small inhaled dose be repeated before infection, sensitization, or toxicity of dental personnel becomes evident? Wells *et al.* (1948) demonstrated that in-herited natural resistance against infection can be overwhelmed by single massive inhaled doses of tubercle bacilli. Conversely, Goldberg *et al.* (1954) have shown that the same numbers of pathogenic microorganisms were required to cause disease, whether administered at one time in a large infective dose, or in smaller time-spaced increments. It is not very difficult to translate these laboratory results in terms of the dentist working upon patient after patient with hidden clinical or sub-clinical infections, inadvertently and repeatedly inoculating himself with micro-organisms of pathogenic potential.

In a survey of respiratory disease in dentists and their assistants conducted at Great Lakes Naval Training Station, the weekly prevalence of acute respiratory disease was documented for the dentists, dental assistants, and patients at 6 dental clinics (Carter and Seal, 1953). One clinic treated only newly arrived recruits; 4 clinics treated recruits who had completed at least 4 weeks of training; the sixth clinic treated only personnel who were permanently stationed at Great Lakes (this last group may be assumed to be "acclimated" to the area). The highest prevalence of respiratory disease was found among dental personnel treating new recruits; a lesser amount of respiratory disease existed among dental personnel treating the other two groups (Table 18-4). In evaluating these results, one should keep in mind that (1) the dental assistants also acted as secretaries for the dentists, and therefore did not spend as much time within the area of aerosol generation as they now do; and (2) this survey was conducted before the advent of high-speed operative techniques, so that aerosol magnitude and dispersal was not as great as it is today.

An interesting study was performed by Bensch, Dominguez, and Liebow (1967) who introduced albumin and globulin molecules into the

Table 18-4   Acute respiratory infections in dentists and their technical assistants, 1951–1952.

Type of patients	Man months of observation						Percent with respiratory infections					
	Nov.	Dec.	Jan.	Feb.	Mar.	Apr.	Nov.	Dec.	Jan.	Feb.	Mar.	Apr.
Dentists Treating:												
Recruits[a]	49	114	94	68	152	41	18.4	13.1	13.8	11.8	9.2	9.7
Non Recruits	31	59	21	28	22	7	16.1	18.6	14.3	3.6	9.1	0.0
New Recruits[b]	23	35	13	15	18	7	26.1	25.7	26.7	23.1	0.0	14.3
Technicians Assisting with:												
Recruits[a]	54	119	94	66	146	38	33.3	16.8	6.4	3.0	5.5	5.3
Non Recruits	34	62	25	30	30	10	20.6	14.5	16.0	13.3	6.7	0.0
New Recruits[b]	20	33	3	12	18	8	45.0	21.2	0.0	25.0	5.5	0.0

[a] Mostly recruits having four or more weeks of service.
[b] Mostly new recruits with less than four weeks of service.
(From Carter and Seal, 1953. Reprinted with permission of the Officer in Charge, Naval Medical Research Unit No. 4.)

alveoli of dogs to determine whether protein molecules could cross the alveolar membrane into the blood supply and yet remain in an antigenically intact state. Human serum protein (15 ml), labeled with I^{131}, was instilled into the lungs of 15 experimental animals; control animals were injected intravenously. Blood samples taken at regular intervals from both groups, beginning 15 minutes post-treatment and continuing for 7 days, showed similar percentages of radioactivity. Exact measurement of instilled protein absorption was complicated by factors such as: (1) loss of lung-instilled material to ciliary action, (2) continuing absorption of the protein with time, (3) individual variation in turnover rate of heterologous serum, and (4) difficulty of achieving maximum precipitation of absorbed intact protein by the precipitation technique. Therefore, the values reported for lung absorption can be considered minimal.

Because these results showed that protein molecules could be introduced into the circulatory system via the alveolar membrane and remain antigenically intact, Bensch *et al.* felt that the lung can absorb large quantities of intact long-chain protein molecules. It is evident that whatever proteinaceous material the dentist breathes, in the form of small particles, will permit the introduction of antigenic substances into the general circulation and produce constant immunogenic challenge.

Booker and co-workers (1967) studied lung elimination of 5-μ particles of Cr^{51}-labeled polystyrene. Each subject inhaled a single breath of aerosol suspension (140 to 500 ml), and the particles were traced by scintillation counters. Elimination of particles was found to occur in two phases. In the first phase, which lasted 10 to 20 hours, elimination of only a fraction of the total inhalant occurred; there was no evidence of a rapid initial removal as reported by Albert and Arnett (1955) and Morrow (1966); in fact, the elimination rate was very slow during the first 3 hours. During the second phase of elimination, half-life values of 150 to 300 days were calculated. [The International Commission of Radiological Protection (1959) adopted 120 days for insoluble particles.] Booker *et al.* concluded that the first phase of elimination was the result of respiratory ciliary action and the second phase represented a much slower alveolar elimination. The length of time for particle clearance is noteworthy and raises the question of the potential for pneumonosis and allergenic effects from inhalation of that fraction of any dental aerosol which contains similarly sized particles.

Green and Kass (1966) have studied the mechanisms of resistance to bacterial infections by exposing mice to aerosols of radioactive bacteria, and comparing the number of viable organisms recovered to the amount of radiation registered. Their data suggest that bacteria were killed before they could be eliminated. Fluorescence techniques showed that phagocytosis had occurred in the alveoli; however, hours later the same phagocytes could be found in the bronchial mucus being carried out of the lungs. They concluded in part that the effectiveness of lung clearance varied with the species of bacteria, the effect of exogenous stress (hypoxia, alcohol, starvation, and cortisone) and the antibacterial (phagocytic) activity of the host.

To summarize, the pertinent points set forth in the preceding pages are:

1. Most important, clinical experimentation to date reports conservative figures for recovery of airborne microorganisms, and for recovery of bacteria from respiratory tracts.
2. Dental aerosols contain bacterial and inanimate particles which run the gamut of size, from the large pieces of debris we see flying out of a patient's mouth to the invisible, submicron particle which can penetrate our upper respiratory tract and ultimately invade our regional lymph nodes or circulatory system.
3. Human and animal experiments have shown that living particles probably enlarge by surrounding themselves with a fluid coat, thus changing their aerodynamic properties; bacteria lodge in the respiratory regions more easily than inanimate particles.
4. Particulate tracer experiments can delineate the dose required for lung penetration and accurately measure the duration of such particles within the tissues.

The following section will offer some suggestions which ultimately may afford us better protection from dental aerosol bombardment and insufflation.

RECOMMENDATIONS FOR REDUCTION OF AEROSOL HAZARDS TO DENTAL PERSONNEL

We shall now examine methods to control potential respiratory hazards within dental operatories. There are two general approaches to solution of the problem: (1) control of the source of particle dissemination, and (2) control of the degree of dissemination as well as length of time particles remain airborne.

The concentration of aerosolized particulates can be reduced at the source by (1) reducing initiation of dental aerosols, or (2) preventing the inhalation of dental debris and microbial particles within the immediate working area. However, since a certain portion of the aerosol will inevitably escape the immediate working area and enter the general office environment, our second approach must be toward development of methods to reduce the microbial population and particulate concentration in the air rapidly enough so that distribution throughout an entire dental suite will not occur.

Mohammed and Manhold (1964) and Manhold and Manhold (1963) advocated that the patient rinse thoroughly with a mouthwash before dental operative procedures, to reduce the number of oral microbes. However, previous work by Crowley and Rickert (1937), Bloomfield (1919–1921), and Brown and Cruickshank (1947) on ecological relationships in the oral cavity showed that microbes return in greater numbers shortly after they have been reduced by various physiological procedures, or after use of a mouthwash. One would have to work rapidly, after an oral rinse, to use this as a feasible way to reduce the microbial count in dentally generated aerosols.

Recently, many investigators have attempted to prevent inhalation of aerosolized particles by shielding either the patient's oral cavity, or the dentist's face. Docking and Amies (1948) stated that "for the mutual protection of the patient and himself, it is most desirable that the dental surgeon should wear a suitable form of face mask." They suggested, as an adjunct to the conventional hospital gauze mask, the dentist wear a clear, plastic face-piece hinged to a metal headband that may be swung upward if desired. Ferguson (1960) advocated wearing spectacles with lenses large enough to adequately protect the eyes from aerosol particles. Kortsch (1964) advocated use of a plastic face shield to "guard

the dentist's eyes" and protect "the nose from dust and droplets."
Travaglini and Larato (1965) designed a disposable dental face-mask
with an attached plastic eye shield that could be worn by both dentist
and patient. The dentist's mask covered both his nose and mouth,
whereas the patient's mask covered only his nose, permitting access to
the oral cavity. Recently, Calderone (1966) designed a kidney-shaped,
transparent plastic shield which an assistant held against the patient's
upper lip just below the nose. This "barrier," 8 inches long and $4\frac{1}{2}$ inches
wide, was intended to stop most of the visible splatter emerging from
the patient's mouth. All of these devices helped reduce the potential
hazard from large particles, but probably were of little value against
true aerosols (1 to 5 μ particles).

Guyton, Buchanan, and Lense (1956) used *Bacillus subtilis var.
globigii* aerosols to determine the filtering efficiency of the following
three types of contagion masks: (1) a thin, absorbent cotton layer
backed by a single-thickness gauze, (2) a typical tie-on surgical mask
composed of four layers of gauze, and (3) a mask made of a single
sheet of wax-impregnated paper and held on the head by elastic ear
loops. Comparing these masks with commercial dust respirators, they
found the typical hospital surgical mask was only 17.6% efficient as
compared to 38% and 39.6% for the other two, whereas commercial
dust respirators evinced filtering efficiencies from 97.35% to 99.993%.
Guyton *et al.* concluded that these three types of contagion masks offered
poor respiratory protection, especially against microorganisms in the particle size range of 1 to 5 μ.

Guyton and Decker (1963), again using *B. subtilis var. globigii* aerosols, evaluated the filtering efficiencies of five newer masks. Four of
the masks were 71% to 95% efficient; the fifth, a nondisposable type
made by placing a glass fiber filter between two layers of 20-mesh screen
(Deseret mask), was 99.1% efficient. An interesting sidelight to their
study included an evaluation of the effects of exhaled CO_2 and moisture
on filtration efficiency. Twelve masks of two different types (which contained filter media representative of all five types) were worn for periods
up to 8 hours. The masks were then evaluated and the "results," when
compared with those of unused masks taken from the same production
lot, showed no significant decrease in efficiency.

Ford and Peterson (1963), in testing eleven masks, almost duplicated
Guyton and Decker's results with the tie-on cotton gauze mask (15.50%
efficient) and nondisposable (Deseret) glass fiber filter mask (99.19%
efficient). Nicholes (1964) also found glass fiber filters to be very efficient—so much so that he suggested a compromise between filter thickness (1.5 to 2.0 mm) and filtering efficiency (95% to 98%) to make

a mask through which one could breathe freely, yet be assured of a high degree of aerosol filtration. Even with this filter thickness, however, some practitioners have encountered breathing difficulties and prefer a pliable styrene foam mask which is inexpensive and disposable.

The glass-fiber filter mask and the styrene foam mask were recently tested by M. T. Hatch (1968) by methods similar to Nicholes'. A known aerosol concentration of *Neisseria meningitidis* was passed through the masks at a flow rate of 1 ft^3/min and into an Andersen sampler. The fiberglass mask was 95% to 98% efficient whereas aerosol particles penetrated the foam mask and some reached the sixth stage of the Andersen sampler (particles < 1 μ diameter). This sampling rate (28.6 liters/min), however, is almost three times the magnitude of the average human breathing rate (10 to 12 liters/min). When the same types of masks were tested at 12 liters/min (three masks in parallel to maintain 1 ft^3/min in the Andersen sampler), the foam mask did impact large particles (larger than 3 μ), but permitted the passage of particles less than 3 μ in size.

Using Andersen samplers operating at a flow rate comparable to the human breathing rate, Miller, Leong, Micik, and Ryge (1968) tested 16 types of face masks for bacterial aerosol filtering efficiency, and found the two most efficient masks to be the Deseret and 3M (Minnesota Mining and Manufacturing Co.) Filtron® masks.

Another way to reduce hazards from aerosols produced by dental operations is to reduce the concentration of particulate matter in the operatory by (1) using sufficient room-air changes per unit time to reduce particle concentration to an acceptable minimum, (2) providing better air filtration and using electrostatic precipitators, (3) adjusting room temperature and relative humidity, as determined by laboratory experimentation, to provide maximal disinfection of circulating microbes, (4 room air sterilization by ultraviolet (UV) irradiation, and (5) spraying the air with various chemical disinfectants.

Elford and van den Ende (1942) introduced ozone (0.05 to 0.06 parts per million) into the atmosphere of a small room (750 ft^3) for the purpose of determining its antibacterial ability. They concluded that ozone concentrations high enough to provide effective protection against airborne bacteria would be destructive to lung tissue.

Williamson and Gotaas (1942) studied the effects of aerosolizing (1) 70% solutions of ethyl alcohol, (2) water solutions of chlorine and chloramine, (3) sodium hypochlorite solutions, and (4) resorcinol-glycerine combinations into a room previously sprayed with bacteria. Their results indicated that the germicidal aerosol is not as satisfactory as UV irradiation.

Elford and van den Ende (1945) evaluated the disinfecting action of sprayed hypochlorous acid gas and sodium hypochlorite solutions against microbial aerosols. The method did not appear to be practical because of the necessity for a high relative humidity (70 to 90%), the possible respiratory irritation from too high an aerial concentration of chemical, and the corrosive effect of the chemical on certain metals.

Ultraviolet radiation has potential for control of airborne microbes. This technique however, is effective only when organisms are directly irradiated, so skillful placement of UV lamps of high intensity is required. Application of UV air sanitation can be made in only three ways: (1) direct irradiation when rooms are unoccupied, (2) indirect irradiation of occupied rooms where occupants can be shielded from direct exposure, and (3) irradiation of air within an air supply duct (Jensen, 1965). In case 3, radiation intensity can be so great that all microbes are exposed while being carried through the ventilating duct. However, radiation inside the duct may not be effective in completely eliminating microbial particulates from the air of rooms being constantly contaminated because the input of sterilized air functions only to dilute the air already in the room and would not influence the propagation rate of microbial particulates.

Hart (1941) used direct UV radiation (2,537 Å) from bulbs placed $10\frac{1}{2}$ ft above the operating-room personnel. Throughout various hospital services, he achieved a reduction of infection rate from $\frac{1}{20}$ to $\frac{1}{100}$ of the previous levels. The report of the Committee on Sanitary Engineering of the National Research Council (1947) and Langmuir et al. (1948) stated that even though UV irradiation had a bactericidal effect on airborne microbes, it should be used in conjunction with other methods of reducing aerial contamination. The possibility that UV irradiation does not permanently injure dry bacteria was suggested by Dimmick (1960).

An effective approach, which can combine any or all of the aforementioned methods, would be the adoption of the "white" or "clean" room design used in industry. Special air-conditioning designs and operating-room techniques have been combined for use in the electronics, optical, pharmaceutical, and aerospace industries, for example, to control airborne microbial and particulate contamination. Aerosol concentration within such a room equilibrates at some (given) level inversely proportional to the circulation rate of air through the room. Normally, 20 air changes per hour are adquate to maintain an industrial "clean" room. Positive pressure is maintained within the room, and to provide optimal removal of particles, the air flow pattern should be "laminar," i.e., non-turbulent and without eddy currents that create "dead air" spaces in corners, nooks, and crannies. To create laminar flow, air enters through a wall or ceiling composed entirely of filters and leaves through the directly opposite entire wall or floor. Study of air changes required and flow patterns would have to be performed in a test operatory before this system could be adopted for the ordinary dental clinic.

Jensen (1967) used T1 *Escherichia coli B* bacteriophage suspensions to challenge the on-site effectiveness of air contamination control systems equipped with both UV irradiation and filtration sterilizers, and operated at flow rates between 600 to 18,000 ft^3/min. Jensen also showed that approximately 81% of the phage particles introduced into the systems were contained in 1- to 6-μ particles and that the 7 systems tested were from 99.8 to 99.99% effective in removing the airborne bacteriophage.

It is obvious from the foregoing that methods for attenuating (indeed, practically eliminating) microbial aerosols do exist. The incorporation of some of these techniques into the design of dental operatories and suites could provide the dentist, his personnel and, probably most important, the patient with high levels of protection from dental aerosols.

SUMMARY AND CONCLUSIONS

1. Dental procedures disseminate airborne particulate matter of varied concentration, size, and composition.
2. These dental aerosols can be assumed to be potentially infective, toxic, or antigenic.
3. Microbes or toxic products rendered airborne by dentistry can remain airborne for prolonged periods.
4. Recent work has shown that aerosols generated by the air turbine handpiece contained as much as 20 times the concentration of particulate matter as aerosols generated by the conventional handpiece. Sixty-five percent of the particles aerosolized by the air-turbine were less than 5 μ in size.
5. The present-day almost universal use of self-contained, recycling air-conditioning systems can convey dental aerosols throughout an entire suite or building serviced with the same ventilating system as the dental operating room.
6. The dentist, his assistant and patients are therefore constantly being exposed to potential health hazards.
7. There is a need to accurately assay the extent of this hazard and then to determine the calculated hygienic risk of ignoring protective measures balanced against the simple factors of time and cost.

REFERENCES

Albert, R. E. A., and Arnett, L. C. 1955. Clearance of radioactive dust from the human lung. *A. M. A. Arch. Ind. Health,* **12:** 99–106.

Ames, A. M., and Nungester, W. J. 1949. The initial distribution of airborne bacteria in the host. *J. Infect. Diseases,* **84:** 56–63.

Belting, C. M., Haberfelde, G. C., and Juhl, L. K. 1964. Spread of organisms from dental air rotor. *J. Amer. Dental Assoc.,* **68:** 648–651.

Bensch, K. G., Dominguez, E., and Liebow, A. A. 1967. Absorption of intact protein molecules across the pulmonary air-tissue barrier. *Science,* **157:** 1204–1206.

Bloomfield, A. L. 1919. The fate of bacteria introduced into the upper air passages. I. *Sarcina lutea. Johns Hopkins Hosp. Bull.,* **30:** 317–322.

Bloomfield, A. L. 1920a. The fate of bacteria introduced into the upper air passages. II. *Bacillus coli* and *Staphylococcus albus. Johns Hopkins Hosp. Bull.,* **31:** 14–19.

Bloomfield, A. L. 1920b. The fate of bacteria introduced into the upper air passages. III. *Bacillus influenzae. Johns Hopkins Hosp. Bull.,* **31:** 85–89.

Bloomfield, A. L. 1920c. The fate of bacteria introduced into the upper air passages. V. The Friendlander Baccilus. *Johns Hopkins Hosp. Bull.,* **31:** 203–206.

Bloomfield, A. L. 1921. The significance of the bacteria found in the throats of healthy people. *Johns Hopkins Hosp. Bull.,* **32:** 33–37.

Bloomfield, A. L., and Huck, J. G. 1920. The fate of bacteria introduced into the upper air passages. IV. The reaction of the saliva. *Johns Hopkins Hosp. Bull.,* **31:** 118–121.

Booker, D. V., Chamberlain, A. C., Rundo, J., Buir, D. C., and Thompson, M. L. 1967. Elimination of 5 μ particles from the human lung. Nature, **215:** 30–33.

Brown, R. V. 1965. Bacterial aerosols generated by ultra high-speed cutting instruments. *J. Dentistry Children,* **32:** 112–117.

Brown, E. A., and Cruickshank, G. A. 1947. A comparative study of the effects of glycerite of hydrogen peroxide and of hexylresorcinol on the bacteria of the normal mouth. *J. Dental Res.,* **26:** 83–90.

Burton, W. E., and Miller, R. L. 1965. The role of aerobiology in dentistry. In *A Symposium on Aerobiology, 1963* (R. L. Dimmick, Ed.), pp. 87–94. Nav. Biol. Lab., Nav. Supply Center, Oakland, Calif.

Calderone, C. V. 1966. A protective shield for high speed equipment. *J. Prosthetic Dentistry,* **16:** 583–584.

Carter, W. J., and Seal, J. R. 1953. Upper respiratory infections in dental personnel and their patients. Naval Med. Res. Unit No. 4, Res. Proj. Rept. NM005 051.14.03, Feb. 1953.

Committee on Sanitary Engineering, National Research Council Division of Medical Science. 1947. Recent studies on disinfection of air in military establishments. *Amer. J. Public Health,* **37:** 189–198.

Crowley, M. C., and Rickert, U. G. 1937. Effect of certain mouthwashes on the number of oral bacteria. *J. Dental Res.,* **16:** 531–535.

Dimmick, R. L. 1960. Delayed recovery of airborne *Serratia marcesens* after short-time exposure to ultraviolet irradiation. *Nature,* **187:** 251–252.

Docking, A. R., and Amies, A. B. D. 1948. The efficacy of dental and surgical face masks. *Australian J. Dentistry,* **52:** 93–99.

duBuy, H., Arnold, F. A., and Olson, B. J. 1947. Studies on the air transmission of microorganisms derived from the respiratory tract. *Public Health Rept.* (U.S.), **62**(39): 1391–1413.

Duguid, J. P. 1945. The numbers and sites of origin of the droplets expelled during expiratory activities. *Edinburgh Med. J.*, **52**: 385–401.

Elford, W. J., and Van Den Ende, J. 1942. An investigation of the merits of ozone as an aerial disinfectant. *J. Hyg.*, **42**: 240–265.

Elford, W. J., and Van Den Ende, J. 1945. Studies on the disinfecting action of hypochlorous acid gas and sprayed solution of hypochlorite against bacterial aerosols. *J. Hyg.*, **44**: 1–14.

Ferguson, G. W. 1960. Eye safety for dental officers. *U.S. Navy Med. Newsletters*, **36**: 7.

Ford, C. R., and Peterson, D. E. 1963. The efficiency of surgical face masks. *Amer. J. Surg.*, **106**: 954–957.

Goldberg, L. J., Watkins, H. M. S., Dolmatz, M. S., and Schlam, N. A. 1954, Studies on the experimental epidemiology of respiratory infections. VI. The relationship between dose of microorganisms and subsequent infection or death of a host. *J. Infect. Diseases*, **94**: 9–21.

Green, G. M., and Kass, E. H. 1966. Proceedings of the Second International Symposium on Inhaled Particles and Vapours. Pergamon Press, London. Reported in *Nature*, **209**: 24–27.

Grundy, J. R. 1967. Enamel aerosols created during use of the air tubine handpiece. *J. Dental Res.*, **46**: 409–415.

Guyton, H. G., Buchanan, L. M., and Lense, F. T. 1956. Evaluation of respiratory protection of contagion masks. *Appl. Microbiol.*, **4**: 141–143.

Guyton, H. G., and Decker, H. M. 1963. Respiratory protection provided by five new contagion masks. *Appl. Microbiol.*, **11**: 66–68.

Hart, D. 1941. The importance of airborne pathogenic bacteria in the operating room. A method of control by sterilization of the air with ultraviolet radiation. *J. Amer. Med. Assoc.*, **117**: 1610–1613.

Hatch, M. T. 1968. Personal Communication. The Naval Biological Laboratory, Oakland, Calif.

Hatch, T. F., and Gross, P. 1964. *Pulmonary Deposition and Retention of Inhaled Aerosols*. Academic Press, New York.

Hausler, W. J., Jr., and Madden, R. M. 1964. Aeromicrobiology of dental handpieces. *J. Dental Res.* (Suppl., #5) abstract #378, **43**: 903.

Hoffman, H. 1958. Air hygiene in dental practice. *Oral Surg., Oral Med., Oral Pathol.*, **11**(9): 1048–1054.

International Commission for Radiology Protection. 1959. Threshold limit values for 1959. *A. M. A. Arch. Ind. Health*, **20**: 226–270.

Jay, P., Crowley, M., Haley, F. P., and Bunting, R. W. 1933. Bacteriologic and immunologic studies on dental caries. *J. Amer. Dental Assoc.*, **20**: 2130–2148.

Jensen, M. M. 1965. Inactivation of virus aerosols by ultraviolet light in a helical baffle chamber. In *A Symposium on Aerobiology, 1963* (R. L. Dimmick, Ed.), pp. 219–226. Nav. Biol. Lab., Nav. Supply Center, Oakland, Calif.

Jensen, M. M. 1967. Bacteriophage aerosol challenge of installed air contamination control systems. *Appl. Microbiol.*, **15**: 1447–1449.

Kazantzis, G. 1963. Air contamination from high-speed dental drills. *Proc. Royal Soc. Med.*, **54**: 242.

Kortsch, W. E. 1964. Recognition and correction of energy robbing sources for the general practitioner. *J. Wisconsin Dental Soc.*, **40**: 301–303.

Langmuir, A. D. 1961. Epidemiology of airborne infection. *Bacteriol. Rev.*, **25**: 173–181.

Langmuir, A. D., Jarrett, E. T., and Hollaender, A. 1948. Studies of the control of acute respiratory diseases among naval recruits. III. The epidemiological pattern and the effect of ultraviolet irradiation during the winter of 1946–47. *Amer. J. Hyg.*, **48:** 240–251.

Madden, R. M., and Hausler, W. J., Jr. 1963. Microbiological comparison of dental handpieces. I. Preliminary report. *J. Dental Res.*, **42:** 1146–1151.

Madden, R. M., and Hausler, W. J., Jr. 1966. Microbiological comparison of dental handpieces. II. Aerosol decay and dispersion. *J. Dental Res.*, **45:** 52–58.

Manhold, J. H., and Manhold, B. S. 1963. Further *in vivo* study of commercial mouth wash efficacy. *N. Y. J. Dentistry*, **33**(10): 383–386.

May, K. R. 1949. An improved spinning top homogeneous spray apparatus. *J. Appl. Phys.*, **20:** 932–938.

Mazzarella, M. A., and Hancock, J. G. 1968. Dissemination of bacterial aerosols by an ultrasonic dental prophylaxis unit. *Bacteriol. Proc.*, M31, p. 70.

Meyers, C. E., Hresko, J., Zippin, C., Coppoletta, J., and Wolochow, H. 1963. Recovery of aerosolized bacteria from humans. III. Recovery from nasal mucus. *A. M. A. Arch Environ. Health*, **6:** 643–648.

Meyers, C. E., James, H. A., and Zippin, C. 1961. The recovery of aerosolized bacteria from humans. I. Effects of varying exposure, sampling times, and subject variability. *A. M. A. Arch Environ. Health*, **2:** 384–390.

Micik, R. E., Miller, R. L., Mazzarella, M. A., and Ryge, G. 1969. Studies on dental aerobiology. I. Bacterial aerosols generated during dental procedures. *Dent. Res.*, **48:** 49–56.

Micik, R. E., Miller, R. L., Ryge, G., and Mazzarella, M. A. 1967. Microbial aerosols discharged from the oral cavity of dental patients. 45th General Meeting of the International Association for Dental Research, March 1967, Washington, D.C., paper #243, p. 97.

Miller, R. L., Burton, W. E., and Spore, R. W. 1965. Aerosols produced by dental instrumentation. In *A Symposium on Aerobiology, 1963* (R. L. Dimmick, Ed.), pp. 97–120. Nav. Biol. Lab., Nav. Supply Center, Oakland, Calif.

Miller, R. L., Leong, A. C., Micik, R. E., and Ryge, G. 1968. An evaluation of surgical masks used to protect dentists from bacterial aerosols. Abstract #483, Int. Assoc. Dental Res. Program and Abstracts.

Mohammed, C. I., and Manhold, J. H. 1964. Efficiency of preoperative oral rinsing to reduce air contamination during use of air turbine handpieces. *J. Amer. Dental Assoc.*, **69:** 715–718.

Morrow, P. E. 1966. Discussion in Symposium on the Structure, Function, and Measurement of Respiratory Cilia. *Amer. Rev. Respirat. Diseases*, **93:** 125–133.

Morrow, P. E., Gibb, F. R., and Gazioglu, K. 1965. *Proceedings of the Second International Symposium on Inhaled Particles and Vapours*. Pergamon Press, London.

Nicholes, P. S. 1964. Comparative evaluation of a new surgical mask medium. *Surg., Gyn., Obstet.*, **118:** 579–583.

Nungester, W. J. 1951. Mechanisms of man's resistance to infectious disease. *Bacteriol. Rev.*, **15:** 105–129.

Ostrom, C. A., Wolochow, H., and James, H. A. 1958. Studies on the experimental epidemiology of respiratory disease. IX. Recovery of airborne bacteria from the oral cavity of humans: The effect of dosage on recovery. *J. Infect. Diseases*, **102:** 251–257.

Palm, P. E., McNermey, J. M., and Hatch, T. 1956. Respiratory dust retention in small animals. *A. M. A. Arch Ind. Health,* 13: 355–365.

Pistocco, L. R., and Bowers, G. M. 1962. Demonstration of an aerosol produced by air-water spray and air-turbine handpiece. *U.S. Navy Med. Newsletter,* 40: 24.

Rayner, A. C., and Hurtig, H. 1954. Apparatus for producing drops of uniform size. *Science,* 120: 672–673.

Shay, D. E., and Clendenin, G. G. 1963. Incidence of coagulase positive *Staphylococci* in the upper respiratory tract of dental students and a study of their transmission during a routine dental prophylaxis. *J. Dental Res.,* 42: 110–122.

Stevens, R. E., Jr. 1963. Preliminary study—air contamination with microorganisms during use of air turbine handpiece. *J. Amer. Dental Assoc.,* 66: 237–239.

Travaglini, E. A., and Larato, D. C. 1965. A disposable dental face mask with a plastic eye shield for operating with the air turbine drill. *J. Prosthetic Dentistry,* 15: 525–527.

Wells, W. F. 1934. On air-borne infection—Study II. Droplets and droplet nuclei. *Amer. J. Hyg.,* 20: 611–618.

Wells, W. F. 1941. An apparatus for study of experimental airborne disease. *Science,* 91: 172–174.

Wells, W. F., Ratcliffe, H. L., and Crumb, C. 1948. On the mechanics of droplet nuclei infection. II. Quantitative experimental air-borne tuberculosis in rabbits. *Amer. J. Hyg.,* 47: 11–28.

Williamson, A. E. and Gotaas, H. B. 1942. Aerosol sterilization of airborne bacteria. *Indust. Med.,* 11: 40–42.

Wright, G. W. 1961. Structure and function of the respiratory tract in relation to infection. *Bacteriol. Rev.,* 25: 219–227.

19

AIRBORNE MICROORGANISMS: THEIR RELEVANCE TO VETERINARY MEDICINE

Denny G. Constantine

U.S. DEPARTMENT OF HEALTH, EDUCATION, AND WELFARE,
PUBLIC HEALTH SERVICE, NATIONAL COMMUNICABLE DISEASE
CENTER, ON ASSIGNMENT TO NAVAL BIOLOGICAL LABORATORY
SCHOOL OF PUBLIC HEALTH
UNIVERSITY OF CALIFORNIA, BERKELEY

Veterinarians, guided by epidemiologic principles that are so much a part of "herd medicine," were among the first scientists to recognize, apply, and benefit from aerobiology. Initially, economic factors demanded preoccupation with prevention of the spread of disease to healthy members of the herd rather than with treatment of the infected individual. Later, these same economic factors sometimes limited the extent to which practical prophylactic techniques could be applied. Now, aerobiological methods are coming into greater use in veterinary medicine; improved environmental control, and diminution of disease spread by other means, leaves the airborne route as one of the principal pathways of transmission.

World-wide losses attributed to animal diseases range from $3 to $4 billion annually, seriously hampering efforts to lessen the effects of hunger and the development of sound national economies and world-wide commerce (Steele, 1962). Man also acquires many diseases from domestic and wild animals, resulting in a further drain on public health resources and the economy in general.

Measures that curtail these losses should be encouraged. Recognition of the principal routes of transmission is prerequisite to exercising control of diseases. Where transmission by the respiratory route is involved, aerobiological research is a necessary approach to recognition, evaluation, and control of certain diseases.

The respiratory route of transmission has been implicated in the genesis of many diseases of veterinary importance. Table 19-1 lists etiologic agents of diseases having the respiratory route as a possible portal of entry. In many instances it supplements other successful routes. Some of these agents also cause disease in man and are known to be transmitted from person to person by the respiratory route.

Of course, the aerobiological approach to the study of the genesis of disease is not unique in that it requires the same study of the pathogen, the environment, the host, the incubation period, etc., as well as the interrelationship of these factors, as do other epidemiological studies. It is unique, however, in that the aerobiological approach requires a specialized and not universally understood methodology. For example, studies of infectivity and viability of specific microbes have shown the great variety of possible responses of the species tested, and indicate the critical role that small changes in environment plays in the over-all potential for contagion.

ENVIRONMENTAL STUDIES

The environment outside the animal body, fortunately, constitutes a serious impediment for most microorganisms as they pass from host to host. The route may be direct, individual to individual, or indirect wherein the organisms accumulate or multiply prior to being inhaled. The fungi, for example, often multiply in soil or other materials where they are saprophytic. Soil that contains animal feces is a particularly favorable growth site for many fungal species.

The animal's environment plays a critical role in airborne transmission of disease agents. Environmental factors that favor infection should be recognized, evaluated, and their modification should be considered in programs of prevention and control.

Modern husbandry methods employed in animal production result in amassing great numbers of animals, and large quantities of feed and supplies, in relatively small spaces. Consequently, a considerable amount of feces, mud, dust, and associated vermin is concentrated as a result of these relatively dense animal populations. Moreover, ruminants, such as cows, sheep, and goats, probably contaminate the atmospheres about

Table 19-1   Disease agents of veterinary medical importance known to be transmitted by respiratory route.

BACTERIA

Actinobacillus mallei	Listeria monocytogenes
Alcaligenes bronchicanis	Leptospira pomona
Bacillus adenitis equi	Mycobacterium tuberculosis
Bacillus anthracis	Mycoplasma gallisepticum
Brucella abortus	Mycoplasma hyorhinus
Brucella suis	Mycoplasma mycoides
Corynebacterium equi	Pasteurella multocida
Erysipelothrix rhusiopathiae	Pasteurella pestis
Escherichia coli	Pasteurella pseudotuberculosis
Francisella tularensis	Salmonella pullorum
Haemophilus gallinarum	Salmonella typhimurium
Haemophilus ovis	Staphylococcus aureus

FUNGI

Absidia ramosa	Histoplasma capsulatum
Aspergillus flavus	Histoplasma farcinosum
Aspergillus fumigatus	Mucor corymbifer
Aspergillus nidulans	Nocardia asteroides
Aspergillus niger	Nocardia brasiliensis
Aspergillus terreus	Rhinosporidium seeberi
Blastomyces dermatitidis	Rhizopus equinus
Coccidioides immitis	Rhizopus suinus
Cryptococcus neoformans	

RICKETTSIAE

Coxiella burneti

PROTOZOA

Toxoplasma gondii

VIRUSES

Avian encephalomyelitis	Infectious bovine rhinotracheitis
Avian leukosis	Infectious bronchitis of fowls
Aleutian disease of mink	Infectious laryngotracheitis of fowls
Borna disease	Infectious nephrosis of fowls
Canine distemper	Infectious porcine encephalomyelitis
Canine herpesvirus	Louping ill
Eastern equine encephalomyelitis	Newcastle disease
Equine infectious arteritis	Ornithosis
Equine influenza	Porcine enterovirus
Equine rhinopneumonitis	Poxviruses
Feline rhinotracheitis	Rabies
Feline pneumonitis	Swine influenza
Foot and mouth disease	Transmissible gastroenteritis of swine
Fowl plague	Western equine encephalomyelitis
Hog cholera	

them by exhaling aerosols of microorganisms from within the rumen, since ruminal microorganisms were found to pass in eructed gas from the animal's rumen to its lungs (Mullenax et al., 1964).

Crowded conditions frequently permit transmission of pathogens by any of several routes. The survival value to any agent of having more than one route of transmission is self-evident. An example is the virus of infectious bovine rhinotracheitis. This agent, long known for the contagious vaginitis syndrome it caused, evidently adapted itself to the respiratory tract, because a new respiratory syndrome emerged after the advent of management practices that precluded venereal transmission, but favored transmission by respiratory route (McKercher, 1964).

Many pathogens accumulate in or on dust in man's environment; when such dust particles are inhaled, they may cause infection. The environment of domestic animals is similarly contaminated with microbes, but usually to a far greater extent. For example, spores of the pathogenic fungus *Aspergillus fumigatus* were found in "outside" air in a maximum concentration of 7 spores per cubic meter compared to 2,300 in a hospital ward and 100,000,000 in a cowshed (Austwick, 1966).

To show the presence of airborne, microbial pathogens of animals in the environment, mechanical air samplers, similar to standard techniques (see Akers and Won, Chapter 4), have been employed. Delay *et al.* (1948) were the first to recover an airborne virus when they collected Newcastle disease virus from the air in poultry houses. In addition to employing mechanical air samplers, they used chicks as "sentinels" to detect airborne virus by serologic conversion; one might consider this as a biological method to amplify the sensitivity of the detection method.

Whereas it is important to establish the simple presence or absence of pathogens in air, air-sampling studies should also yield quantitative data. The study of concentrations of microbial pathogens in air or dust of appropriate environments has received little attention in recent years, and much of the early work was too general to find practical application today. Knowledge of concentrations of the agents can confirm the importance of suspect sources, particularly when dose-time responses of the host are known. Thus, data on concentration are important as one proceeds from problem detection to problem evaluation.

Hanson and Anderson (1967) studied the significance of environmental factors in poultry disease. They exposed chickens to Newcastle disease virus in rooms kept at temperatures ranging from 2 degrees above freezing to that of a hot summer day. Mortality in birds kept at 90°F was 100%, but the mortality rate was lower as birds were kept at lower temperatures, the rate being lowest (55%) in birds maintained at 34°F.

They observed that ammonia is produced as litter decomposes in poultry houses, and found concentrations can reach levels of 100 parts per million (ppm). They measured the effect of this gas on the infection rate of birds exposed to airborne Newcastle disease virus. After only 3 days of exposure to an ammonia level of 20 ppm, the rate of infection was doubled compared to controls.

Certain disease problems such as the ones just noted may be controlled by modifying the environment. In addition to such measures as source removal, improved hygiene, or addition of chemicals to soil, control may be approached by ventilation, air filtration, ultraviolet irradiation, or by addition of chemicals to the air (Bourdillon et al., 1948). It is possible that airborne transmission of microbes can be attenuated within rather narrow ranges of temperature or relative humidity (Hatch and Dimmick, 1966)—factors that can be modified in programs of control.

HOST STUDIES

Diseases acquired by inhalation may be characterized either by involvment of particular organs or tissues, such as the respiratory tract, or by generalized infection. They may be manifested as acute or chronic infections, and sometimes only detected by immunological techniques.

Surveys for the pathogen may be made in air or dust arising from sites of deposition by an infected animal, or by seeking the pathogen in swabbings from the upper respiratory tract, noting clinical signs or other specific or nonspecific factors that influence spread. Evidence of specific microbial activity may be sought in serologic surveys. Pathology techniques can be employed to study the course of the disease. Animals may be experimentally exposed to infectious aerosols to study the epidemiology or the genesis of disease for the purpose of evaluating, or better understanding, methods of prevention or control. Finally, exposure of animals to aerosols of avirulent agents may be undertaken to effect immunization.

Surveys

An example of problem detection and analysis involving survey techniques is provided by recent studies of staphylococcal nasal carriage in dogs. In man, the area of skin just inside the anterior orifice of the nose is well known as a site in which *Staphylococcus aureus* can commonly be found. This association is of particular importance in hospitals, because the establishment of the healthy carrier seems to be the method by which antibiotic-resistant staphylococci maintain themselves in the everchanging community of patients within a hospital.

A similar association has been observed in dogs. Up to 71% of all dogs sampled were found to be carriers of staphylococci, the organisms being recovered most frequently from the nares by swab (Blouse *et al.,* 1964). Floors had the highest rates of staphylococcic contamination of all environmental areas sampled. Kennel personnel had staphylococci phage patterns analogous to those found in dogs. Infection rates of kennel personnel were comparable to those reported for the normal adult population, so no conclusions regarding interspecies transfer could be made.

Experimental Exposure of Animals to Airborne Agents in the Field

Domestic and laboratory animals have been deliberately exposed to airborne agents as they occur naturally in the field. In some instances the method was used to demonstrate the presence of pathogens in the air when other methods either were unavailable or had failed. In other instances the method was used in epidemiologic studies to investigate the infectivity of atmospheres for natural hosts, and to investigate the pathogenesis of the disease in a host as it influences further dissemination of the pathogen. In still other applications this method has been employed to test the effectiveness of control methods.

Coccidioidomycosis is caused by the dimorphic fungus *Coccidioides immitis,* limited mainly to the desert regions of southwestern North America and the Grand Chaco-Pampa region of South America. Respiratory exposure results from inhalation of arthrospores of the saprophytic phase, which grow in the soil and are disseminated by wind during dust storms. The disease is of great economic importance in endemic areas from the standpoint of human and animal infections.

To study this kind of problem or to evaluate the results of control methods it is necessary to determine the infectivity of the atmosphere at different times. Converse and Reed (1966) were unable to recover the agent from air by mechanical sampling techniques. Observing that man probably was similar to dogs and monkeys in susceptibility to infection by the fungus, they successfully employed the lower animal species as biological air samplers to detect amounts infectious for man.

Vampire bat-borne rabies is not only the foremost livestock disease problem in Latin America, but also a disease to which man is susceptible. Possibly the migratory habits of bats makes the problem of our southern neighbor related to the problem of rabies in bats and other susceptible species in the United States, where each year many livestock die and some 20,000 persons must take the antirabies treatment.

An instance where two men died from rabies led to efforts to demonstrate the infectivity of air in bat caves (Constantine, 1962, 1967). These men had been exposed to the atmosphere in a bat cave in Texas and to no other known source of the virus. Initial efforts to detect airborne virus by use of mechanical air samplers, or by stationing laboratory rodents in the cave, were unsuccessful, but sentinel carnivores (foxes, coyotes, and a ringtail) that were exposed only to the cave air became infected. Subsequently it was shown that opossums could be similarly infected. Thus, air in populous bat caves had to be acknowledged as sources of infection for man, and as potential sources of infection for bat predators as well as for bats.

Animals were similarly exposed in caves to investigate the effect of ventilation in the control of airborne rabies. Cave ventilation may have lowered the infection rate in carnivores and lengthened the incubation period in those that succumbed. Bats became infected despite ventilation—perhaps because they became infected by bite, or because, in ventilated caves, they frequently retired into poorly ventilated crevices.

Experimental Exposure of Animals to Airborne Agents in the Laboratory

Basic Considerations

The established principles and techniques used in experimental exposure of laboratory animals to microbiological aerosols, as discussed elsewhere in this book, apply as well to veterinary research except for variations in methods of animal restraint and modifications of apparatus to accommodate the larger species. The Henderson aerosol apparatus was modified for the exposure of large and small domestic animals by the addition of an interchangeable small-animal exposure box and a large-animal exposure helmet (Beard and Easterday, 1965a). Various other exposure methods were used in earlier research with domestic species.

The species of animal selected for experimentation of this kind is selected according to its intended use. Frequently, species such as mice or guinea pigs are employed as samplers for the mere detection of agents in air or as assay systems of droplet infectivity, and often are also used to study aerobiological models of the genesis of disease. However, we are concerned here with the use of the domestic animal host in studies of the genesis, evaluation, and control of disease problems known to occur naturally in these species.

Pappagianis (Chapter 16) has described the structure and function of the respiratory tract in relation to respiratory infections in man. The same principles apply to domestic animals and laboratory animals. However, tremendous differences exist between these species with respect to respiratory systems (e.g., turbinates, mucociliary patterns and directional flow, sinuses, lung functional units), breathing habits (such as mouth breathing and panting), and air volumes inhaled per time unit. These differences are important to the measured extent of susceptibility, and as such they are realities to be directly dealt with and to not be by-passed by substituting unrealistic laboratory models.

Unfortunately, considerably less is known about domestic animals than about laboratory species concerning respiratory system functions that are important to achieving quantitatively significant aerobiological studies. For instance, where the effective route of exposure is through the lung, one should know the percentage of a noninfectious aerosol (of given particle shape, size, chemical composition, and physical consistency) that will be retained and the time required for its clearance from the lung prior to exposing the species to similar infectious aerosols. Although certain preliminary data, i.e., minute respiratory volumes, have been obtained for several domestic species such as dogs, goats, and burros (Jemski and Phillips, 1965), baseline data (percent retention and clearance times) have been obtained only for certain laboratory animal species. In the absence of this kind of information, the dose received must be estimated on the basis of air volume inspired during exposure.

The possibility that experimental animals may have previously acquired, latent or overt infections should be considered. "Silent infections," or undetected disease, may be converted to clinical disease by experimental stress or be mistaken for experimental pathology at autopsy. For example, chronic murine pneumonia affects virtually all rats not reared in germ-free or specific pathogen-free (SPF) conditions; 96% of standard laboratory rats had lung disease compared to 19% of SPF rats (Paget and Lemon, 1965). McKercher (1964) cited roles of stress, and single or multiple microbial agents, in the genesis of respiratory diseases of cattle. Care in the selection of experimental animals and the use of adequate controls help to minimize errors of this kind. Special methodologies have been discussed at length by Jemski and Phillips (1965).

Experimental Pathology

The majority of laboratory exposures of domestic animals to infectious aerosols have been done to determine their susceptibility to infection by that route or to challenge the immune status of vaccinated animals. Little attention has been paid to the pathogenesis of disease following experimental exposure by inhalation. Several examples will be cited.

Sinha *et al.* (1954) demonstrated that chickens could be infected with Newcastle disease virus by exposure to clouds of virus particles produced in the laboratory. They employed air-sampling techniques to estimate the concentration of virus in exhaled air from the experimentally infected birds. Although the air appeared to contain few infective particles per liter, it proved to be infective for other chickens.

Beard and Easterday (1967a, 1967b) nebulized the same virus to challenge the immunity of chickens. Notably, vaccinated birds became infected when exposed by this natural route although they were refractory to exposure by intramuscular inoculation of virus; the virus was no longer a pathogen by that route.

Pigs were exposed to nebulized hog cholera virus in order to challenge the effectiveness of vaccinations (Beard and Easterday, 1965b). In this instance, aerosol and intramuscular challenges with virus yielded similar results.

Hyslop (1965) experimentally generated aerosols of foot-and-mouth disease virus that proved infective for cattle, and recovered virus from the air in enclosures containing experimentally infected cattle. These results supported field observations indicating that cattle may occasionally be infected by airborne virus.

Six strains of Newcastle disease virus, some of high and some of low pathogenicity for chickens, were studied quantitatively for their tissue predilection in chickens infected by aerosols of small quantities of virus (Sinha et al., 1952). All 6 strains of virus multiplied to high titers in the lung. The blood contained the virus of 5 strains and the spleen contained the virus of 4 strains; 4 of the 6 strains were isolated from the brain. The 2 not found in the brain were the least pathogenic for chickens. These authors noted that although chickens could be experimentally infected by other routes, the course of the infection and sometimes the outcome differed from that observed in chickens following exposure by the simulated natural route.

Immunization by Aerosolized Vaccines

This subject is reviewed by Hottle in Chapter 15. It will be discussed briefly here as it pertains to veterinary medicine.

The feasibility of immunization by inhalation exposure to nebulized microbial vaccines has been explored in relation to several diseases of veterinary medical importance, and in several instances vaccination is now routinely achieved by this method.

Hitchner and his co-workers (1948, 1952) demonstrated that when poultry inhaled aerosolized, live Newcastle disease virus, many birds became immune.

Refining the approach to vaccination by airborne route, Bankowski and Hill (1954) showed that aerosols of Newcastle disease virus could be successfully used for vaccination of poultry, provided (a) the strain of virus employed was properly attenuated, or was avirulent; (b) the virus dosage was controlled; (c) the particle size was controlled; and (d) the depth of deposition within the respiratory system was controlled. The optimal dose of virus was determined by adjusting viral titer and duration of exposure, then measuring host response by antibody development and the ability to withstand challenge. Optimal particle size was determined by measuring sizes of the particles after they were impinged on slides coated with alkyd resin. The optimal depth of deposition in the respiratory tract was determined by use of a radioactive marker.

Over 90% of Newcastle disease vaccine produced in licensed establishments is of the type administered as aerosol, dust, or by addition to drinking water (Agriculture Research Service, 1964). The same methods are employed to immunize poultry against the virus of infectious bronchitis. The bronchitis vaccines and Newcastle disease vaccines have been combined as an added convenience, but interference between the two viruses has been demonstrated in certain combinations (Hofstad, 1965).

Attempts to immunize pigs by aerosolized attenuated hog cholera virus were unsuccessful, although complete protection was achieved when the vaccine was inoculated by intramuscular route. It was suspected that the difference was due to the low titer of the vaccine virus or to the instability of the vaccine when aerosolized (Beard and Easterday, 1965b).

Gorham et al. (1954a, 1954b) immunized mink and ferrets against the virus of canine distemper by exposing them to aerosols or sprays of egg-adapted virus. Cabasso et al. (1957) showed that the success of immunization of these species with egg-adapted virus depends on the size of the aerosolized particle. A properly operated nebulizer created particles that consistently afforded 100 percent protection, whereas when larger particles were employed only 50% of the animals became immunized. This finding is not surprising if one considers where, on a distribution plot of particle sizes emitted from various atomizers, the size for maximal retention might lie. Johnson et al. (1957) demonstrated that 68 to 100% of mink on ranches responded after a single exposure to the nebulized vaccine.

DISCUSSION

Mankind must increase his husbandry effectiveness if he is to overtake and master the current meat shortage and its associated deleterious effect on the health, welfare, and economy of mankind. Specific problems must be recognized, analyzed, and subjected to control; the effectiveness of the latter must be evaluated. In the course of this pursuit, it is clear that man will acquire fewer artificial impediments the closer he adheres to reality. The methods he employs should be as precise in effect as they are in intent if he is to avoid misdirection of energy. These principles are as true in reference to dealing with animal diseases as they are in reference to animal breeding or feeding.

Many studies of diseases acquired by airborne route have been handicapped because experimental animals were exposed by unnatural routes or because other convenient, though unrealistic, techniques were used. The investigator should be acutely aware that the more he removes his problem from nature, the less realistic is his conception of the problem. Of course, there are times when individual variables must be experimentally studied in isolation from the influences of other phenomena, and when uncontrollable circumstances impose the use of unrealistic methods—hence, laboratory studies. The departure from reality, however, should be recognized with respect to its possible significance. In some instances, unrealistic impediments may be assumed unwittingly, or because they are attractively convenient, and there is a great temptation to underemphasize deviations when results are interpreted. Much information can be gained by observing samples of the unproductive or misleading outcome of an unwitting misapplication of "standard techniques."

Various investigators failed to recover naturally airborne organisms (*Histoplasma capsulatum*, *Coccidioides immitis*, *Mycobacterium tuberculosis*, rabies virus, etc.) by the convenient mechanical air sampling methods. However, opposite results were obtained when susceptible animals, exposed to contaminated air, were used as samplers, i.e., when the more realistic or natural event was approximated. Either mechanical methods failed to collect the airborne pathogens (because of lack of sensitivity or efficiency), or the collection technique "killed" the microbes. I know of no one investigating this intriguing and most important problem.

As has been noted, differences in responses of experimental animals to a given pathogen administered by various routes may be as different as life and death. It has been common procedure in experimental pathology and epidemiology to infect animals by unnatural routes. Although circumstances frequently justified the act, the responses of the animals were of questionable significance relative to what happens in nature. The practice of inoculating infective organisms by syringe to study the pathogenesis of an infection normally acquired by inhalation may be likened to stoning an animal to study the pathogensis of disease caused by ingestion of gravel. Use of artificially produced agents is probably in the same category. Effective doses differ quantitatively and qualitatively, and responses usually differ in kind, degree, and in time of occurrence. The use of an unnatural route of exposure may serve no practical purpose—the results may only obscure the course of natural events.

Knowledge of the natural route of infection, and of techniques for the experimental reproduction of the event, has found remarkably practical application in the techniques of administering aerosolized vaccines and in effectively challenging the resistance of vaccinated animals. Knowledge of natural events was prerequisite to experimental reproduction of disease, mimicking natures methods, methods that later proved effective for immunization of animals with avirulent organisms. The same natural method, used to challenge the resistance of vaccinated animals, produced meaningful results that differed from results obtained when unnatural routes of exposure were used.

This is not to belittle the experimental approach to the study of airborne contagion, but rather to view laboratory studies in their proper perspective. Field studies by veterinarians, or air-hygienists, for that matter, whether infections studied are natural or induced, usually are not concerned with "why" or "how" (i.e., basic principles), but rather with "what" and "how much." We need, however, better tools and increased knowledge of host-pathogen relationships and of environmental effects which can be provided selectively only by the experimental approach. What is needed is greater communication between the two allied fields.

REFERENCES

Agriculture Research Service, USDA. 1964. Biological Products Notice 128, Activities of Licensed Establishments Supervised by the Animal Inspection and Quarantine Division.

Austwick, P. K. C. 1966. The role of spores in the allergies and mycoses of man and animals. In *The Fungus Spore* (M. F. Madelin, Ed.), pp. 321–337. Butterworths, London.

Bankowski, R. A., and Hill, R. W. 1954. Factors influencing the efficiency of vaccination of chickens against Newcastle disease by the air-borne route. *Proc. Am. Vet. Med. Assoc. 91st Ann. Meeting*, pp. 317–327. Seattle, August 23–26, 1954.

Beard, C. W., and Easterday, B. C. 1965a. An aerosol apparatus for the exposure of large and small animals: Description and operating characteristics. *Am. J. Vet. Res.*, **26:** 174–182.

Beard, C. W., and Easterday, B. C. 1965b. Aerosol transmission of hog cholera. *Cornell Vet.*, **55:** 630–636.

Beard, C. W., and Easterday, B. C. 1967a. The influence of the route of administration of Newcastle disease virus on host response: I. Serological and virus isolation studies. *J. Infect. Diseases,* **117:** 55–61.

Beard, C. W., and Easterday, B. C. 1967b. The influence of the route of administration of Newcastle disease virus on host response: II. Studies on artificial passive immunity. *J. Infect. Diseases,* 117: 62–65.

Blouse, L., Husted, P., McKee, A., and Gonzalez, J. 1964. Epizootiology of staphylococci in dogs. *Am. J. Vet. Res.,* 25: 1195–1199.

Bourdillon, R. B., Lidwell, O. M., and Lovelock, J. E. 1948. Studies in Air Hygiene. Med. Res. Council (Brit.); Spec. Rep. Ser. No. 262, 356 p. His Majesty's Stationery Office, London.

Cabasso, V. J., Johnson, D. W., Stebbins, M. R, and Cox, H. R. 1957. "Atomized" distemper vaccine of avian origin. I. Experimental immunization of ferrets and mink. *Am. J. Vet. Res.,* 18: 414–418.

Constantine, D. G. 1962. Rabies transmission by nonbite route. *Public Health Rept.* (U.S.) 77: 287–289.

Constantine, D. G. 1967. Rabies Transmission by Air in Bat Caves. Public Health Service Publication No. 1617. U.S. Gov. Printing Office, Washington, D.C.

Converse, J. L., and Reed, R. E. 1966. Experimental epidemiology of coccidioidomycosis. *Bacteriol. Rev.,* 30: 678–694.

Delay, P. D., DeOme, K. B., and Bankowski, R. A. 1948. Recovery of pneumoencephalitis (Newcastle Disease) virus from the air of poultry houses containing infected birds. *Science,* 107: 474–475.

Gorham, J. R., Leader, R. W., and Gutierrez, J. C. 1954a. Distemper immunization of mink by air-borne infection with egg-adapted virus. *J. Am. Vet. Med. Assoc.,* 125: 134–136.

Gorham, J. R., Leader, R. W., and Gutierrez, J. C. 1954b. Distemper immunization of ferrets by nebulization with egg-adapted virus. *Science,* 119: 125–126.

Hanson, R. P., and Anderson, D. 1967. Significance of environmental factors in poultry disease. *Proceedings Western Poultry Disease Conference,* pp 1–3, Univ. of Calif., Davis, Calif.

Hatch, M. T., and Dimmick, R. L. 1966. Physiological responses of airborne bacteria to shifts in relative humidity. *Bacteriol. Rev.,* 30: 597–602.

Hitchner, S. B., and Johnson, E. P. 1948. A virus of low virulence for immunizing fowls against Newcastle disease (Avian pneumoencephalitis). *Vet. Med.,* 43: 525–530.

Hitchner, S. B., and Reising, M. S. 1952. Flock vaccination for Newcastle disease by atomization of the B_1 strain of virus. *Proc. Am. Vet. Med. Assoc. 89th Ann. Meeting,* pp. 258–264. Atlantic City, June 23–26, 1952.

Hofstad, M. S. 1965. Infectious Bronchitis. In *Diseases of Poultry* (H. E. Biester and L. H. Schwarte, Eds.), pp. 605–620. Iowa State Univ. Press, Ames, Iowa.

Hyslop, N. St. G. 1965. Airborne infection with the virus of foot-and-mouth disease. *J. Comp. Pathol. Therap.,* 75: 119–126.

Jemski, J. V., and Phillips, G. B. 1965. Aerosol challenge of animals. In *Methods of Animal Experimentation* (W. I. Gay, Ed.), Vol. I, pp. 273–341. Academic Press, New York and London.

Johnson, D. W., Cabasso, V. J., Huffman, K., and Stebbins, M. R. 1957. "Atomized" distemper vaccine of avian origin. II. Field experience in mink. *Am. J. Vet. Res.,* 18: 668–671.

McKercher, D. G. 1964. Some viruses associated with the respiratory tract of cattle and the syndromes which they cause. III. Intern. Meeting Diseases of Cattle, pp. 3–26. Copenhagen, Denmark.

Mullenax, C. H., Allison, M. J., and Songer, J. R. 1964. Transport of aerosolized microorganisms from the rumen to the respiratory system during eructation. *Am. J. Vet. Res.*, **25**: 1583–1593.

Paget, G. E., and Lemon, P. G. 1965. The interpretation of pathology data. In *The Pathology of Laboratory Animals,* pp. 382–405. Charles C Thomas, Springfield, Ill.

Sinha, S. K., Hanson, R. P., and Brandly, C. A. 1952. Comparison of the tropisms of six strains of Newcastle disease virus in chickens following aerosol infection. *J. Infect. Diseases,* **91**: 276–282.

Sinha, S. K., Hanson, R. P., and Brandly, C. A. 1954. Aerosol transmission of Newcastle disease in chickens. *Am. J. Vet. Res.*, **15**: 287–292.

Steele, J. H. 1962. Animal Disease and Human Health. Freedom from Hunger Campaign Basic Study No. 3, 50 pages. Food and Agriculture Organization of the United Nations. Rome, Italy.

GLOSSARY

James C. Warren / D. D. Flynn

NAVAL MEDICAL RESEARCH UNIT NO. 1,
UNIVERSITY OF CALIFORNIA, BERKELEY

Writings from each branch of science contain a "jargon"; so it is
with aerobiology. Some words are used more or less in the sense defined in standard dictionaries, some bear little resemblence to the
English language.

Following is a partial list of words found either in the literature,
or in this book, that might confuse the uninitiated reader. We have
tried to define the words as they are used in the aerobiology literature,
although we must admit that even our most knowledgeable colleagues
do not agree with our interpretations in all instances, nor among themselves—another indication that aerobiology is currently an art, not a
science.

AEROBIOLOGIC PROPERTIES: Those properties of an organism that either hinder or
enhance survival in the airborne state.

AERODYNAMIC DIAMETER: *See* Aerodynamic particle size.

AERODYNAMIC PARTICLE SIZE: Size of particles measured by their inertial parameter
and assuming a spherical particle, usually of unit density. Sometimes termed
"Stokes diameter" or equivalent mean diameter.

AEROGENIC IMMUNIZATION: *See* Aerosol immunization.

AEROSOL: A collection of particles in air with special reference to those that remain
suspended for appreciable times (e.g., particles less than 100 μ diameter). An
aerosol is not a "sol" in the sense of "hydrosol"; i.e., an aerosol is not indefinitely
stable.

AEROSOL ADDITIVES: *See* Aerosol stablizers.

AEROSOL AGE: The interval of time between the creation of an aerosol and some
other event used to characterize the aerosol.

AEROSOL BEHAVIOR: The observed, time-dependent, properties of an aerosol, which
include changes in particle concentration, size distribution and, for microbial
particles, the number remaining viable.

477

AEROSOL CHAMBER: A vessel (box, flask, barrel, etc.) to contain aerosols.

AEROSOL IMMUNIZATION: Exposure of animals to aerosols of antigens for purposes of inducing immunity.

AEROSOL MONITOR: *See* Light-scatter aerosol monitor; Penetrometer.

AEROSOL STABILITY: 1. The ability of any organism to retain its viability or infectivity while suspended in an aerosol. 2. Measure of tendency of a collection of particles to remain airborne.

AEROSOL STABILIZERS: Also called "aerosol additives" and "protective agents"; substances added to a slurry prior to aerosolization to enhance the survival of the microorganisms in the aerosol.

AEROSOL TRANSPORT APPARATUS, ATA: Any aerosol apparatus transporting aerosols from a point of input to distant sampling ports, or to disposal apparatus.

AEROSOL TRANSIT TIME: In an aerosol transport apparatus, the amount of time required for an organism to travel from its point of entrance into the apparatus to any given sample port.

AEROSOLIZATION: The act of generating an aerosol.

AEROSOLIZER: Any device for generating an aerosol.

AGAR: A complex organic material derived from sea weed. Forms a sol in water when heated to 95 to 100°C, but gels at 42 to 45°C. Used as semi-solid matrix for growth media.

AGENT: 1. A specific microbe: 2. A specific chemical.

AGGREGATE: (*n.*) A mass of individual particles. (*v.*) To form a mass.

AGI-30: A standard glass impinger in general use today. AGI = "all glass impinger"; 30 refers to distance in millimeters from the critical orifice to the bottom of the collection tube. Commonly operated at 12.5 liters per minute, air flow.

AIRBORNE DECAY: *See* aerosol stability.

AIRBORNE INFECTION: (*v.*) Act of transmitting disease from one person to another by airborne particles laden with microbes. (*n.*) Capacity of air to infect; should properly be "airborne infectivity" or "airborne contagion."

AIRBORNE TRANSMISSION: *See* airborne infection.

AIRBORNE PARTICLES: Particles suspended in a gaseous medium; the cause of particle suspension may be natural, i.e., wind-generated, or artificially generated by laboratory apparatus.

AIRBORNE STABILITY: Capacity of microbes (or measurement of that capacity) to remain both airborne and viable.

ALVEOLUS: The terminal gas-exchange sac at the end of a respiratory tree; a human lung has over 150 million such sacs, the sacs having a mean diamenter of 15×10^{-4} cm.

ANTIBODY: A protein that appears in the blood of an animal in response to certain materials (antigens, e.g., microbes or their constituents) entering the body.

ANTIGEN: Any material that elicits an antibody response and can be demonstrated to react specifically with the antibody.

ASPIRATION: 1. Movement of fluids by means of a vacuum. 2. The act of moving particulate material out of the respiratory tract.

ATOMIZATION CHAMBER: 1. Chamber in which aerosol is generated. 2. Chamber in which a newly generated aerosol is added to a stream of pre-conditioned air.

ATOMIZER: *See* Aerosolizer.

ATOMIZER FLUID: A liquid, which may contain dissolved or suspended material, to be aerosolized by an atomizer or nebulizer.

BIOLOGICAL AEROSOL: A suspension of viable particles in a gaseous medium. *See also* Viable particles.

BIOLOGICAL DECAY: Difference between total decay (loss of particles and loss by death) and physical decay (loss of particles).

BIOLOGICAL EFFICIENCY: Efficiency of a sampling device as a collector of airborne microbes in a viable state. Should be used only in a comparative sense.

CAPILLARY IMPINGER: Device to collect airborne particles. It is composed of a jet through which the air sample is moved at sonic air-velocity, and a liquid collection menstruum. See AGI-30.

CASCADE IMPACTOR: A sampler for aerosol particles, having two or more impaction stages in series.

CLOUD: A visible aerosol with defined boundaries in free air.

CLOUD CHAMBER: Any apparatus for enclosing an aerosol. See Aerosol chamber.

COAGULATION: A spontaneous and continuous process of particles joining into larger masses.

COLLISON ATOMIZER: A three-nozzle, two-fluid, right-angle, refluxing spray unit contained in a housing. The liquid spray is directed against the housing wall or other baffle.

COLLECTION EFFICIENCY: The degree to which any sampling device is capable of stopping and collecting particles from an aerosol.

CONDENSATION NUCLEI: Very small particles on which water vapor can condense when the air mass becomes saturated or supersaturated with water vapor.

CONTAMINANT: Any material (substance), not deliberately introduced into an experimental situation, which interferes in an unpredictable manner with experimental aims.

CRITICAL ORIFICE: An orifice placed in a system to control the volume of gas passing through the system per unit time. For calibration, and in use, the pressure drop (as a ratio, P_{out}/P_{in}) across the orifice must be less than 0.5.

CUNNINGHAM'S CORRECTION: A factor applied to velocity of fall of small particles to correct for slippage between air molecules.

DEATH RATE: The apparent number (percent, ratio) of microorganisms dying per unit of time. An exponential rate is usually assumed, e.g., percent per minute or logs per minute. See Survival rate; D-value.

DECAY RATE: The rate (percent or fraction per unit of time) at which viable organisms or inert particles become unrecoverable from the aerosol state.

DEW-POINT (SATURATION TEMPERATURE): The temperature to which air must be cooled, at constant pressure, to become saturated.

DIE-AWAY: "Death rate" when used as indefinite quality.

DROPLET NUCLEUS(I): Airborne particles containing viable microbes. Originally referred to particles from mouth or nose that dried to a size small enough to remain airborne for appreciable times. Does not have connotations of either a central core or of point of assembly, as in "condensation nuclei."

DROPPING PIPETTE: A pipette with a tip calibrated to deliver a known volume of liquid per drop. Volume is dependent on outer diameter of the tip (which must be grease-free) and the surface tension of the liquid.

D-VALUE: Time interval for a tenfold population reduction to occur.

$$D = \frac{\Delta t}{(\log N_0 - \text{Log } N)}$$

N_0 = concentration at an initial time t_0
N = concentration at time t when $\Delta t = t - t_0$
$1/D = K$ = decay constant or death rate

DYNAMIC AEROSOL: An aerosol that is being continuously generated or in a state of continuous motion from a point of origin to a point of disposal while being studied.

DYNAMIC CHAMBER: An aerosol container where air and particles are being continuously introduced and removed—as in a duct or pipe.

EDGEWOOD BUBBLER: Device wherein air is bubbled through liquid, usually water, to remove larger particles.

FOMITES: Substances or things capable of carrying and disseminating microbes.

GASEOUS IONS: Molecules, or small aggregates, of gaseous elements or compounds which may have an excess, or lack of, one or more electrons.

HOMOGENEOUS AEROSOL: 1. An aerosol in which any sample is equivalent to any other sample of that aerosol. 2. An aerosol having particles of uniform properties; e.g., size.

IMMUNE SERUM: The liquid portion of blood containing one or more specific protein antibodies.

IMPACTION: The entrapment of aerosol particles onto a solid surface within a sampling device after particles have been accelerated to high velocity by passing through an orifice. (Impaction and impingement are synonymous; "impact" is the past participle of "impinge.") In aerobiology studies, impinge refers to collection in a liquid.

IMPACTION PARAMETER: A dimensionless measure of the collection characteristic of an impactor stage. Determined in part by the size and density of particles collected.

IMPINGEMENT: *See* Impaction.

IMPINGER: Any device used for the collection of aerosol particles in liquid.

INACTIVATION: The apparent "kill," or loss of viability, of microorganisms by exposure to hostile elements: e.g., osmotic shock, relative humidity shifts, temperature changes, mechanical impaction, etc.

INERTIAL PARAMETER: A measure of particle inertia. Usually determined by the efficiency of collection in a calibrated impactor stage, or by Stokes' Law.

INFECTIVITY: Extent of ability to cause infection. May be used in a comparative sense, i.e., comparing numbers of the infecting agent required to achieve a given number of infected animals, etc.

LAG PERIOD: The interval between planting a culture and evidence of growth or reproduction.

LAMINAR FLOW: Fluid flow in which streams do not intermingle.

"LECTRODRYER": A trademarked name for a commercial air-drying device. Manufactured by Pittsburgh Lectrodryer Corp., Pittsburgh, U.S.A.

LIGHT-SCATTER AREA: In a collection of particles, an assumed, total area available to reflect (scatter) incident light. *See* RLS.

LIGHT-SCATTER AEROSOL MONITOR: A photo-electronic device used to detect the presence of an aerosol by scattered light, and to measure aerosol concentration. Measurements may be in absolute units (e.g., mass/unit volume) or comparative (e.g., change of units with time). If single particles are detected, it is termed a particle counter.

LIGHT-SCATTER PHOTOMETER: *See,* Light-scatter aerosol monitor.

LOG PHASE: Interval when bacteria are growing at an exponential rate.

LOSS OF VIABILITY: The decrease in numbers of organisms able to reproduce in some observable manner.

MASS MEAN DIAMETER, d_m: A calculated diameter, used to characterize a collection of particles of heterogeneous size. If the total *mass* of material in all the particles were equally distributed to N particles, each particle would have the mass mean diameter, d_m.

$$d_m = \sqrt[3]{\frac{\Sigma(n \times d^3)}{N}}$$

where n = number of particles of diameter, d, and
$\quad\quad N$ = total number of particles of all diameters.

MASS MEDIAN DIAMETER, MMD: A calculated diameter, used to characterize a collection of particles of heterogeneous size. Fifty percent of the total mass of the collected particles is in the size range *smaller* than the MMD. Determined by plotting the cumulative mass versus particle diameter.

MECHANICAL EFFICIENCY: Ratio of numbers of particles trapped (collected) versus total (true) number in an aerosol sampler. Depends on way true total is measured.

MEDIUM: 1. Any material, of known or unknown composition, which will support growth of a microorganism. 2. Any porous material which traps particles in a moving fluid (liquid or gaseous).

MIXING CHAMBER: 1. A chamber in which streams of wet and dry air may be thoroughly mixed to gain a desired relative humidity. 2. A chamber used to mix and dilute concentrated aerosol.

MONODISPERSE AEROSOL: Aerosols composed of particles of uniform size.

MULTIPLE-STAGE DEATH RATE, MSDR: More than one death rate, each apparently occurring in sequence, during an experiment with a given aerosol preparation.

NBL ROTATING DRUM; DYNAMIC AEROSOL TOROID, DAT: *See* Rotating drum.

NEBULIZER: A two-fluid, right-angle refluxing spray nozzle contained in a small housing. The liquid spray is directed against a closely spaced baffle. Sometimes used as a synonym for atomizer.

PARTICLE COUNTER: *See* Light-scatter aerosol monitor.

PARTICULATE: In the form of fine particles. Often used redundantly as "particulate aerosols."

PENETROMETER: A light-scatter monitor (or other device) used to assess the penetration of an aerosol through a filter medium.

PHYSICAL DECAY: Decrease in numbers of particles over a time period, usually defined as percent/min or logs/min. *See* Decay rate.

PLATE COUNT: Number of colonies, or plaques, on petri plate. *Not* number of plates.

PRE-CONDITIONED AIR: Air which has been either dryed, heated, or moistened by various pieces of apparatus before being used elsewhere. Process air.

PRE-IMPACTOR: Any device which may be used to selectively remove particles, of a size *larger* than those of interest, from an airstream prior to collection of these particles. Synonymous with pre-impinger.

PRIMORDIAL PARTICLES: The individual particles which collectively make up an aggregate. *See* Aggregation; Satellite particles.

PROTECTIVE AGENTS: *See* Aerosol stabilizers.

REACTIVATION: The apparent revitalization, or return to viability of part of a population of cells, as a result of a changed environment (e.g., light reactivation of cells damaged by UV).

RECOVERY: 1. The number of living microbes in a sample after an interval of experimentation. 2. Ability of a microbe to be alive after apparently being dead.

RECUPERATION: An increase in the apparent number of survivors that occurs while cells are *in* a condition of stress (i.e., without a change of environment; spontaneous). Assumed to be caused by internal, cellular processes, but not uniformly initiated or maintained by cells in a population.

REFLUX ATOMIZER: An aerosol generator that creates a continuous aerosol in a reservoir within an atomizer bulb such that liquid, not escaping as an aerosol, returns to the reservoir to be re-aerosolized.

REHUMIDIFICATION: The process of increasing the moisture content of air, hence of particles suspended in that air.

RELATIVE HUMIDITY, RH: The ratio of the quantity of vapor actually present to the greatest amount possible at the given temperature; usually expressed as percent.

RELATIVE LIGHT SCATTER, RLS: The amount of light scattered by particles; measured by arbitrary but standardized photoelectronic techniques, usually in reference to some initial value.

ROTATING DRUM: Cylinder-shaped aerosol container rotated at speeds from 3 to 10 rpm on a central axis.

"S" UNIT: Contraction for a ventilated storage unit in which infected animals are housed.

SA: Abbreviation for "slurry count after"; refers to the numbers of viable organisms in a slurry remaining just after the interval of atomization.

SAMPLE PORT: Any hole, entrance, or access to an aerosol chamber from which samples of an aerosol may be withdrawn for assay or analysis.

SATELLITE PARTICLES: Particles formed from a larger aggregate particle by impaction, impinging, or centrifugal forces. Satellite particles are usually smaller than the bulk of the particles of interest.

SB: Abbreviation for "slurry count before"; refers to the number of viable organisms in a slurry just before the atomization period.

SCFM: Standard cubic feet per minute; i.e., at 68°F and 29.9 in Hg.

SECONDARY AIR STREAM: In introducing any aerosol into a chamber, the airstream supplied to dilute the primary aerosol.

SENSITIVE CELLS: In a population, a fraction of cells assumed to be incapable of surviving a given stress.

SHIFT-DOWN (OR SHIFT-UP): A term denoting an abrupt change from a higher to a lower relative humidity, or the reverse, within a dual aerosol transport apparatus. Also used in microbiology with reference to temperature change.

SHIPE IMPINGER: Modification of a capillary impinger.

SLIPPAGE: That number or percentage of particles in an aerosol not trapped by a collection device.

SLIT SAMPLER: A particle collector using an orifice in the shape of a slit. Usually particles are impacted onto a rotating, agar plate.

SLURRY: A suspension of a microbial culture which has been prepared for aerosolization. Sometimes an antifoaming agent is added to the suspension.

SLURRY COUNT: The number of viable organisms existing in a slurry (atomizer fluid) either before or after aerosolization.

"SORBED DEATH": Death of an organism following the uptake of moisture.

STABILIZING ADDITIVES: *See* Aerosol stabilizers.

STAIRMAND BAFFLES: Baffles (of $\frac{1}{2}$ duct area) placed within a duct system that induce mixing of the air stream for at least a distance of 6 duct-diameters.

STATIC AEROSOL: Aerosols confined to a single aerosol chamber and not continuously generated throughout an observation or study period.

STATIC CHAMBER: A container wherein aerosols are held with no disturbances other than those imparted by gravity, thermal diffusion, convection or by deliberate, gentle, stirring.

STATIC ENVIRONMENT: That environment within a static aerosol chamber.

STATIONARY PHASE: Interval directly following a growth phase during which the number of viable bacteria remains constant.

STIRRED SETTLING CHAMBER: An aerosol apparatus (i.e., a sealed box, barrel, or drum) with some method of stirring the aerosol to maintain uniform concentration throughout the chamber.

STORAGE STABILITY: Ability of an organism to retain its viability and infectivity during storage under a given condition.

SURVIVAL: The ability of a microorganism to remain viable or retain its infectivity. Often used to describe qualitatively the number of surviving cells in a population, as "high" or "better" survival. *See* Survival curve.

SURVIVAL CURVE: A curve describing survival of microorganisms over a period of time.

SURVIVAL PATTERN: An abstract representation (usually graphical) of changes in survival of populations studied under different conditions; e.g., survival rates plotted versus humidity; surviving fractions at constant humidity and aerosol time plotted versus age of culture.

SURVIVAL RATE: The change in apparent number (per unit volume, or ratio to an inert tracer) of microorganisms remaining viable over a unit time-span. Usually expressed as percent per minute or logs per minute.

SUSPENSION FLUIDS: Various liquid media containing, or destined to contain, microorganisms (growth broths, impinger fluid, diluent fluid).

THERMAL PRECIPITATION: Collection by impacting particles gently on cooled surfaces by means of a thermal gradient.

TOROID DRUMS: *See* Rotating drum.

TOTAL DECAY: The *observed* loss in viable-cell concentration in an aerosol. This loss is the sum of that attributable to physical losses (settling, impaction) and that resulting from death of the microbes.

TYNDALL EFFECT: Scattering of a beam of light by small particles; the basis for operation of the light scatter aerosol monitor.

VENTURI DILUTION UNIT: A nozzle held within an air stream such that an aerosol can be drawn through the nozzle to mix with the air stream.

VENTURI SCRUBBER: Device based on venturi principle to mix air and water at high efficiencies.

VIABLE PARTICLE: An airborne particle containing at least one living microbe. Term arises from fact that some airborne particles (e.g., in an aerosol or as collected by a sampler) contain no microbes, the microbes are dead, or one or more living microbes per particle are present. Shorthand jargon, then, for "airborne particles containing viable microbes" and used, for example, to distinguish between a "viable particle sampler" that collects gently enough not to kill, and one that might shatter or disrupt a collected microbe. Reasonably synonomous with droplet nucleus.

VIABLE STATE OR STATE OF VIABILITY: 1. State of a microbe being alive. 2. Number of viable microbes in a population.

VIABILITY: With respect to microbes, the ability to reproduce, or to increase in protoplasmic mass; extent of that ability in a population.

VIABLES: A number of viable cells.

WAGNER SAMPLER: An in-line air sampler using a filter pad.

WELLS ATOMIZER: A twin-fluid, jet atomizer developed by William F. Wells to generate aerosols.

WET-DRY BULB: The simplest, most common instrument for measuring relative humidity. A wetted thermometer bulb in an air stream (> 15 ft/sec) indicates a decrease in temperature compared to an adjacent dry bulb thermometer; the lower the humidity the greater the difference in temperature.

INDEX